Galileo's Telescope

Galileo's Telescope

A EUROPEAN STORY

Massimo Bucciantini
Michele Camerota
Franco Giudice

Translated by Catherine Bolton

Harvard University Press
CAMBRIDGE, MASSACHUSETTS
LONDON, ENGLAND
2015

Printed in the United States of America

First printing

First published as *Il telescopio di Galileo: Una storia europea,* © 2012 by Giulio Einaudi editore s.p.a., Torino

The translation of this work has been funded by SEPS
SEGRETARIATO EUROPEO PER LE PUBBLICAZIONI SCIENTIFICHE

Via Val d'Aposa 7—40123 Bologna—Italy
seps@seps.it—www.seps.it

Library of Congress Cataloging-in-Publication Data
Bucciantini, Massimo.
[Telescopio di Galileo. English]
Galileo's telescope : a European story / Massimo Bucciantini, Michele Camerota, Franco Giudice ; translation by Catherine Bolton.
pages cm
Includes bibliographical references and index.
ISBN 978-0-674-73691-7 (alk. paper)
1. Telescopes—Europe—History—17th century.
2. Astronomical instruments—Europe—History—17th century.
3. Galilei, Galileo, 1564–1642.
4. Astronomy—History—17th century.
I. Camerota, Michele. II. Giudice, Franco. III. Title.
QB88.B8313 2015
522'.209409031—dc23
2014033443

Book designed by Dean Bornstein

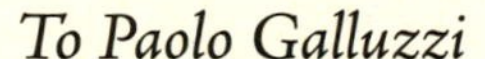

To Paolo Galluzzi

Contents

Illustrations, Plates, and Maps

Illustrations

Plates

Color plates follow page 180

Maps

Galileo's Telescope

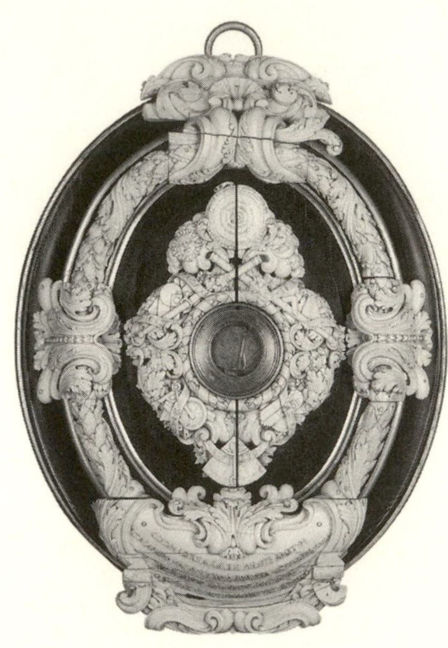

Figure 1. Galileo Galilei, objective lens, late 1609. Inserted in an ebony and ivory frame built by Vittorio Crosten in 1677

Figure 2. Galileo Galilei, objective lens, late 1609

Prologue

The students who flock to Florence to visit the Galileo Museum along the Arno, close to the Uffizi and the Ponte Vecchio, gaze at the object with great anticipation and wonder. Then, disappointed by its rather unattractive appearance, they wander around looking for something that might better pique their curiosity. A few—the most motivated, of course—persevere and take a closer look, trying to glean some mysterious secret.

Yet even upon closer inspection, it holds little appeal. Opaque and dotted with tiny flaws, it has a wide ridged band around it and is a milky, dusty gray color that almost makes you want to pick it up, blow on it, and then polish it gently to make it clearer and shinier. If it weren't so well made and, like a saint's relic, imprisoned in an exquisite frame of ebony, ivory, and gilt brass, it would be little more than an ordinary piece of glass—and one that is broken to boot, crossed by three clearly visible cracks (Figures 1 and 2).

That piece of glass is Galileo's only extant lens. At the very least, it is the only one that, beyond the shadow of a doubt, has survived to our day of the many he made, first in Padua and then in Florence: of the hundreds that, starting in the summer of 1609, he worked ceaselessly to grind himself or with the help of expert spectacle makers, in order to fit out the "cannon" he wanted to turn skyward to seek new stars and planets.

Optics experts describe it as a biconvex objective lens measuring 58 mm in diameter, with an effective aperture of 38 mm. There is little we don't know about it. We know, for example, that Galileo gave it to Grand Duke Ferdinando II; that it was broken accidentally a few years later; that after Galileo died it was kept in the Guardaroba of the prince (later cardinal) Leopoldo de' Medici; that in 1675 it entered the Medici collection and was jealously guarded at the Uffizi, where it remained until 1793, when it was transferred to the Museum of Physics and Natural History; and that in the mid-nineteenth century it was placed in the Tribuna di Galileo with other memorabilia that had belonged to the scientist. We also know the exact year—1677—that the Medici commissioned Vittorio Crosten, a skilled Dutch carver who worked at the court of Cosimo III, to make the frame in which the lens was set.

Moreover, we have accurate information about Galileo's only two extant telescopes, which are displayed close to the framed lens at the Florentine museum. One is composed of a tube made of two fluted wooden shells held together by copper hoops and covered in paper; it is 1,273 mm long. The lenses and diaphragms (possibly Galileo's originals) are secured with copper rings. The objective lens measures 50 mm in diameter and the eyepiece 40 mm, and it can magnify up to fourteen times. The other is composed of a tube made of wooden slats covered in brown leather overlaid with strips of red leather, and it is finished with exquisite decorations in gold leaf. It is 927 mm long, the objective is 37 mm in diameter, and the eyepiece—not original and added in the nineteenth century—measures 22 mm; it has a magnification of about twenty times (Figures 3–6).

We could go on, of course, perhaps listing the radius of curvature, the positive and negative focal length, the breadth of visual field, the center thickness of each lens. Yet ultimately nothing would change, because this type of information, useful as it may be, gives us little insight into one of the most fascinating chapters in human history.

There is something about these tiny lenses that surprises and captivates us, though no expert on optics and instruments seems able to explain it. How is it possible that, of the many secrets guarded by European courts between the late sixteenth and early seventeenth centuries, such an ordinary and essentially ugly object so radically transformed the world? Amid the countless physicians hawking miraculous formulas and ointments, astrologers compiling almanacs and horoscopes, alchemists seeking a philosopher's stone of some sort, artisans, mechanics, and self-styled inventors of the oddest and most peculiar things, the secret of the *occhiale*—the spyglass—must have seemed like one of the countless creations designed specifically to wrangle privileges and benefits: just another way to earn a bit of money and make ends meet.

To learn more we must thus be willing to read this story, the story of the telescope, with new eyes and try to recount it with other languages that

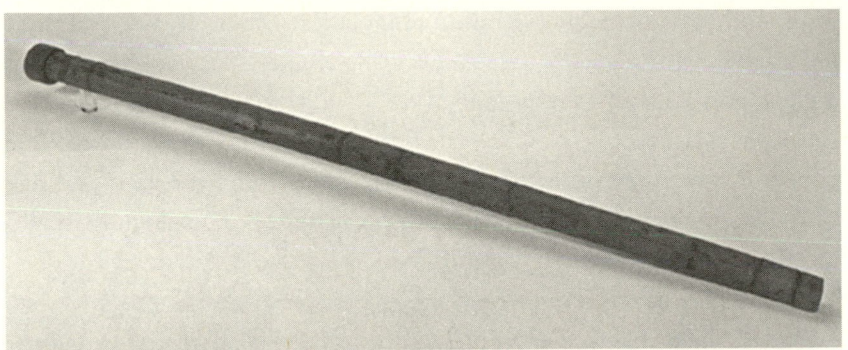

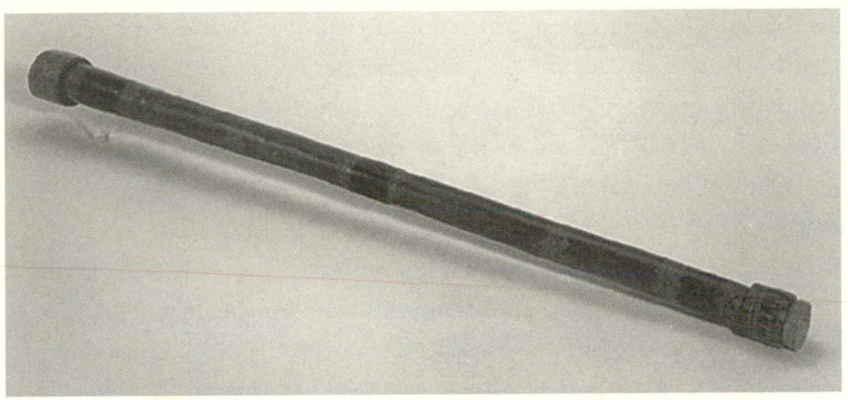

Figures 3 and 4. Galileo's telescopes, 1609–1610

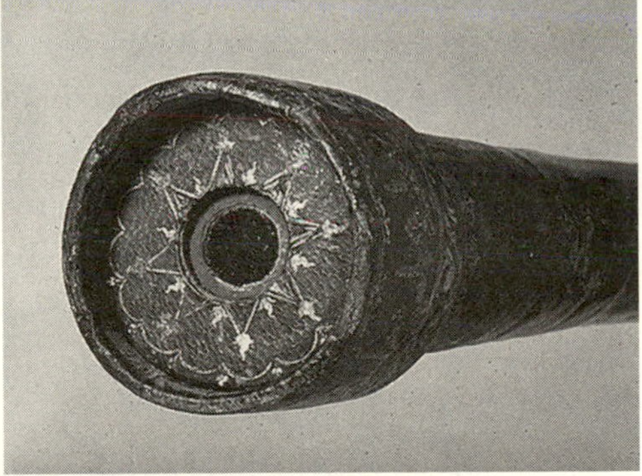

Figures 5 and 6. Galileo's telescope: eyepiece and objective lens, 1609–1610

go beyond those of technology and the applied sciences. We need far more impure eyes and languages once we decide to set our "most excellent" lenses in the context of their own space and time, within broader realms than those generally ascribed to them. Here it is pointless to try to lay down clear boundaries between different areas of knowledge, and the clear and distinct ideas of the present day ultimately prove to be poor and empty shells.

• •

This is a crime story. Everything began between 1608 and 1610, and then exploded when the sky, so familiar until then, was shattered, when the heavens contemplated by Homer and Ovid, Aristotle and Ptolemy, Dante and Thomas Aquinas were wiped out forever. It is the story of a journey that starts in the middle of 1608 in the Netherlands with the invention of an instrument able to make distant things seem close, winds its way across Europe, and ends in Rome in the spring of 1611, when the Jesuit mathematicians of the Collegio Romano acknowledged that Galileo's telescopic discoveries were well grounded. In between lies the transformation of a toy, a Dutch spyglass that could magnify slightly more than three times, into an astronomical instrument. And therein lies the ensuing discovery of a new sky, announced on March 13, 1610, through the publication of a short but immensely powerful book: the *Sidereus nuncius*, the sidereal message, the starry message or messenger (*nuncius* means both "message" and "messenger").

Couched in these terms, however, the story seems far too simple, unfolding so clearly that it would not be worth exploring by either writer or reader. Yet it has often been recounted in precisely these terms: that, of course, there was initially some resistance (but minimal), as is always the case with any valuable innovation; that everything went smoothly; and that those extraordinary discoveries were universally recognized and admired.

But this was not quite the case. It takes little to cast doubt on this seemingly straightforward story—complete with a happy ending—and shatter its entire construction. Suffice it to recall a detail often overlooked. During Galileo's triumphal journey to Rome on May 17, 1611, the cardinals of the Holy Office ordered Padua to inform them if the scientist was somehow involved in the trial of the philosopher Cesare Cremonini, a professor at the university where Galileo had taught for nearly twenty years, who had been accused of denying the immortality of the soul. This was the first time Galileo was mentioned by the Roman Inquisition, and it came at the height of his fame for his astronomical discoveries, well before the start of the exegetic-scriptural battle that led to the Index's anti-Copernican decree of March 5, 1616.

Why was this information requested? Was Galileo's Roman sojourn really as successful as he hastened to tell the grand duke upon his return? It is in the light (and reevaluation) of documents such as this one that Galileo's celestial novelties acquire far different weight. Their impact was much more sudden than we might imagine, just as the discussions and debates that broke out after March 13, 1610, have a much deeper meaning than may appear.

And then, of course, we come to the subtitle, the fact that it is a European story. In fact, if we hadn't been afraid that it might seem like an exaggeration, we would have added "Another World" to the title, though "world" would be so much more than a synonym for the sky. For Galileo, the discovery of a new sky immediately triggered a conflict going beyond the boundaries of astronomy and cosmology. The use of a telescope would lead him not only to reform astronomy but also to establish a new philosophy destined to overturn the traditional relationship between man and nature (and thus between man and God). By abandoning all teleological and anthropocentric views, he distanced himself not only from Copernicus but also from Kepler. Indeed, 1610 is much more than year zero of telescopic astronomy. It is the starting point of a revolution that is at once cosmological and anthropological. The sixteenth century—the long European sixteenth century—did not end with the year 1600, in which Giordano Bruno was burned at the stake, but with 1610 and the *Sidereus*. It was through the telescopic discovery of a new sky that Galileo revealed another world, and this is what so terrified a poet such as John Donne and a theologian as eminent as Bellarmine.

* *

It all began exactly four centuries ago. This is an enormous span of time, above all for the young people grappling with that broken lens in the

Florentine museum. It was then that, for the very first time, we realized the heavens were no longer what they appeared to be, nor what we could see with our eyes; that the word "reality" no longer corresponded to something in front of us, something we always thought we knew. For the first time, we realized that we cannot discover the truth about nature through the senses alone. We grasped that there is another world beyond what had appeared to us for two thousand years (and soon, with the invention of the microscope, we would learn of yet another world beyond what we can touch).

The act of seeing was itself being redefined and no longer coincided with our natural organ of sight. For this new way of seeing to be transformed into an act of knowing, however, observational techniques in astronomy also had to change, for gazing at heavenly bodies as we had always done no longer sufficed. Far more than in the past, it became necessary never to lose sight of those luminous points, never to stop observing them, day in and day out. To understand them, it was necessary to learn how to recount the sky.

Yes, perhaps these are the right words, the right caption, to set under our little piece of cracked glass in order to restore the living splendor of a great thing. But even if this were the case, these words alone cannot fully reveal the secret of the spyglass: namely, that it inaugurated a new way of seeing. To grasp this watershed in human history, there is something we must do first. We must take our new artificial eyes, those minuscule, transparent lenses inserted at the ends of wooden or cardboard tubes, and place them in their original location, in Galileo's workshop in Padua, filled with his mathematical, mechanical, and military instruments. We need to set them on his worktable with its penetrating smell of glue and putty; amid wood shavings and glass dust; alongside wooden and brass compasses, armed magnets, sheets of glass and crystal to be cut and ground, prepared wooden cylinders already drilled at the ends, and pieces of wet cardboard left in a corner to dry.

Then we must try to understand the man, Galileo. At a certain point in his life, why did he feel compelled to throw himself heart and soul into an undertaking that few were able to comprehend? Why did he decide to interrupt a project that, while daring in its own right, happened to be logical—writing a Copernican work starting with the examination of familiar phenomena such as the tides—and instead invest all his energy into making an instrument that might even prove to be inadequate, given the enormous distances between Earth and the heavenly bodies, and which would contribute little to improving the concrete work of the astronomers calculating ephemerides?

In those years virtually no one questioned the ontological difference between heaven and earth. This was taught in the university classrooms of northern Europe as well as of Italy and Spain, and was unconditionally accepted by the lower classes. The difference between Earth and the heavenly bodies was an absolute truth for astrologers and astronomers, theologians and philosophers of every ilk and school, and both high and low culture. It was also the first crucial objection raised against anyone who wanted to follow in Copernicus's footsteps and insisted on tossing the Earth up amid the heavenly bodies. It was a theological and philosophical objection, but above all it was a consideration based on common sense. Thinking of Earth as a heavenly body would destroy not only the image of nature and the cosmos but also the accepted and shared way of thinking.

Starting in the spring of 1609, when the Dutch spyglass began to circulate, Galileo's project for Copernican physics and cosmology reached a radical turning point. It unexpectedly entered a new phase, as he oriented his research toward observing the sky rather than attempting to explain terrestrial phenomena such as the tides. If one could use a powerful spyglass to observe the surface of the Moon and compare it with Earth's, then what would prevent someone from considering Earth to be a heavenly body like the Moon? If the Moon turned out to be covered with mountains, just like Earth, a millenary representation of the sky would be shattered. And this would occur not in the realm of abstract principles, in positing a new and different world system, but in the far more tangible and substantial realm of facts. These were the very facts that forty years earlier had allowed Tycho Brahe to demonstrate the fluidity of the heavens and the presence of new stars in the firmament. The challenge would no longer lie between books and books, between commentators and commentators; it would no longer be an exercise conducted behind the closed doors of university classrooms. No, it would be between philosophers *in libris* and new producers of knowledge, who were at once mathematicians *and* philosophers, but at the same time also the skilled builders of new objects: men capable of working not only with their minds but also with their hands.

• •

At this point, we might be tempted to think that our picture is almost complete; its contours have been sketched out and we need only begin. But this is not actually the case, because the birth of a new sky, and of the ensuing relationships between humans, God, and nature, is much more complex than it seems. In our attempt to understand one of the watersheds in the transition to modernity—along with the discovery of America—we cannot merely study Galileo's most famous astronomical work as if it were an island and there were nothing else left to say. It takes much more to understand the importance of those spare little pages. If this were the aim of our book, then why bother, given that it has already been done countless times? Wouldn't that be a pointless rhetorical exercise? A repetition of universally known words and events?

Our starting point was different. This book was inspired by the conviction that often it is places alone that can tell stories properly and that, as a result, this extraordinarily renowned book—the *Sidereus nuncius*—can be examined differently. We might even be tempted to eliminate the adverb "often" and say that if we do not start where men have acted and interacted, where certain objects were juxtaposed with others, and where particular words and ideas were uttered or developed, then we will grasp little of this story. That said, we have no intention of venturing into abstract and often futile discussions on historical method. We are not interested in doing that, because there are always cases that cannot be reduced to any kind of theoretical framework.

In short, what worked for us, for our project, was the idea that places are archives of the truth they have witnessed. It was only by starting with the close relationship between these new, far-seeing objects and the different places in which they were crafted, donated, purchased, compared, and coveted that we have been able to reconstruct such an important part of European history.

There is no such thing as *the* history of the telescope unfolding vertically around a single figure at the top. But there is one punctuated by countless horizontal histories that intersect in a number of places and that have never been fully recounted before. This is why the book you're reading drew on various strengths and areas of expertise. We felt that the best way to voice this synchronic spatial dimension was to let the many (and often divergent) ways in which Galileo's contemporaries viewed his telescopic undertaking speak for themselves: the various ways in which this new object was grasped, used, and reinvented as it changed hands and cities.

The reader will travel with us across Europe—this is our invitation—because our pages focus not only on Galileo but also on a far more irregular universe, punctuated with words, images, and objects pertaining to different worlds and cultures. This is why Galileo's point of view alone cannot suffice. Nor was his way of thinking about the universe enough to give us insight into what actually happened during those extraordinary years. His projects tied to perfecting the spyglass and constructing a heliocentric cosmology were thus interwoven with those of other men who viewed things differently, and they spawned the enthralling social history of one of the most important scientific discoveries of modernity, outside of which Galileo's own story would be impoverished and perhaps not fully understood.

Recounting the birth of a new sky meant that we had to immerse ourselves in a broader spatial perspective than has been done so far. Everything unfolded in Europe, as we have noted, but it ultimately spread as far as India and China, and the leading players in this story—alongside Galileo—were not only mathematicians, astronomers, philosophers, and theologians such as Sarpi, Kepler, Harriot, Bellarmine, and Magini, but also princes and sovereigns such as Rudolph II, Henry IV, and James I, along with ambassadors, papal nuncios, court dignitaries, artisans, travelers, cardinals including Scipione Borghese and Federico Borromeo, poets and writers of the caliber of John Donne, and painters such as Jan Brueghel, Peter Paul Rubens, and Ludovico Cigoli. And alongside them there was a world of *botteghe* and workshops, post houses and seaports, court banquets and assemblies of cardinals, in which different bodies of knowledge converged and clashed, and where certain places, with their boundaries and religious beliefs, played a critical role in the circulation and even the production of knowledge.

It was in places such as these that the news of first the Dutch spyglass and then Galileo's telescope spread like wildfire, and through the most disparate and unusual channels. Ambassadors' dispatches, travel diaries, and secondhand information in private letters wrought a climate of feverish curiosity, so by the summer of 1609 demand was soaring in every class of European society for this strange object that, before being referred to as a spyglass or telescope, was christened with an array of highly imaginative names: *fistula dioptrica* and *perspicillum* in Latin; *lunettes* in French; *instrument om verre te sien* in Dutch; *cylinder, perspective cylinder,* and *perspective trunckes* in English; *cannone, cannone dalla vista lunga, trombetta, visorio,* and, in some cases, simply *occhiale grande, occhialone,* or *occhiale di canna* in Italian. But some called it the *occhiale di Galileo,* almost as if it had its own brand name.

This may be the first case in which a secret—and a scientific invention—immediately gained such vast public and international prominence that it crossed every geographical barrier, was projected throughout the known world, and did not merely remain within the confines of the Republic of Letters. It is also because of this that our book can be defined as belonging to the social history of science.

• •

We must nevertheless observe that the Galileo who emerges from these pages is quite different from the one portrayed in *Galileo, Courtier,* by Mario Biagioli (1993), a book indubitably packed with information and original ideas but in our opinion unsupported in its underlying thesis. This is not only because we are convinced it would be simplistic to interpret Galileo's Copernican commitment and his proposal for a new philosophy of nature as opportunism, dictated by the cultural system of patronage and his desire to attain a social and professional standing as a philosopher and mathematician. It is above all because we think it would be misleading to assume that his undertaking was centered solely in Florence, as if the time he spent in Padua—nearly eighteen years—were a random digression from the long-coveted climb up the social ladder unexpectedly offered to him by the Dutch spyglass. In fact, according to Biagioli's recent *Galileo's Instruments of Credit* (2006), Galileo's astronomical discoveries were merely the chance to construct his own scientific authoritativeness and personal credibility, thanks to an astute plan to maximize his credit by dedicating the satellites to the Medici so he could acquire a fully legitimate social status.

We are instead convinced that it all began *before* that, in Padua and Venice, where Galileo's endorsement of the Copernican cosmology was consolidated, and where his close rapport with Sarpi and the latter's circle played a decisive role. The watershed was not Florence but Padua and Venice in the years 1609–1610. And it was the telescope, the arduous construction of this new instrument, that transformed what until then had been his "silent Copernicanism" into a great and daring public project. This wholly unexpected path would forever change his life.

⁂

As we worked on our book, there naturally emerged elements that demanded further investigation, leading us to probe things in greater detail and trace such analytical paths that maps and chronologies ended up being more and more important. No longer tools intended to supplement the text, they became an integral part of this work, essential not only in order to understand how the individual circumstances unfolded but also to revisit books whose drafting cannot be separated from a dense network of spatial relationships.

Readers will thus find we have taken an approach to the history of science that avoids the contrast between intellectual history that privileges the world of ideas (scientific, philosophical, and theological alike) and another history played out in a political and social context. This, in any case, has been our intent. Likewise, the unique and concrete nature of the thousands of threads running through these stories made it impossible to take an overly rigid approach and belied historiographical theories that, intriguing though they may be, have ultimately proven to be sterile and baseless. In attempting to understand how certain theories arose and were interpreted, we felt the need to be in several places at once. Allowing these places to speak meant giving voice to the tangible actions of men and ideas in the realms of politics, literature, and art. What has emerged is an integrated history of the telescope composed of faint and sporadic traces, but it is a very rich and multiform history. It is a reality composed of many elements that do not always entwine to form a common thread but that, in some cases, end up overlapping chaotically and, in turn, become the center and origin of something else: a wonderful challenge for those of our métier.

Figure 7. Jan Brueghel the Elder,
Landscape with View of the Castle of Mariemont, 1611

ONE
From the Low Countries

This is the very first image of a Dutch spyglass (Figure 7 and Plate 1).[1] Despite the fact that the scene is dominated by the lush landscape around the castle, the most striking aspect is the detail with the instrument. Although at first glance it seems to hold little significance, it is impossible not to notice it, above all when we close in on the figures at the lower left to examine the man holding it up to his eye. Along with the instrument, he is the true focal point of the painting (Figure 8).

The man with the spyglass is Archduke Albert of Austria, brother of Emperor Rudolf II, and the artist who portrayed him in this pose was Jan Brueghel the Elder, nicknamed "Velvet Brueghel." The castle is Mariemont, a few miles south of Brussels, boasting a park full of luxuriant flora and wild animals, crossed by brooks and dotted with impressive fountains.

The instant we understand the focus of this image, the whole picture emerges in a different and unexpected light. The landscape simply becomes an evocative setting observed by the protagonist, yet this is not a natural gaze of the naked eye. In an interplay of allusions, the painting also flaunts a status symbol, the unequivocal sign of power, conveyed by the fact that the archduke possesses the latest technological invention.

Brueghel depicted the castle of Mariemont a number of times, and the spyglass also appears in several of his paintings, often executed in collaboration with Peter Paul Rubens, such as the famous *Allegory of Sight* (1617), now at the Prado in Madrid (Plate 2).

But this is the only work in which Mariemont and the spyglass are depicted together. It was executed between 1609 and 1612, shortly after a twelve-year truce was reached between the southern provinces, controlled by the Habsburg dynasty, and the rebellious provinces of the north. Starting in 1598, Albert and his consort, Infanta Isabella of Spain and Portugal, ruled

Figure 8. Jan Brueghel the Elder, *Landscape with View of the Castle of Mariemont,* 1611, detail

over all the Low Countries, although this sovereignty was effectively limited to the Belgian provinces. That truce had been pursued zealously, as it acknowledged the independence of the United Provinces but also put an end to the religious and civil wars that had lacerated the Low Countries for over forty years.

Taking advantage of the truce, the archdukes wanted to bring new prosperity to their territories, rebuild the cities that were in ruins, restore dignity to the churches devastated by iconoclasm, and renovate the palaces in which they lived. These works reflected their status and prestige as sovereigns, and were aimed at reviving the lost splendor and ancient grandeur of the Burgundian court.[2] This was certainly an ambitious plan, and to implement it they decided to call in artists and architects of the caliber of Otto Van Veen, Wenceslas Coebergher, Brueghel, and Rubens.[3]

The royal palace in Brussels was renovated, and one of its wings was completely rebuilt and filled with lavish tapestries. Then it was the turn of the castle of Tervuren, the archdukes' hunting lodge, followed by their summer residence, Mariemont. Albert and Isabella transformed the three residences into microcosms, a living encyclopedia, decorating them with hundreds of paintings by Flemish artists as well as Italian masters, and filling them with rare plants, precious stones, and exotic animals. The vast collection of scientific instruments was equally important, with priceless astrolabes, solar quadrants, theodolites, and celestial globes, all of which were exquisitely crafted at the workshops of Michel Coignet in Antwerp and Juan Cocart in Madrid.[4] These were tangible signs of a court that expressed its political power in part through the magnificence of its venues and the furnishings chosen to adorn them. Just as we now associate the name of Louis XIV with Versailles, the people of the era connected Albert and Isabella with their castles in Brussels, Tervuren, and Mariemont.[5]

Brueghel's painting, now in the United States at the Virginia Museum of Fine Arts in Richmond, unquestionably reflects this cultural policy. The painting shows the castle of Mariemont as it appeared between 1608 and 1611, by which time most of the initial expansion and renovation work had been completed.[6] To the right of the main palace we can clearly see the new ramparts and the red-brick clock tower, built between 1608 and 1610 (Plate 1).[7]

Moreover, Brueghel, appointed court painter in 1609, reproduced this very image in two other paintings, one now at the Prado in Madrid and datable around 1611–1612, and the other at the Musée des Beaux-Arts in Dijon, signed by the author and dated 1612.[8]

The castle of Mariemont became one of the archdukes' favorite residences the moment they set foot in the Low Countries, and their court was officially headquartered there in the summer.[9] When the truce was signed with the United Provinces in 1609, Mariemont became the symbol of renewed peace and the legitimate sovereignty of Albert and Isabella over the Low Countries of the south. In effect, Brueghel's paintings so masterfully illustrated their full splendor that one of them—now at the Prado—was sent to the king of Spain for this very purpose.[10]

Yet there is something unique about our painting: the spyglass through which the archduke observes the birds flitting around the park. Albert obtained it at the end of March 1609, just a few days before the truce was formally signed (April 9 of that year). It was during the decisive stages of these negotiations—the last week of September 1608—that the spyglass officially debuted at The Hague, the headquarters of the States-General of the Republic of the United Provinces. In a certain sense, this too was a political symbol, set within the context of the grueling and stormy negotiations that would culminate in a temporary agreement.

Underlying the painting is the troubled management of the truce, yet the artwork's light colors and orderly presentation of objects intentionally convey the archduke's palpable aura of serenity while, out for a stroll, he enjoys the scenery using an instrument from the place where the negotiations had been held. Few people had such an instrument at their disposal in March 1609. Yet just a few months later, these odd new spectacles reached Europe's most important cities and, one way or another, were tied to what would occur in The Hague in September and October 1608.

• •

The history of the Dutch spyglass clearly went through Mariemont. This story has been narrated time and again, but it has always been sketched out in the vaguest of terms, so it merits retelling. Here we will investigate some of the key details that, for some reason, have been overlooked, but we will also probe others that have never been discussed before.

Toward the end of September 1608, Hans Lipperhey, a German-born spectacle maker, set out for The Hague from his adopted home, Middelburg, the capital of Zeeland, which was one of the seven United Provinces of the Low Countries. Lipperhey hoped to be granted an audience with Count Maurice of Nassau, stadtholder and commander of the armed forces of the United Provinces, and one of Europe's most adroit military strategists.[11] He brought with him a letter of presentation from the councilors of his province, dated September 25, 1608, saying that he wanted to show Maurice a new invention: "a certain device through which all things at a very great distance can be seen as if they were nearby."[12]

The Middelburg spectacle maker reached The Hague at the height of a political crisis that had yet to be resolved. A terrible clash was unfolding, sparked not only by the Spanish Habsburgs' acknowledgment of the independence of the United Provinces but also by the right of the latter to continue trade with the East Indies. This was a crucial point, given the huge financial interests involved. It was so important, in fact, that it led to a breakdown in negotiations at the end of August 1608. In effect, just a few weeks later the United Provinces proved they had no intention of relinquishing their contacts with the Far East when Maurice of Nassau received two Siamese ambassadors on their first visit to a European country.[13] These emissaries arrived with the retinue of Admiral Corneille Matelief, who had left three years earlier at the helm of a fleet of eleven ships owned by the Dutch East India Company, with the aim of establishing commercial relations with China. Since the admiral's mission had ultimately failed, however, the United Provinces hoped to achieve their goal through the good offices of the king of Siam.[14]

It was during this course of events that Lipperhey traveled to The Hague and entered its difficult climate, bogged down by the hustle and bustle of delegates participating in the umpteenth series of negotiations in an attempt to resolve this impasse. The United Provinces were represented by Maurice of Nassau and Grand Pensionary Johan Van Oldenbarneveldt, and the archdukes by the Genoese Ambrogio Spinola, who had led the Spanish Armada in the Southern Low Countries since 1605. Other leading European diplomats were also seated around the negotiating table, including Pierre Jeannin, the special envoy of Henry IV, and the ambassadors of James I, Sir Ralph Winwood and Sir Richard Spenser.

Discussions revolved around the proposal advanced by England and, above all, France to stipulate an extended truce with Brussels.[15] News about the developments of these complex negotiations reached the main European courts through the usual confidential diplomatic channels—in other words, dispatches and reports sent by ambassadors stationed in Holland. At the same time, however, the news began to circulate more widely thanks to handwritten notices and the first printed folios, which arose in this very period.

One of these is especially important to our story. This document is just twelve pages long and does not cite the printing venue; its title—in ungrammatical French—reads *Ambassades du Roy de Siam envoyé à l'Excellence du Prince Maurice, arrivé à la Haye le 10. Septemb. 1608*. Even though the only indication on the title page is a generic "L'an de grace 1608," it is very likely that it was published in October of that year.

The anonymous author, indubitably a partisan from the United Provinces, gives us a rather singular opinion of the negotiations. He does not refer directly to the difficulties and perplexities the motion for a truce raised among delegates. Instead, he was struck by two episodes that, during the days so decisive for the fate of the Low Countries, may have seemed irrelevant or marginal: the visit of the ambassadors of the king of Siam and the presentation of Lipperhey's "new invention" to Maurice of Nassau.

This well-informed chronicler can be credited with the vivid account of the curious outcome of the Middelburg spectacle maker's journey to The Hague. This is how things went in his reconstruction.

> A few days before the departure of Spinola from The Hague, a spectacle maker from Middelburg, a humble, very religious and God-fearing man, presented to His Excellency [Prince Maurice] certain glasses by means of which one can detect and see distinctly things three or four miles removed

> from us as if we were seeing them from a hundred paces. From the tower of The Hague, one clearly sees, with the said glasses, the clock of Delft and the windows of the church of Leiden, despite the fact that these cities are distant from The Hague one and a half and three and a half hours by road, respectively. When the States[-General] heard about them, they asked His Excellency to see them, and he sent them to them, saying that with these glasses they would see the tricks of the enemy. Spinola too saw them with great amazement and said to Prince Henry [Frederick Henry of Nassau, Maurice's younger brother]: "From now on I could no longer be safe, for you will see me from afar." To which the said prince replied: "We shall forbid our men to shoot at you." The master [spectacle] maker of the said glasses was given three hundred guilders, and was promised more for making others, with the command not to teach the said art to anyone. This he promised willingly, not wishing that the enemies would be able to avail themselves of them against us.[16]

The events cited in this report spread across Europe with astonishing speed, and at virtually the same time the instrument became an overnight success. But why did Lipperhey's glasses create such a stir? Obviously, the novelty was not glasses used to correct poor vision. After all, spectacles fitted with convex lenses for presbyopia and concave lenses for myopia had been around for centuries.[17] They were used regularly by this time, and it is unlikely that they would have been newsworthy in 1608. No, they attracted attention for another reason. What was so intriguing was the way the Middelburg craftsman had put these two types of lenses together to build a satisfactory magnifying instrument.

Yet there was little new about this either. Efforts to understand how a combination of two lenses, or of a lens and a mirror, could make distant objects seem larger went back to the second half of the sixteenth century. It was not a secret; in fact, the practical effect was rather well known in various parts of Europe. There were numerous allusions to it in the literature of the era, in which there was also fanciful speculation on legendary and "telescopic" burning glasses such as those of Archimedes and the Lighthouse of Alexandria.[18] One had merely to read the second edition of Giovan Battista Della Porta's *Magia naturalis* (1589), a very popular scientific

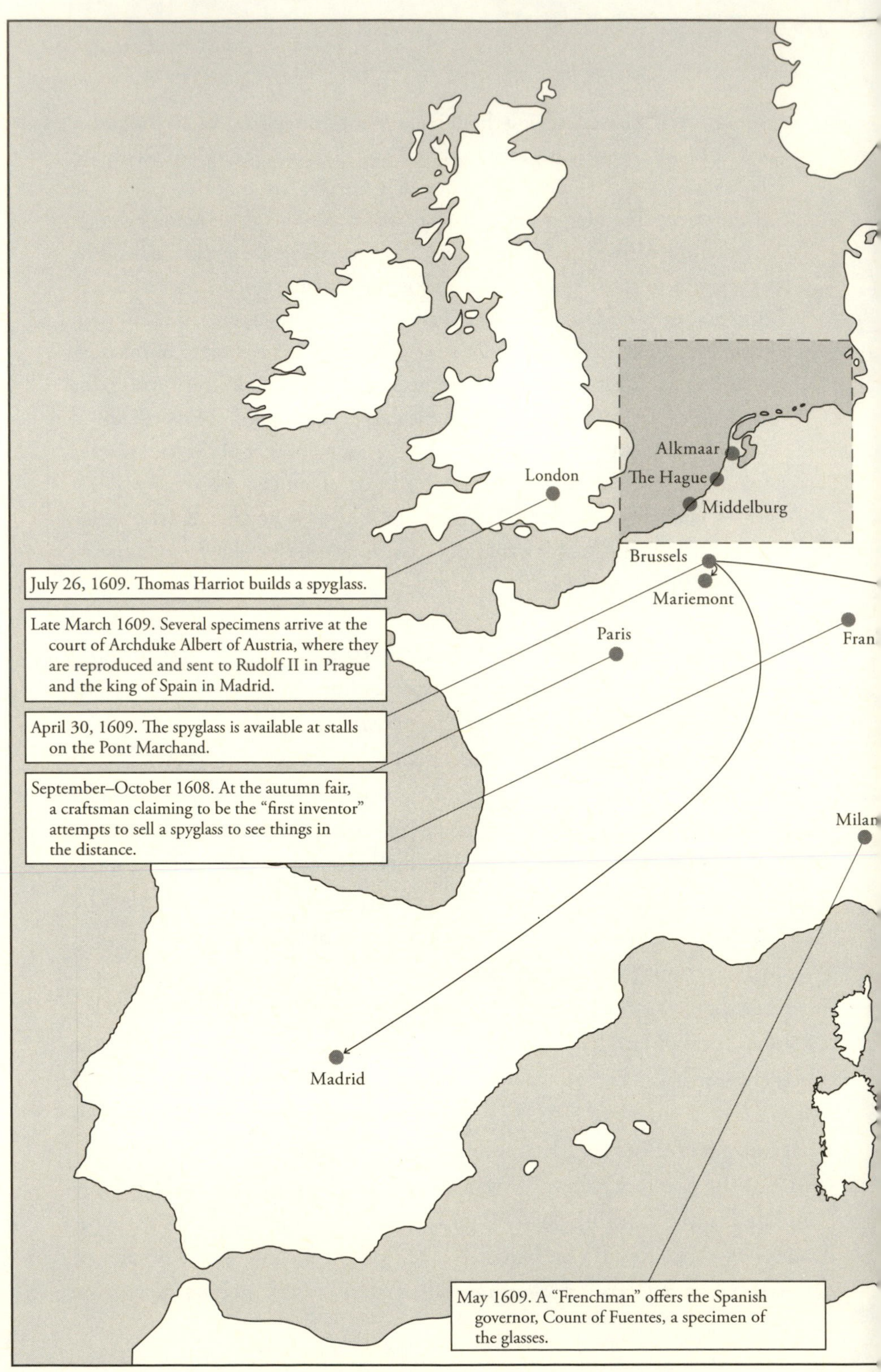

Alkmaar
London
The Hague
Middelburg
Brussels
July 26, 1609. Thomas Harriot builds a spyglass.
Mariemont
Late March 1609. Several specimens arrive at the court of Archduke Albert of Austria, where they are reproduced and sent to Rudolf II in Prague and the king of Spain in Madrid.
Paris
April 30, 1609. The spyglass is available at stalls on the Pont Marchand.
September–October 1608. At the autumn fair, a craftsman claiming to be the "first inventor" attempts to sell a spyglass to see things in the distance.
Madrid
May 1609. A "Frenchman" offers the Spanish governor, Count of Fuentes, a specimen of the glasses.

Map 1. Circulation of the Dutch spyglass: September 1608–August 1609

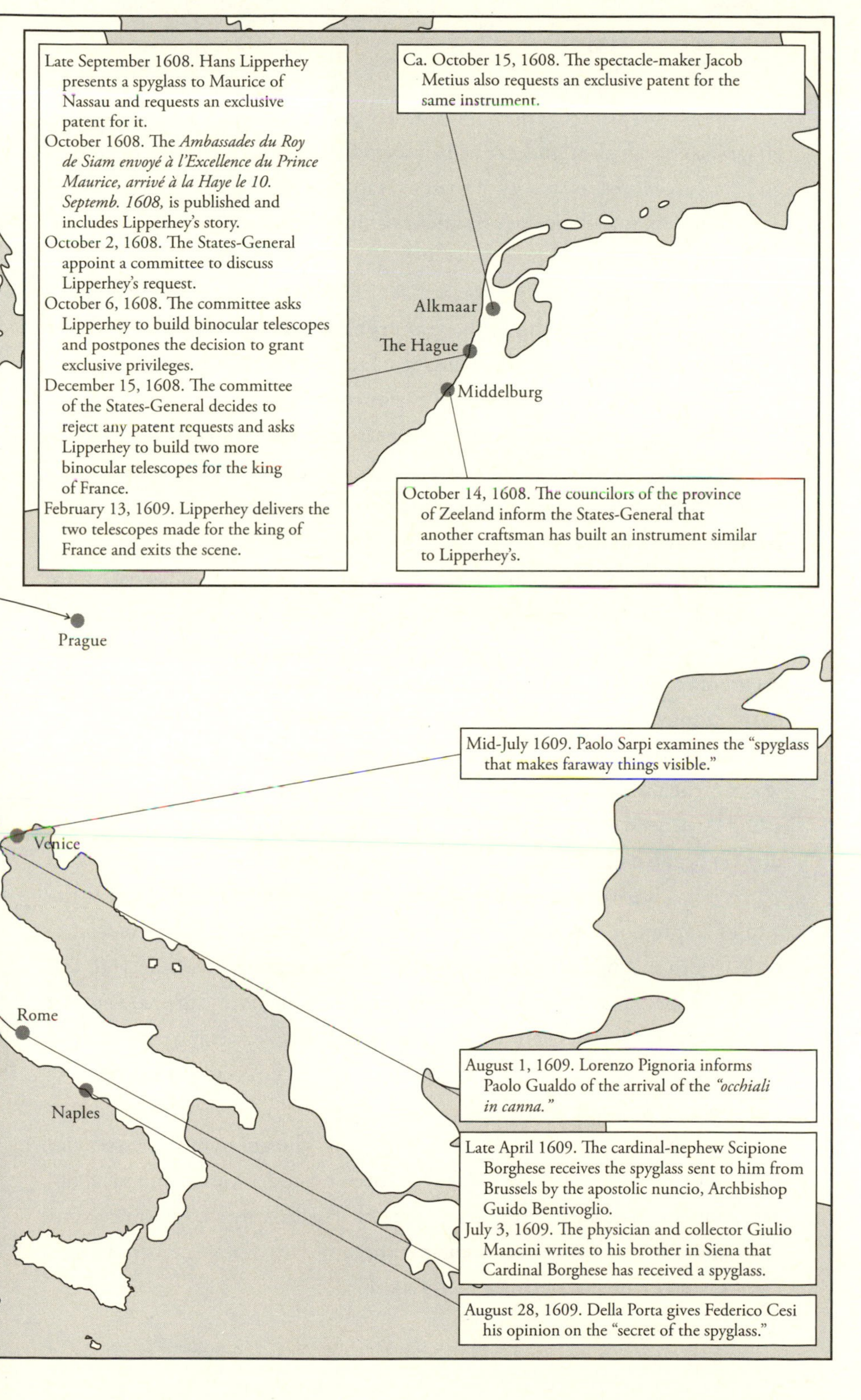

Late September 1608. Hans Lipperhey presents a spyglass to Maurice of Nassau and requests an exclusive patent for it.
October 1608. The *Ambassades du Roy de Siam envoyé à l'Excellence du Prince Maurice, arrivé à la Haye le 10. Septemb. 1608,* is published and includes Lipperhey's story.
October 2, 1608. The States-General appoint a committee to discuss Lipperhey's request.
October 6, 1608. The committee asks Lipperhey to build binocular telescopes and postpones the decision to grant exclusive privileges.
December 15, 1608. The committee of the States-General decides to reject any patent requests and asks Lipperhey to build two more binocular telescopes for the king of France.
February 13, 1609. Lipperhey delivers the two telescopes made for the king of France and exits the scene.
Ca. October 15, 1608. The spectacle-maker Jacob Metius also requests an exclusive patent for the same instrument.
Alkmaar
The Hague
Middelburg
October 14, 1608. The councilors of the province of Zeeland inform the States-General that another craftsman has built an instrument similar to Lipperhey's.
Prague
Mid-July 1609. Paolo Sarpi examines the "spyglass that makes faraway things visible."
Venice
Rome
Naples
August 1, 1609. Lorenzo Pignoria informs Paolo Gualdo of the arrival of the *"occhiali in canna."*
Late April 1609. The cardinal-nephew Scipione Borghese receives the spyglass sent to him from Brussels by the apostolic nuncio, Archbishop Guido Bentivoglio.
July 3, 1609. The physician and collector Giulio Mancini writes to his brother in Siena that Cardinal Borghese has received a spyglass.
August 28, 1609. Della Porta gives Federico Cesi his opinion on the "secret of the spyglass."

encyclopedia reprinted a number of times and translated into various languages, to find a clear and explicit description of the properties of convex and concave lenses, as well as of their extraordinary magnification power.[19]

Therefore, it took no special talent or unique inventiveness to come up with the idea that combining two different lenses—for example, inserted at the ends of a tube—would create a device allowing people to see faraway objects enlarged. If anything, the real problem was that this combination of lenses failed to produce sharp images. Despite the fact that the progress made in the production of glass and the introduction of new grinding techniques had improved the sphericity of concave or convex surfaces, lens quality still left much to be desired.[20] As a result, the game was played out on a material level: transforming a widely known idea into an instrument that could perform well.

Therein lay the secret to Lipperhey's success. Although his lenses were no better than those of other craftsmen, he must have used an approach that allowed him to remedy the problem of the lenses' poor quality. Although no proof remains today, the most plausible theory is that he applied a paper diaphragm with a small aperture in front of the lenses, thereby mitigating the optical distortions caused by inadequate grinding of their edges.[21] It was a rather simple expedient—in fact, it was promptly noted and understood by others, including Galileo—thanks to which the images of distant objects became much clearer.[22]

When Lipperhey showed his instrument to Maurice of Nassau at the end of September 1608, the innovation was not the concept of a system of lenses to magnify things, but the fact that it actually worked. This was the sensational news that immediately spread throughout Europe. And the *Ambassades du Roy de Siam* simply announced this, disclosing that some of Europe's most skilled men-at-arms had witnessed the extraordinary results of Lipperhey's instrument.

There was more. With growing emphasis, the anonymous chronicler seemed to suggest that these results had been extended to astronomical observation, reporting that "even the stars which ordinarily are invisible to our sight and our eyes, because of their smallness and the weakness of our sight, can be seen by means of this instrument."[23]

Aside from this vague and minimal reference, there are no other documents attesting to the use of Lipperhey's spyglass for astronomical purposes. In short, we do not know if someone in Holland turned it skyward to discern new things or details invisible to the naked eye. In any case, it is highly unlikely that such a rudimentary instrument would have been up to the task. What is instead certain is that what aroused such enormous interest was its obvious military applications, and it is equally certain that it proved so easy to build that, after merely hearing about it and without even needing to examine it, a number of spectacle makers were able to reproduce it. In fact, within nine months the spyglass could be purchased in all the most important European cities.

◆ ◆

At this point, we must closely follow the channels through which the news spread, and with it the instrument itself. We need to understand the features of the device Lipperhey brought to The Hague.

Although the *Ambassades* offers no technical details about its structure or magnification, other sources provide a rather accurate idea. For example, we find that the spyglasses Pierre de L'Estoile, auditor of the Royal Chancellery, examined in Paris in April 1609 were "composed of a tube about a foot long" and that "at each end there [was] a glass."[24] In August of that year, Della Porta wrote from Naples to describe "a silver-plated tin tube measuring a *palm*" that had "a convex lens at one end," inserted into which was "another tube measuring *four fingers long*" attached to a lens "concave at the end."[25] This tells us that the first specimens measured a palm and four digits, or about 30 to 35 cm. These figures tally with Johannes Walchius's detailed description in his *Decas fabularum*, published in the fall of 1609, in which we find another important detail: that the tube measured "about three digits" in diameter.[26] Moreover, this also matches what Galileo wrote in the *Sidereus nuncius*, where he recounted that his first spyglass magnified objects three times over.[27] This was the kind of instrument that Lipperhey showed Maurice of Nassau: a spyglass that was not very powerful but was unquestionably useful for sightings on land, as documented by the *Ambassades*.

Lipperhey demonstrated his creation at Maurice of Nassau's palace toward the end of September 1608. At the same time, the first news of the spectacular performance of this new invention reached The Hague, and naturally, its promising military applications did not escape two experts of the caliber of Ambrogio Spinola and Pierre Jeannin. In fact, the two soon procured specimens for their respective sovereigns.

What happened at this point—when Count Maurice realized that Lipperhey's instrument actually worked—clearly emerges from the minutes of the States-General. Everything transpired in the space of a month or so.

On October 2, 1608, discussions were held regarding the spectacle maker's request for an exclusive patent in exchange for his promise to build the instrument solely for the United Provinces. Instead of making an immediate decision, however, the assembly chose to appoint a committee to examine the matter in greater depth and negotiate with the Middelburg craftsman.[28] An agreement was reached on October 6, just four days later. As requested by the committee, Lipperhey pledged to craft an improved version of his device, which had to make it possible to see "with both eyes" and envisioned the use of lenses made of "rock crystal," telling us just how poor the quality of the glass used for his first model must have been. He was given a down payment of 300 guilders and would receive the balance of 600 as soon as he finished the work—but only if it was satisfactory. At that point, the States-General would hand down a decision. In any case, Lipperhey was asked not to disclose his secret.[29]

Within a matter of days, however, the scenario was upended by news that came, yet again, from Middelburg. The town was home to the oldest glassworks of the Northern Low Countries. Founded in 1581 with the help of several Italian craftsmen, it had rapidly become the region's leading glassmaking industry, a prime position it shared with Amsterdam. In the early seventeenth century, these cities had become known for glass of such exceptional quality that it was even exported to Venice.[30] Therefore, it is unsurprising that on October 14, 1608, while Lipperhey's patent application was still pending, the councilors of the province of Zeeland informed the States-General that in Middelburg there was "a young man who says he too knows the art, and who has demonstrated the same thing with a similar instrument." Moreover, they expressed the valid suspicion that there were several around and that "the art could not remain a secret," given how easy it was to reproduce, particularly once "the form of the tube" was seen and it became clear that there was a lens at both ends.[31]

As if this were not enough, the arrival of this information virtually overlapped another patent request submitted to the States-General by a craftsman named Jacob Metius, from the town of Alkmaar in the province of Holland. After two years of studying the effects that could be obtained with "the use of glass," he too stated that he had "invented" an instrument to see distant things "more clearly." He had devoted most of his time to improving it, and once he achieved a satisfactory product, he offered it to Maurice of Nassau, who observed that it was just as sound as the one that had recently been shown to him by the spectacle maker from Middelburg. Consequently, as soon as Metius heard about Lipperhey's patent application, he claimed the priority of his discovery and requested that he not be robbed of the legitimate fruits of his labor.[32]

• •

Less than three months after its official debut, a battle was already raging over paternity of the spyglass. Not two but three claimants asserted that they were the inventors, and two of them—Lipperhey and Metius—even demanded that the States-General give them an exclusive patent. The situation was further complicated by the fact that, outside the boundaries of the Low Countries, there was a "Dutchman" who, according to the testimony of the astronomer Simon Mayr, considered himself the "first inventor" and was trying to sell a spyglass at the autumn fair in Frankfurt, at essentially the same time as Lipperhey was setting out for The Hague.[33]

In short, at the beginning of October 1608, in three different cities well connected with The Hague—Middelburg and Alkmaar are about forty-five miles away, respectively to the south and north, while Frankfurt is just over three hundred miles to the southeast—there were at least four craftsman who owned a spyglass. Regardless of any decision reached by the States-General, there was no doubt about one thing: that there were no longer any secrets about the instrument, as its mechanisms were known by this time and could easily be reproduced.[34]

The minutes from the subsequent sessions of the States-General illustrate how this complicated and somewhat embarrassing situation unfolded.

Metius's case was examined during the meeting of October 17. The delegates decided to reward him with 100 guilders, urging him to "further his invention for seeing far, in order to bring the same to perfection," and promising him that they would hand down a statement regarding his patent application.[35] In other words, aside from the financial aspect, Metius was being treated in exactly the same way as Lipperhey had been around ten days earlier. As to the latter, we can easily imagine that he spent the months of October and November building his binocular spyglass with rock-crystal lenses.

On December 15, 1608, "the instrument for seeing far with both eyes" was on the table of the committee appointed by the States-General. At this point, a decision was finally made: Lipperhey's patent application was rejected because it was obvious that "several others have knowledge of the invention." This provision applied to all other requests, past and future, and put an end once and for all to a situation that had spun out for about three months. Nevertheless, given that his instrument was deemed to be of good quality, Lipperhey was ordered to build two more, for which he received 300 guilders, with the agreement that he would receive the balance of 300, as established, when the work was finished.[36]

We do not know what use the States-General made of Lipperhey's first binocular telescope. We do know, however, why they wanted two more: to give them to a valuable and powerful ally, King Henry IV of France, who, with his minister Jeannin, was playing a decisive role in negotiating the conflict with the Southern Low Countries.[37]

According to records, Lipperhey delivered the two instruments intended for the French king on February 13, 1609. That same day he received the 300 guilders he was owed, but from then on his name fails to appear in any official Dutch document. So we have no way of knowing if he built other spyglasses for private customers—nor, if so, how many—before he passed away in Middelburg later that year, on September 29.

• •

While Jeannin was in a position to request the spyglass directly from the States-General, General Spinola, as delegate for the archdukes and thus the counterpart in the dispute, certainly did not enjoy the same privilege.

We do know, however, that he was still in The Hague when Lipperhey presented his "glasses" to Count Maurice and that he managed to procure an instrument for himself. He left the city on September 30, 1608, and stopped in Brussels, where he announced the invention of the spyglass to at least two high-ranking figures. One was none other than Archduke Albert of Austria, governor of the Spanish Netherlands, a man with a passion for science, and a knowledgeable collector of instruments.[38] The other was Archbishop Guido Bentivoglio, who had arrived in Brussels in the summer of 1607 to serve as apostolic nuncio for Pope Paul V.

As an authoritative representative of the Church of Rome at a Catholic court, Bentivoglio was welcomed into the most exclusive aristocratic circles and was privy to information off-limits to other diplomats.[39] From the moment he arrived, he sent a continuous flurry of dispatches on the difficult political situation in the Low Countries. He did not always write about politics, however, and one letter in particular discusses the spyglass. Sent from Brussels on April 2, 1609, and addressed to the cardinal-nephew Scipione Borghese, it is a valuable document because it tells us how the instrument reached Brussels from the United Provinces. We can read it in its entirety:

> When Marquis Spinola returned from Holland, among other things he reported to the Most Serene Archduke that Count Maurice had an instrument, with which he could see the most distant things as if they were right before his eyes; and the Marquis added that he thought the Count had procured this instrument to see in the distance on the occasion of wars or of strongholds he wished to besiege, or encampments, or enemies marching in the countryside or similar things, which could benefit him. The Archduke and the Marquis himself then had the very great desire to have such an instrument and several came into their hands, although not as fine as Count Maurice's. I recently had the chance to test that of His Royal Highness, as I was with him, and with the Most Serene Infanta, hunting herons outside the gates of Brussels. From there the Tower at Malines, which with the naked eye can barely be discerned, appearing very small and indeed blurred, seen through the aforesaid instrument was revealed with its very

> large and distinct form, and one could make out the orders of the Tower, and the windows with other even smaller parts still under construction. Yet this is a distance of about ten Italian miles.[40]

This means that at the end of March 1609 Albert of Austria owned instruments like the one shown to Maurice of Nassau the previous September. Moreover, it was with one of these very instruments that the archduke had Brueghel portray him in the park of his castle at Mariemont. Shortly thereafter, spyglasses started to be built in the Southern Low Countries as well, as documented by a deed dated May 5, 1609, in which, from Brussels, Albert ordered a payment of 90 guilders "for two artificial tubes to see far away, made for us."[41]

Yet Bentivoglio's letter also tells us more. In it, we learn that the apostolic nuncio found "this instrument quite intriguing," thought it could "provide great enjoyment," and went to great lengths to "procure one." His zeal was motivated not by his own satisfaction but, rather, by the pleasure it might give "to His Holiness [Paul V] and Your Excellency." Consequently, with the same courier he had sent "the instrument to the hands of Lord Entio, my brother, so that he may deliver it to Your Excellency in my name. If the pleasure of Your Excellency is not as great as my opinion of the instrument, I hope that at least my diligence will be appreciated."[42] On April 2, 1609, another spyglass was shipped to Rome, where it was probably delivered toward the end of the month.[43] In any event, it reached the city before May 23, as documented by a draft penned by Scipione Borghese that same day and addressed to Bentivoglio: "The precious gift and the instrument [was] given to me in the name of Your Lordship by Lord Entio, your brother, which I greatly appreciate, as befits the perfection and powerful virtue enclosed in a small body, and the new device unknown to past centuries cannot help but show us infinite marvels. I myself saw the effect, and with the greatest pleasure. Therefore, Your Lordship can believe that not only the gift but also the thought of obtaining it for me has been and is most dear to me. I thank you most sincerely and wish you all happiness."[44]

That the news immediately spread through Rome is confirmed by a letter the doctor and refined art collector Giulio Mancini, future chief physician to Urban VIII, sent his brother Deifebo in Siena on July 3.[45] "A glass arrived here, said to have belonged to Count Maurice, which shows very distant things, they say in Grottaferrata or in Frascati from here in Rome," he wrote. "I have not seen it, but someone who tested it and whose sight is not poor but quite imperfect told me that from Monte Cavallo he saw a friend and recognized him as he went to San Pietro in Montorio and it is a most wondrous thing, especially because from the other end it makes very nearby things most tiny though they are very large. The glass is in the hands of His Excellency Borghese."[46]

As we can see, although less than a year had elapsed since the new spectacles had first appeared, they had already sparked a great deal of interest. We need merely consider the specimens discussed so far to grasp not only their widespread circulation but also their social prestige. At The Hague, the States-General and the head of their armed forces owned more than one; in Paris, two others were in the hands of the king of France; in Brussels, Archduke Albert and General Spinola had managed to procure and then manufacture them on their own; and it is also certain that they were sent to Rudolf II in Prague and the Spanish king in Madrid.[47] Lastly, in Rome one was at the pope's complete disposal.[48]

What we can sketch out here is a map that, aside from its geographical significance, is characterized above all by the political standing of the instrument's recipients. It is a map of those in power, who almost immediately grasped the importance of the new invention and were attracted by its promising possibilities in the political and military arena.

• •

The complicated story of the Dutch spyglass does not end here. There are still two key details to recount, allowing us to leave the Low Countries to go first to Paris and then on to London and Venice.

Jeannin informed the French king of Lipperhey's "new invention" on December 28, 1608, but the *Ambassades du Roy de Siam* had already reached Paris about a month earlier.[49] On November 18, Pierre de L'Estoile noted in his journal that a friend had shown him "two manuscript notices from The Hague, in Holland." The first had been printed in Sedan, while the latter, still in manuscript form and "much shorter and more significant," had to do with "certain glasses presented to Count Maurice." L'Estoile had given a copy of both notices to a printer, who would publish them shortly thereafter.[50] This promptly came to pass just a few days later, as documented on November 23, 1608: "C. B. [Claude Berion, the printer] has delivered four printed copies of the small notices I was shown in manuscript form last Tuesday." He gave copies to two of his friends, kept one for himself, and offered the other to the English ambassador in Paris.[51] This contact was unquestionably one of the ways—perhaps the most important—through which information about the spyglass and its possible applications reached England by the end of the year.[52]

But L'Estoile was not the only one in Paris with a copy of the *Ambassades*. An Italian-born Huguenot by the name of Francesco Castrino read the notice and immediately informed Paolo Sarpi in Venice, who replied that he had already received it in November 1608 directly from The Hague.[53] Five months later, in April 1609, it wasn't just the news that circulated but the instrument itself. It could be found and purchased virtually everywhere, confirming what Jeannin had already grasped: that it was "by no means difficult to imitate this first invention."[54] Again thanks to L'Estoile's testimony, we know that on April 30 a spectacle maker was even selling it on the Pont Marchand.[55] The following month, "a Frenchman" claiming to be a partner of the Dutch inventor "rushed into Milan" to offer just such an instrument to Pedro Enríquez de Acevedo, Count of Fuentes, who was serving as governor and general there.[56] It is possible that this Frenchman was the "ultramontane" who, as the erudite antiquarian Lorenzo Pignoria wrote on August 1 to the clergyman Paolo Gualdo in Rome, had brought to Padua "one of the spyglasses [*occhiali in canna*], about which you have already written to me."[57] This sentence bears emphasizing, because it tells us that a few months earlier Gualdo must have informed Pignoria that the "instrument" Bentivoglio sent Cardinal Scipione Borghese on April 2 had arrived, but also that, by the end of July at the very latest, a similar specimen had made its appearance in Padua.

Yet it was not in Padua that the *occhiali in canna* made their debut in the territories of the Venetian Republic. In mid-July 1609 Sarpi had already had the chance to see them in Venice, almost certainly thanks to the "ultramontane," who, before heading to Padua, had attempted to sell them to the doge as if they were something highly confidential. Yet in other Italian cities the new instrument was neither secret nor new by this time. It could be found in Naples, where Della Porta, writing to Prince Federico Cesi on August 28, 1609, poked fun at it ("I've seen the secret of the spyglass and it's a hoax"), though he had no problem claiming paternity or at least the original concept behind the invention.[58] By this time, it was also available in Rome and Milan.

It had all started about nine months earlier in the small city of The Hague and, through the most diverse channels, had rapidly spread across most of Europe. As a result, the spyglass was easily copied in many places. After creating a six-power instrument, on July 26, 1609, the English astronomer and mathematician Thomas Harriot used it at Syon House, near London, to observe the surface of the Moon.[59] Nevertheless, it was not from Syon House that the spyglass came to the world's attention as the ideal stargazing instrument, but from Venice and Padua. This marked the beginning of yet another season for the Dutch spyglass: one that would prove to be completely new and unpredictable.

TWO
The Venetian Archipelago

The news arrived in autumn—in November 1608, to be precise—and Paolo Sarpi was one of the first to hear it. In his letter of December 9, he wrote to the Huguenot Francesco Castrino: "A month ago, I received from The Hague your lordship's report regarding the embassy sent to Count Maurice by the Hindu king of Siam, and about the new spyglasses made by that gifted man, and this has given me much to ponder. However, because these philosophers instruct us that we should not speculate about the cause before seeing its effects with our own senses, I have resigned myself to waiting for such a noble thing to spread throughout Europe."[1]

The news of the spyglass was already circulating in Europe and certainly would not have gone unnoticed by a man such as Sarpi, who had long cultivated an immense passion for any sort of device and human invention. He did not add anything else beyond "this has given me much to ponder," as he was well aware that it would be premature to embark on gratuitous overthinking without even having laid eyes on the instrument. But on January 6, less than a month later, he returned to the subject:

> When I was young, I thought about such a thing, and it occurred to me that a glass made in the shape of a parabola could produce such an effect. There were demonstrative arguments, but because they are abstract and fail to consider material constraints, I hesitated. So I did not pursue that work much, as it would have been time-consuming and, as a result, I did not confirm or refute my thinking through experience. I don't know if perhaps that artisan has matched my ideas; if, as rumors tend to do, the thing was perhaps magnified along the way.[2]

That device, which for Sarpi was still hearsay, intrigued him and piqued his curiosity. He continued to mull over this news and couldn't wait to examine one of these instruments to probe its secrets. Perhaps, though, he also wanted to see if he might actually have been able to implement his youthful idea about how one or more parabolic lenses could work, an idea he had set aside years earlier but which was now uppermost in his mind once more. At the time, the difficulty of putting his idea into practice had convinced him to let it go, as had been the case with many other of his "thoughts," often mere outlines of experiments that remained on paper but which nevertheless testify to his immeasurable interest in any kind of natural phenomenon.

Unfortunately, the terrible fire of 1769 that destroyed the library of the Venetian convent of Santa Maria dei Servi, where his documents were preserved, has had devastating consequences for our knowledge of Sarpi as a natural philosopher with a keen interest in technology. Of the five tomes holding his writings, only the fifth—with the *Pensieri naturali, metafisici e matematici*—is known. None of the others have survived. The so-called *Schedae Sarpianae*, which contained notes and comments on the period following the *Pensieri* (and which thus would have been very important for us), has been lost.[3] On July 22, 1608, he wrote: "Even before the events in the world beckoned me to think about serious things and not pastimes, about the considerations in which your lordship has seen me immersed, all my tastes were for natural and mathematical things."[4] *This* is the Sarpi who intrigues us: the man who, before becoming *totus historicus* and *politicus*—director and prime actor in the battle against the Papal Interdict (1606–1607) and then the famous condemned author of the *Istoria del Concilio Tridentino* (1619)—tirelessly probed scientific matters and who, in the story of the spyglass, was destined to play a key role that has often been underestimated in studies on the history of the telescope and in some cases overlooked entirely.

Only if we take this premise into account can we grasp the importance of the letter of March 30, 1609, from Sarpi to Giacomo Badoer (Jacques Badouere), a young Venetian-born diplomat in the service of the French king and once Galileo's pupil in Padua. Perhaps Badoer, well aware of the Servite theologian's scientific interests, had turned to him in the hope of obtaining a statement on the merits and quality of the widely discussed

Dutch spyglasses. If he did, his curiosity was indubitably left unsated. None had appeared in Venice so far, and Sarpi himself was anxious for more information ("If you learn anything further, I would like to know what the thinking is there").[5] While waiting to see one, he preferred not to get his hopes up only to have them dashed. He even warned his correspondent not to expect much from him, as he had "almost abandoned the thought of natural and mathematical things, and, truth be told, due to old age or out of habit, [his] brain [had] become somewhat obtuse in such contemplations."[6]

This is a rather odd conclusion, particularly if we consider the fact that just a short time earlier, when compiling a list of some of the figures who best represented the "new century," Sarpi had focused entirely on the world of naturalists and mathematicians. In his letter of June 12, 1608, he wrote to Alvise Lollino, "In this century I have not seen a man who has written something of his own, with the exception of Vieta [François Viète] in France and [William] Gilbert in England."[7] On July 8, he wrote, "I would very much like to have Vieta's *Harmonicon coeleste*."[8] Moreover, just a few years earlier he had extolled Gilbert as a man worthy of "eternal memory."[9] In turn, Gilbert had said how much he praised and admired Sarpi, specifically for his gifts as a scientist and naturalist, but rather than saying this in a private letter, he chose to make it public, including it in the first chapter of Book I of *De Magnete*, his most famous work.[10]

Therefore, it is hard to believe that Sarpi would have been uninterested or turned his back on a discovery that was nothing short of sensational. Nevertheless, several months would pass before he finally offered his opinion, a few days (or at most a few weeks) before July 21, 1609.

• •

That day, Sarpi took quill in hand and wrote to Castrino, whom he regularly informed of the latest news on Venetian and Italian politics. This time, however, he had little to report, except that "a spyglass has arrived that makes faraway things visible. I greatly admire it because of the beauty of the invention and its skillful craftsmanship, but I find it worthless for military uses, either on land or at sea."[11] More than seven months after first hearing about the device, Sarpi was finally in a position to make a statement. In his opinion, the idea was nothing short of marvelous, but its execution was entirely unsatisfactory. The specimen he had had the chance to examine was quite disappointing and unsuited to any military use.

That's all he said. Yet if we take a closer look, his words are quite telling. The letter to Castrino is fundamental for at least two reasons. First of all, we learn that between July 7 (the date of the previous letter) and July 21, Sarpi saw the spyglass for the first time. Second, it gives us grounds to theorize that Sarpi immediately expressed his doubts to Galileo.

Unfortunately, this is an isolated document. It would take an entire month before further information about the spyglass finally reached Venice, in the form of Giovanni Bartoli's letter to Belisario Vinta, dated August 22. We are talking about thirty long days, all of which are crucial for our story. This means that unless new documents are unearthed, any attempt at a detailed reconstruction of the chronology of Galileo's invention is destined to remain in the "domain of pure conjecture."[12] Mario Biagioli's recent effort also—and predictably—leaves a number of unresolved issues.[13] When did Galileo start to work on his first spyglass? Who worked with him? To what extent are the narrative passages in the *Sidereus nuncius* and *Il Saggiatore* aimed at making him appear to be the sole "hero" and "inventor" of the telescope? What—and how much—did he omit? We can offer only vague responses to all of these questions.

There is no doubt that the collaboration of Sarpi and his entourage was decisive, although Galileo never publicly acknowledged this. There are too many elements, albeit seemingly divergent, that point in this direction.[14] Likewise, it seems all too evident that Galileo tried to create a wholly theoretical image of his work based on the "doctrine of refraction," in order to emphasize that his mathematical *inventio* was far removed from that of spectacle makers, who worked on a purely empirical basis.[15]

But let us get back to our story. At the end of July, one of the new *occhiali in canna* also appeared in Padua, brought there by an anonymous "ultramontane."[16] Twenty days later, on August 22, Giovanni Bartoli, secretary to Asdrubale Barbolani, the resident of the Grand Duke of Tuscany in Venice, informed the court that "a person came here who wanted to give his lordship the secret to a spyglass or a cannon or some other instrument, with which one can see even twenty-five and thirty miles away so clearly that they say it seems to be nearby."[17] Moreover, Bartoli reported that "in France and elsewhere this secret is known to all, and that it can be purchased cheaply; and many say that they have seen and obtained it." He then added a detail found in no other document: that many had "seen and tested [the spyglass] from the Campanile of St. Mark's."[18]

In short, in the span of just over two months, thanks to the work of master craftsmen who instantly grasped that there was money to be made, the *occhiali in canna* or *trombette* became an open secret and could be purchased cheaply in many cities, proving not only their widespread circulation but also the poor quality of many specimens.

• •

Let us turn once more to Venice, and specifically to the home of the Tuscan ambassador, which, as is the case with every respectable embassy, was a veritable message factory. Ordinary dispatches, meaning those sent regularly to Florence, tended to go out weekly. As we have seen, on August 22 Bartoli sent one to inform the court that a certain "foreigner" had brought a spyglass to Venice and that it instantly became a public attraction, handed from person to person, who took turns peering at the sea from the city's tallest bell tower. A week later, on August 29, Bartoli sent additional news about it, and his next report was even more richly detailed. For the first time, he mentioned Galileo and, along with him, Sarpi.

> What has given us most cause for discussion this week is Messer Galileo Galilei, mathematician from Padua, with the invention of the spyglass or cannon to see far away. And it is said that a certain foreigner who came here with this secret, having learned from I don't know whom (supposedly the Servite theologian, Brother Paolo) that he would not earn anything here, as he demanded 100 sequins, left without making any further attempts; so that, since Brother Paolo and Galileo were friends, and having explained to him the secret he had seen, they say that Galileo himself, with his mind and with the aid of a similar instrument from France, though not of excellent quality, investigated and found the secret; and having implemented it, with the aura and approval of several senators, he obtained from these sirs a raise in his stipend of up to 1,000 florins, but with the proviso that he serve as professor there perpetually.[19]

The week of September 22 to 29 proved fateful for Galileo. It would mark the beginning not only of a new trade, that of a lens maker admired and sought after throughout Europe, but also of a new life. At forty-six years of age, he was about to see his existence change dramatically. The perfection of the Dutch spyglass and its repurposing into an astronomical instrument would completely transform his routine and work, starting with projects he had commenced much earlier but would soon suddenly abandon, namely, the great works of cosmology and mechanics that he had planned to complete and publish, and to which he had devoted all his energy over the previous two decades.

People in town spoke of nothing else. Everyone was abuzz with talk about Galileo and his lifetime appointment as professor at the University of Padua, with a yearly stipend of 1,000 florins generously offered as recompense for building a spyglass that would "greatly surpass the fame of the one from Flanders."[20] This was an extraordinary salary for a mathematician, far higher than what many of his colleagues at the university could ever hope to get, and it even exceeded the payment (800 scudi) the Senate had proposed to Sarpi for his work as a theologian in the service of the Republic.[21]

In the meantime, news of the new Venetian spyglass was spreading like wildfire. While the previous September an unknown craftsman had been the one to show diplomats and soldiers the surprising qualities of this new invention, choosing as his vantage point the tower in The Hague and training his spyglass on the clock in Delft and the windows of the Leiden cathedral, now the scene had shifted to Venice. This time, the mathematician from one of Europe's most prestigious universities was the one to present to political powers the extraordinary effects of an instrument he had perfected, with which, from the top of the Campanile of St. Mark's, one could see the bell tower in Chioggia and ships far off in the open sea, demonstrating an object "fifty miles away as large and near as if it were just five miles away."[22]

Galileo reminisced about the atmosphere of those frenetic days in a document that reads more like the screenplay for a Chaplin comedy than an actual report—and it was written four centuries ago. Venice was a city in the thrall of odd and unexpected euphoria, in which the most popular sport was clambering up and down towers and campaniles. Not only young men but also portly middle-aged lords vied to be the first to get to the top, determined not to miss a spectacle that had become the social event of the year. The spyglass of that certain foreigner was also carried up the stairs of St. Mark's, but there was no rivalry there. "And since news reached Venice that I had built one, six days ago I was called before the Most Serene Signoria [the Doge] to show it to him and the entire Senate, to everyone's boundless astonishment; and there were countless gentlemen and senators who, though elderly, climbed the stairs of the tallest bell towers of Venice more than once."[23]

For just a moment, let's try to imagine this scene, never losing sight of Galileo. How amused he must have been to watch all those eminent Venetian aristocrats sweating and panting as, one by one, they climbed the narrow steps of the bell tower of St. Mark's and then, at the top, crowded around that odd device, kneeling in front of it and, with one eye open and the other squeezed shut, "discovering sails and vessels at sea so far away that when they headed to port under full sail, it took them two hours or more before they could be seen without [his] spyglass."[24]

◆ ◆

In just days the news swept through Italy as well as outside, ricocheting between Rome and Florence, and spreading from France to Germany. Galileo was immediately inundated with requests, starting with those from his native city and, first and foremost, Grand Duke Cosimo II and Don Antonio de' Medici.[25] However, it was not just the reports sent by the leading player in this turn of events that sparked such astonishment. The letters sent by a figure of the caliber of Bartoli, who was by no means tenderhearted toward Galileo, acknowledged the merits of the new spyglass, considered the finest in circulation—better than those sold in Venice by "a certain Frenchman" who made them there "like a secret of France" even though they were worth "only a few sequins."[26] Neither the "secret" of Holland nor that of France could match "Galilei's secret." None of the other spyglasses could create

> the marvels and singular effects of the cannon of said Galileo . . . ; with which, keeping one eye open and the other shut, each of us clearly saw,

> beyond Liza Fusina and Marghera, also Chioggia, Treviso, and even Conegliano, and the bell tower and domes with the façade of the church of Santa Giustina in Padua: we could make out those entering and leaving the church of San Giacomo in Murano; we could see people getting in and out of the gondolas by the column of the ferry crossing at the beginning of Rio de' Verieri, with many other details of the lagoon and the city that were truly admirable.[27]

This is one of the first descriptions of a landscape viewed with a telescope, and we owe it to the diarist Gerolamo Priuli, a witness to this exceptional event, who also provided a brief description of the instrument itself: "a cannon with a diameter slightly larger than a silver scudo and less than a *braccio* long, with a lens at each end, and that, when set in front of the eye, multiplies sight nine times over."[28] By the end of August 1609, people no longer turned to The Hague for news about the spyglass, but to Venice. Here as well, the dates are fundamental and must be borne clearly in mind, above all considering the extraordinarily rapid spread of a discovery that, due to its military and commercial potential, piqued the interest of the entire European political class.

There is no doubt that the solemn and public approach Venice chose to reward *its* mathematician should also be interpreted as a self-celebration of La Serenissima's openness to technological innovation. Moreover, it is easy to surmise that Sarpi wholeheartedly approved of this recognition, not only because of his prestigious standing in the city's political and cultural life but also because of the leading role he himself played in this turn of events from a scientific standpoint. That Sarpi was the one to persuade the Senate not to take seriously "that foreigner," a man who had demanded 1,000 sequins for his secret, is confirmed not only by Bartoli's aforementioned letter of August 29 but also by Fulgenzio Micanzio's *Vita di Paolo Sarpi*. "He was the one to discover the device when, presenting one [spyglass] to the Most Serene Signoria [the Doge] with the request of a reward of 1,000 sequins, the friar [Sarpi] was asked to test its uses and offer an opinion; because he was not allowed to open it to inspect it, he guessed what it could be and then conferred with Messer Galileo, who thought the friar was right."[29]

As we can see, Micanzio's words and Bartoli's testimony do not fully coincide. Micanzio underscores the role Sarpi played in discovering how the spyglass worked, almost as if to credit him with paternity of the instrument; by contrast, after reporting the information Sarpi conveyed to Galileo ("telling him about the secret he had observed"), Bartoli dwells on the attempt made by Galileo, who, "with his mind and with the help of another similar instrument, but not as high in quality, which came from France," had managed to build a far better one. Despite these different accents, however, both versions agree on one point: that at the end of August 1609 Sarpi and Galileo were still close friends.

As in the good old days, before the fiery years of the Interdict when Galileo's cautiousness clashed with the theologian's stance, the two were constantly in touch, and there is no question that the unexpected arrival of the Dutch spyglass in Venice became yet another opportunity for the two men to meet and collaborate. In March 1610, before the *Sidereus* was published, some people even sent letters for Galileo to Sarpi's address.[30] Bartoli's words ("as Brother Paolo and Galilei are friends"), crafty as they may have been because they were sent to the secretary of a state that had always been loyal to Rome, merely confirmed what everyone in Venice already knew.

In autumn—around September or October—Venice continued to convey new information on the "long-sighted cannon." Moreover, people's attention had by no means dwindled; if anything, it became mingled with other equally curious facts, such as what Lorenzo Pignoria described from Padua: "Here we are dealing with [long-sighted] cannons and some excellent ones have been seen; but so far only a few people have its secret, which is much esteemed. A certain marvelous little lantern is also circulating, which is no less of an invention than the spyglass, as, with a light inside, at night it spreads so much splendor [light] that a letter can be read 500 paces away."[31]

In the meantime, from Venice Bartoli continued to correspond with Secretary Vinta, giving him the latest updates. After all, the "cannons" were no longer new and could be purchased at different prices. Some were made with ordinary glass, others with Murano crystal, and yet others with quartz or "mountain" crystal—the last were the most sought-after and expensive.

Yet they all had the same flaw: they were hard to handle and very unstable. A skeptical Bartoli wrote, "As far as I am concerned, and I have seen several, particularly one sold for three sequins to the postmaster of Prague, I must confess that I have never fully been satisfied because it is longer than an arm's length, so it takes a great deal of time for the eye to find the thing you would like to see and, once you've found it, the instrument must be kept so still that if it wavers even slightly, the thing vanishes once more."[32]

There remained Galileo's spyglasses, but there was no trace of them on the Venetian market. Word had it that they were the best, although nothing leaked out as to how they were made. Only two things were sure: the first had to do with the construction of twelve spyglasses for the Republic, and the second was that Galileo had been ordered "not to teach others" the secret.[33] Consequently, when the Medici rulers insisted, Bartoli had no choice but to buy a "French" one. In his disgruntled letter of October 24, he commented, "If your illustrious lordship had not specifically ordered me to do so, I would not have purchased it." Nonetheless, he could not gloss over what, hither and yon, had proven to be news that was more than reliable. "As to me, I do not find as many miracles as I had heard these instruments can work, although I understand that those of Galilei truly offer great advantages and benefits, supposedly with a magnifying power of ten, meaning that they magnify tenfold what one would see without them."[34]

Nevertheless, no gossip emerged about Galileo as a lens maker, nor can more detailed information be gleaned from the few letters the scientist wrote during that time. Furthermore, we have no way of knowing if—or how—his collaboration with Sarpi was proceeding. What we can say is that Sarpi is never mentioned in Galileo's letters and writings from this period, starting with the *Sidereus nuncius*. The same thing cannot be said of Sarpi, however, who cited Galileo's works on several occasions.

• •

We know next to nothing about the months following August 1609, which are crucial for our story. Very few of Galileo's letters from the period of September through December remain. There are four, to be precise, and only one—dated December 4—mentions the spyglass.[35] If we compare it to the letter written a month later, on January 7, 1610, in which he first announced his celestial discoveries, this is a considerable gap.[36] As a result, we can surmise that he must have attained his best results as a grinder and builder of lenses in December.

When it comes to Sarpi, things are even worse. It would take the *Sidereus* to force the two to resume talking about the spyglass. On March 16, 1610, just three days after the book was printed, Sarpi sent a copy to Jacques Leschassier (a Parisian attorney professing the Catholic Gallican faith), along with a far more analytical description of the instrument than is found in the *Sidereus*. While Galileo merely discussed a lead tube with a lens at either end but specified nothing else, Sarpi not only proved that he was well informed but also implied that he had actively been involved in building the first spyglasses.[37] Let's read his first letter.

> As you know, this instrument is composed of two lenses (which you call *lunetes*), both of which spherical, one with a convex surface and the other concave. We made the convex one from a sphere with a diameter of six *piedi*, and the convex one from another sphere a digit smaller in diameter. These are used to create an instrument with a length of about four *piedi*, through which one can see only part of the object that, if viewed with the naked eye, would subtend six arc-minutes, whereas with the instrument it can be viewed under an angle of more than three degrees. On the Moon, the star of Jupiter and the constellations of the fixed stars we saw those things you will read in the booklet that the lord ambassador will present to you in my name, and a number of other even more marvelous things, about which I shall write to you later.[38]

Sarpi's description shows macroscopic differences with respect to the *Sidereus*. Galileo insisted on the need to dispose of a "perspicillum exactissimum," but then gave the reader no advice as to how to make one, circumventing the matter with a few well-pondered but reticent words. Galileo spoke vaguely about *rationes* to be sought, *media* to be devised, and *inventio* founded on the doctrine of refraction. The only thing he underscored was the exceptional merit of his undertaking. Aside from his initial comments on the news about the Dutch spyglass he had received from Giacomo

Badoer, the rest is hazy and imprecise, including suggestions on how to perfect the instrument, intentionally left secret, although everything is written in the first person.

On this point, Sarpi's position was by no means attuned to Galileo's. His letter to Leschassier, dated March 16, counters the idea of a heroic challenge waged one-on-one to develop an instrument in the utmost isolation: "Our mathematician from Padua and others among us who are not unacquainted with those arts started to use this invention in order to observe celestial things and, once instructed in its use, embraced and further perfected it."[39] Credit for the invention goes not only to the mathematician from Padua but also to others (*alii ex nostris*). As opposed to what the *Sidereus* conveys, Galileo did not do it all on his own. Others collaborated with him in transforming the Dutch spyglass into an astronomical instrument that could magnify no less than thirtyfold.

Sarpi mentions no names. Nonetheless, he was clearly referring to people from his coterie, one of whom was his old friend Agostino Da Mula, a man Galileo too had known for a number of years. Moreover, we cannot reject the idea that the words *alii ex nostris* conceal an act of modesty, "given that, at the time, and as far as we know, Galileo's friends intent on such things included him alone and Agostino Da Mula."[40] After all, that Sarpi was quite knowledgeable about optics and the use of the telescope is amply demonstrated by the subsequent correspondence with the Parisian attorney and by the fact that he conducted his celestial observations at the monastery of Santa Maria dei Servi.[41]

All of this transpired on March 16. On April 5, Leschassier sent Sarpi a long letter informing him that someone in Paris had built an instrument two or three *piedi* long and fitted with two lenses, which could be used to observe "the spots on the Moon that cannot be seen with the naked eye, and a greater number of fixed stars."[42] Above all, however, he told Sarpi that he had immediately read the *Sidereus* and that, despite a few doubts about the shape of the Moon, he found it admirable: "artem novam et naturae miraculum continentem." In other words, a gift of providence that would bring humankind closer to heaven, a new instrument to help celebrate the glory and greatness of God.[43]

Sarpi's reply to Leschassier is surprising, to say the least. In response to the man who lavished such words of praise on the book he had just received, the friar admitted that he had not actually read the work. "I will now talk to you about the Moon. To be honest, I have not read what our mathematician has written, but I have spoken often to him about it, and many things were said on both ends. I will tell you what I think about the Moon and, as I am accustomed to do, will speak only about the things I have personally observed."[44]

The letter is dated April 27, 1610, and if it weren't authentic, it would almost appear to be a joke (and a bad one to boot). The *Sidereus* had been published a month and a half earlier. In Prague Kepler had already written his *Dissertatio cum Nuncio sidereo,* confirming the celestial news, while in Florence the grand duke had given Galileo a gold necklace to thank him for dedicating the work to the Medicean planets. Yet in Venice, where the book had been printed in the first place, Sarpi was declaring that he hadn't even read it yet.

This defies the imagination. Given that the *Sidereus* was in demand across Europe, why would the person who, more than anyone else, had followed every step of the arrival of the spyglass in Venice and unquestionably discussed it with Galileo—possibly even collaborating with him to perfect it—choose at some point to "forget" it and virtually expunge it from his horizon? How else can we interpret the phrase "I have not read what our mathematician has written," immediately followed by "I will speak only about the things I have personally observed"? What pushed Sarpi to take this stance?

One answer surely has to do with the publication of this unusual book that changed the way of seeing and picturing the sky. That said, we must carefully reread each and every page, trying to bear in mind that, rightly or wrongly, the interminable chain of reactions, accusations, and claims is a key part of the story and that without them we cannot fully understand this book.

We could certainly get around this by saying that Sarpi didn't need to read it in the first place, since he had already been apprised of all the improvements made on the spyglass and was fully aware of the instrument's performance even before the discoveries were set down on paper. If this were the case, however, why would he have said that he would only discuss his personal observations? Why would he have drawn such a fine distinction between his own work (and that of his friends) and Galileo's publication? Not for astronomical reasons. In fact, in another passage of his letter to Leschassier, Sarpi dwelled on his telescopic observations, which differed little from Galileo's. The description of the Moon illuminated by the Sun or light reflected by the Earth, with its hills and hollows, its interplay of light and shadow, is perfectly in keeping with what the scientist had written in his *Sidereus*.

The reasons lie elsewhere and are entirely different. First of all, the celestial novelties were part of a manifesto that was certainly not neutral, an announcement characterized by a marked political orientation that surely would not have escaped an expert in "human things" such as Paolo Sarpi. The announcement from its author that this was not only a book on astronomical discoveries but *the* book launching a new astronomy and a new philosophy contains another level of meaning.

The dedication to the grand duke and the tribute to the Medici family, naming the four satellites of Jupiter after them, were clearly signs that Galileo wanted to leave Venice and return to Florence. The Venetian theologian would have opposed such a decision, not only because of the ancient ties between Tuscany and Rome but also because it would have given the grand duchy—the political recipient of those discoveries—wholly undeserved cultural prestige, downplaying the active role and generous support the Venetian government had provided time and again.

In short, Sarpi was clearly distancing himself from the scientist's ingratitude toward the Republic. The "Patrician of Florence and Public Mathematician of the University of Padua" (this is how Galileo described himself on the title page of his work) did not write one thing about his years in Padua or the benefits he had received, nor—more important—the fact that he had been given a lifetime position as professor of mathematics at the university. He spent nary a word on the Venetian craftsmen who had labored with him to perfect his lenses. Nor did he mention Sarpi and the circle of friends who, at least for a time, had shared his same passion to build a new and more powerful spyglass.

• •

Yet there is even more. The *Sidereus* is also a work in which intentional gaps and voids stand out. Maurizio Torrini is right when he notes that "the silence of the *Sidereus* with respect to tradition and coeval discussions must have sounded odd to Kepler's ears."[45] When we read it with disenchanted eyes and, so to speak, *ex parte veneta,* what emerges is an equally systematic and conscious attempt to conceal people, words, and events.

There are rare exceptions, such as Copernicus and the aforesaid Badoer, who converted to Catholicism (supposedly through the Jesuits' efforts) after the abjuration of Henry IV and who, in the spring of 1609, had informed Galileo and Sarpi of the invention of the spyglass. Nevertheless, the presence of this ex-Protestant diplomat in a context with so few other references is telling. What better way for Galileo to distance himself from Venice than by citing his friendship with figures acceptable to the Roman Catholic entourage? The dispatches of the Tuscan ambassadors residing in Venice are invaluable here, as they allow us to understand the immense hostility toward the Republic's violently anti-Roman policies. Even after the "war" of the Interdict, tension remained high. As has been observed, "These were the years of the correspondence between Brother Paolo and the cultural and religious world of Northern Europe, of the dialogues with the German Protestant and Calvinist agents in Venice, in view of the great political design of forming a front against the Habsburgs, but also of bringing Venice into the Reformation, and—lastly—of launching profitable relations with the English Puritan world."[46] We need merely read some of the dispatches from this period to understand just how tense this climate must have been. For example:

> On Monday the Inquisitor was called to the Collegio and was vigorously admonished and scolded against harsh treatment, particularly of the subjects of the books written in favor of the Republic during the past turmoil, and writings that should not in any way be prohibited or be worthy of

> suspension or abolition any more than those [writings] in favor of the opposing party.[47]

According to another letter,

> Now dead is Messer Alessandro Malipiero, of the same age and a close friend of His Serene Highness and Father Master Paolo, the excommunicated Servite, with whom he was when [Sarpi] was stabbed, and he pulled from his face the dagger that had been plunged there, and so also saved him from death, crying out against those who had wounded him. The square is full and this gentleman wanted to pass away without any of the most holy sacraments, and particularly extreme unction, but I do not know if this was because he did not believe there were any nor that he was dying, or because he spurned them, although many say it is the latter.[48]

The image that—rightly or wrongly—most of the world had of the city coincided with the one described by Giovanni Diodati in a letter dated August 1, 1608, to the "pope of the [French] Huguenots" and Sarpi's correspondent, Philippe Duplessis-Mornay: "Venice seems like a new world. . . . Men preach there as they might preach in Geneva, but with such ardor that they attract enormous crowds, and one must arrive well in advance to find a spot."[49] This is also confirmed by the words of Cardinal Scipione Borghese, who said that Sarpi's impiety extended "so far beyond that not only can we consider him a heretic, but a heresiarch who strives for nothing other than to infect that entire Republic, assuming that something sound remains there in the first place."[50]

We could continue with other testimony, but it would be superfluous. The events we have just recounted are enough to allow us to understand the Tuscan diplomats' aversion toward Sarpi and the Venetian government.

Consequently, it behooved Galileo to distance himself from the Servite theologian. The irrevocable decision to leave Venice and return to religiose Florence to enjoy the leisure necessary to complete his two great works (the *Dialogo* and the *Discorsi intorno a due nuove scienze*) called for precisely this type of "sacrifice." After all, what chance would he possibly have had if he had publicly admitted his collaboration with Sarpi's camp, if he had acknowledged his debt of gratitude toward the heretical theologian instead of naming the far too politically correct Giacomo Badoer (whom Pierre de L'Estoile defined a "faciendaire et espion des Jésuites"[51])? What was the likelihood of successfully bringing his project back to Florence if emissaries and officials faithful to the Church of Rome managed to intercept one of the many letters his friend Sarpi wrote to his Protestant correspondents? We can read one at random, such as the missive Sarpi wrote to the Huguenot Castrino on August 17, 1610:

> The most important news we have is about the celebrations of the Jesuit fathers everywhere, for the beatification of Father Ignatius. The comedy they staged in Seville is old, bringing out a life-sized relief image dressed in velvet, with a Jesus on his chest made of diamonds and pearls, and with eleven pages, the founders of other religions. But in Prague for this effect they again held a most solemn feast, to which they invited the Elector of Mainz, Archduke Leopold, the Spanish ambassador, and many Catholic nobles, and they all drank very devout toasts to Blessed Ignatius. In Rome, Naples, Milan, and other places in Italy they set up lavish displays in their churches for this purpose. . . . We will let them be, praying to God that they soon reach the climax, so that the abyss will also be nigh.[52]

For Sarpi, 1610 was a disastrous year to be written off entirely. In addition to the European triumph of the Society of Jesus in regard to the recent beatification of Ignatius of Loyola, the murder of Henry IV, and the appointment of the ultra-Catholic Maria de' Medici as regent of France, now he also had to face the loss of a dear friend who had chosen the pomp and honor of Jesuit Rome over the success and freedom he would have had in Venice for the rest of his life. In short, it was an *annus horribilis* not only because of the changing international political scene but also because of Galileo's decision to publish his news about the telescope in that form. It was a terrible blow to Sarpi and the Republic. To make matters worse, the publishing success of the *Sidereus* would wholly benefit the Roman Curia and, indirectly, reinforce the cultural prestige of the fiercest enemies of the "real"

Church, those whom the Venetian theologian felt represented evil and the "plague" infecting the whole world.[53]

• •

In his letter of April 27 to Leschassier, Sarpi also emphasized another aspect. Work to perfect the spyglass was moving ahead, and the results attained thanks to Galileo's slender volume were merely the beginning of more sweeping research. Above all, he was anxious to announce that the most important news would arrive shortly, because "here our people are making great progress in both building and using the instrument, and I have no doubt that the entire celestial philosophy [*tota philosophia caelestis*] will advance enormously as a result."[54] The *Sidereus nuncius* was the starting point for research still under way but progressing rapidly. In fact, the objections being raised against astronomical observations would fall by the wayside once there was a theoretical explanation for how the spyglass worked. On May 10, Sarpi wrote to Jerome Groslot de l'Isle:

> Despite what one might expect from the new spectacles, touching heavenly things, nothing else has been observed so far, except that by having built one with such craftsmanship, so that one can see only a hundredth of the Moon at a time but as large as what could be seen entirely with the first telescope, the cavities are so conspicuous and seen so exactly that it is astonishing; and the star of Jupiter, observed many times, thus appears to have the size and features of the Sun, when sometimes one can see beneath the haze. But the marvels discovered with this device lie in the profession of perspective, as from this we understand how vision works, and the explanation of spectacles for both weak and short sight: they are things that require a proper volume to explain them.[55]

Perhaps we should read this passage several times, because the most striking thing here is that Galileo is never mentioned. To be more precise, he is referred to indirectly, but no longer plays a leading role. Work on the "new spectacles" continued, but it was no longer spearheaded by the author of the *Sidereus nuncius*. To understand the real "explanation of spectacles," we need to look past the pages of his book, surprising as they may have been. Solid underpinnings were essential to counter the criticism already being leveled against the celestial discoveries, considered tricks of the lenses and not real phenomena. A treatise on optics had to be written to explain "how vision works" and the properties arising from putting two lenses together. But nothing of the kind could be found in the book that had just been published.

In short, much remained to be done to arrive at a full understanding of the new instrument and the knowledge it could provide. This was Sarpi's response following publication of the *Sidereus:* that it would be a fragile and short-lived success unless the delicate theoretical issues underpinning the invention of the spyglass were tackled. Only after constructing a rigorous science of vision would it be possible to build a new astronomy, a new *philosophia caelestis*. Sarpi's answer closely resembles what Kepler said in his *Dissertatio* at the same time, but in this case it put an end to a friendship and scientific collaboration that had spanned nearly twenty years.

Sarpi would never again mention Galileo with regard to the telescope, not in this letter nor in any of his later correspondence. In Venice, however, work proceeded briskly. Just one month later, he sent Leschassier more precise information. "Regarding the spyglasses, I can tell you that here several scholars [*viri aliquot eruditi*] are working on a short treatise on vision [*commentariolum de visione*], in which they explain the reason and cause of the instrument invented in Holland, and at the same time expound the entire theory behind it."[56]

We have no way of knowing exactly who these *viri aliquot eruditi* were. However, if we consider the news arriving from his friends north of the Alps, we cannot reject the possibility that Sarpi himself was part of that group. In fact, on June 29, in thanking him for sending the *commentariolum de visione*, if it should be published, Leschassier congratulated him on the works he had received as a gift, "such as your short comment on the phases of the Moon, about which you wrote to me in your previous letters, and also about the methods to build the spyglass, which you sent me a short time ago."[57] Even if he was not personally involved, there is no question that Sarpi was well informed about the work of his circle of spectacle-making and optician friends, who included one of Sarpi's leading supporters, namely, Agostino Da Mula.

Also a friend of Galileo, the aristocrat Da Mula was a well-known figure in Venetian political and cultural circles. First podesta and captain in Belluno, and then elected several times as *savio di terraferma*, along with Nicolo Contarini, Sebastiano Venier, and Giovan Francesco Sagredo, he was one of the men closest to Sarpi, but he was also one of the most uncompromising when it came to anti-Spanish policies and championing the defense of the Republic's rights against the Holy See.[58] Giovanni Camillo Gloriosi specifically referred to the nobleman's interests in optics and astronomy in his letter of May 29, 1610, to Giovanni Terrenzio, in which he also insinuated that the first person to observe Jupiter's satellites was none other than Da Mula. He wrote: "It is said publicly that Agostino Da Mula, a Venetian aristocrat, was the first to observe these stars and that he informed Galileo."[59] Two years later, it was Sagredo who reminded Galileo that Da Mula had shown him "a very large number of wooden panels carved with different demonstrations, which were to serve for one of his treatises, written in his own hand, in a manuscript of perhaps 100 sheets, but he would not allow me to read a single thing."[60]

Therefore, many clues seem to confirm that "the book of spectacles" Sarpi mentioned to Castrino and Leschassier in two letters dated August 3, 1610, and which had not been printed yet due to difficulties in engraving the panels, must have been Da Mula's work. It is plausible that it was never printed because Da Mula himself considered it obsolete after Kepler published his *Dioptrice*. The fact remains that the history of the spyglass does not travel along a single vector or one-way path but, as Venice demonstrates, is an archipelago of overlapping human and intellectual events: a collective story that must be recounted in its simultaneity in order to be reconstructed and reassembled piece by piece.

THREE

Breaking News: Glass and Envelopes

To emerge from the Venetian labyrinth we entered, we need to talk about lenses and envelopes. And to do so, we need to go back to what Paolo Sarpi wrote to Francesco Castrino on July 21, 1609.

As we already noted, in his letter Sarpi did little to hide his disappointment over the mediocre performance of the spyglass he had had the chance to examine.[1] Nevertheless, his admiration for the invention shows that he felt the idea was worthwhile and improvable in terms of application. If we are to believe Giovanni Bartoli (dispatch of August 29), it was precisely this type of information that Sarpi shared with Galileo.[2] Regardless of how things may have gone between the two men, there is no question that everything started with the spyglass that had arrived in Venice.

As the turn of events in the Netherlands demonstrates, from its advent the spyglass was considered such a rudimentary and easy-to-make object that the States-General decided to reject all patent applications. Yet if it was so simple, a device almost immediately imitated in various parts of Europe, why did Galileo wait until the summer of 1609 to build one?

Nine months—that is how much time elapsed after the news of the Dutch spyglass reached Venice in autumn of 1608. It may seem far too long a time, but this is not altogether surprising, considering how warily both Sarpi and Galileo had greeted the device in the first place. This was also because the news landed in a milieu rife with all sorts of secrets and stories about futuristic optical devices promising prodigious effects. A veritable torrent of such items had reached the squares of many European cities, where unscrupulous charlatans attracted immense throngs of people, hawking a variety of products as if they had come from the most recondite recesses of nature. In Venice these figures had become a real attraction and, as recounted by Thomas Coryat, an English traveler who stayed there in 1608, no visit to the city could be considered complete without a stroll through St. Mark's Square to watch their performances.[3] In such a scenario, it is unlikely that news about the secret of the Dutch spyglass would not have been viewed suspiciously. The fact that Sarpi was not the only one to harbor doubts about it can be deduced from a passage in the *Sidereus,* in which Galileo observes, "About this truly wonderful effect [the instrument's ability to make distant things look close] some accounts were spread abroad, to which some gave credence while others denied them."[4]

Therefore, it is unsurprising that Galileo would have waited so long before finally deciding to commit himself to building a device surrounded by such great incredulity, which diminished when a specimen finally reached Venice in July 1609 and provided tangible proof that the rumors circulating for months were grounded. This new circumstance—unquestionably the most significant to be gleaned from Sarpi's letter to Francesco Castrino—sparked Galileo's involvement. Establishing whether, through Sarpi, he too had the chance to inspect the "cannon to see far away" which "that foreigner" was trying to sell the Venetian Senate is a rather minor consideration.[5] Reproducing the device was not an overly complicated undertaking for many craftsmen, so there is no reason to think it would have been difficult for Galileo either. In fact, as he recounts in *Il Saggiatore* (1623), it took him just a single day to make one for himself.[6]

The real difficulty actually lay elsewhere: not in imitating the Dutch spyglass, but in improving its performance. In the months that followed, Galileo decided to devote his energy to achieving precisely this goal. However, he worked so discreetly that he didn't let anyone know what he was doing, an approach that allowed him to protect the results of his work but indubitably aroused suspicion and resentment. In fact, immediately after he published the *Sidereus,* one of the most frequent criticisms—not only on a private level, as in Sarpi's case, but also publicly, as in Kepler's *Dissertatio*—was that its author had no theoretical understanding of the optics of the instrument that was making him famous around the world.[7]

But was an adequate and detailed theory of the optical properties of lenses really necessary to make a good spyglass? Moreover, does the fact that, despite his statements and promises, Galileo never developed a mathematical theory of the instrument actually mean that he did not understand how it worked or had no optical knowledge whatsoever? These questions have spawned an endless stream of literature, fueling a debate that has yet to subside.[8] Yet various details allow us to recount a virtually unknown story that sheds unexpected light on his work as a lens maker, and it gives us further insight into his first observations of the Jovian satellites.

These documents are sui generis: envelopes used like modern Post-it notes, on which he would hastily jot down an unexpected observation or write detailed lists of items he wished to purchase, including glass and everything needed to work it. It is through this invaluable testimony that we will attempt to follow the new activity Galileo undertook in the fall of 1609.

• •

In the *Sidereus nuncius* Galileo described the path that led him to perfect the Dutch spyglass, in a rather spare account virtually devoid of technical details and without any chronological references. Through a crescendo of digits, he emphasized his rapid and considerable progress. First he wrote about an instrument that showed objects "three times closer"; then "another more perfect one . . . that showed objects more than sixty times larger"; lastly, one "so excellent that things seen through it appear about a thousand times larger and more than thirty times closer." The information about the instrument was cursory: "a lead tube" at the ends of which were "two glasses, both plane on one side while the other side of one was spherically convex and the other concave."[9]

Although his account was intentionally vague, it is not difficult to establish an approximate time frame. Galileo made his first telescope, which could reach a magnification of three, toward the end of July 1609 or, in other words, after the appearance in Venice of the telescope Sarpi described to Castrino in his letter of July 21. It was a simple imitation of the Dutch spyglass and was immediately replaced by another "more exact" model that could reach a magnification of eight or nine, which he made before August 21. This was the "cannon" described by Gerolamo Priuli in his *Cronaca* and, namely, the one solemnly offered to the Doge of Venice on August 24.[10]

Moreover, based on the terse clues offered in the *Sidereus* it seems that, in both cases, Galileo used the same type of lens employed to make spectacles: ordinary lenses with a plano-convex surface for presbyopia and a plano-concave one for myopia. By this time, they had been on the market for over a century[11] and were easy to procure through shops in Venice or Padua. It was through a combination of these lenses—a weak plano-convex one, which served as an objective, and a strong plano-concave one, used as an eyepiece—that he managed to make an eight-power instrument.

Anything further was impossible. Higher-quality lenses were needed and, above all, convex lenses with a uniform curvature and longer focal length, ground more carefully and precisely.[12] Spectacle makers were unable to satisfy these requirements, however, as revealed by two different occurrences during this period.

The first was in Germany in the fall and winter of 1608, and was recounted by the astronomer Simon Mayr. He had learned about the Dutch spyglass from a friend, a certain Johannes Philip Fuchs, who had seen one at the fair in Frankfurt. They had managed to reproduce it easily and rather quickly by taking "glasses out of common spectacles, a concave and a convex." Realizing that "the convexity of the magnifying-glass was too great," however, Fuchs made "a correct mold in plaster of the convex glass and sent it to Nuremburg, to the makers of ordinary spectacles that they might prepare him glasses like it; but it was no good, as they had no suitable tools."[13]

The second took place in Venice toward the end of May 1609, a couple of months before Galileo came on the scene. In this case, the leading player was the Milanese Girolamo Sirtori, author of the first history of the telescope published in 1618, but written in 1612 and, naturally, entitled *Telescopium*. Sirtori reported that he had examined the Dutch spyglass that an unnamed "Frenchman" had offered the Count of Fuentes in Milan,[14] and that he had managed to make "similar ones." Promptly realizing that "many of the inconveniences [of the telescope] occurred because of the glass," he "went to Venice in order to obtain a supply [of lenses] from artisans there," but noted that he had spent "some money uselessly" and had learned "nothing more than that the business is to be perfected by chance and by the laborious selection of lenses."[15]

The situation in Nuremberg and Venice was thus essentially the same. Although the cities were renowned as centers for the production of spectacles,[16] the quality of the lenses made there left much to be desired, as documented by Mayr and Sirtori. In the case of spectacles, of course, flaws in the grinding of the lenses were negligible and did not affect their principal purpose of correcting vision problems. In the case of the telescope, however, those imperfections made it impossible to improve performance, so there was only one answer: Galileo had to learn to make the lenses himself.

There is no point scouring Galileo's correspondence for information about his working conditions or how he managed to master the processes for grinding glass to obtain lenses with good optical properties. The few extant letters between September and December 1609 ignore the subject almost entirely. Yet there is no question that during those months he was engrossed with the idea of improving his eight-power telescope. And he also intensified his visits to the spectacle makers and glass shops lining the alleys of Venice, of which he had long been a habitué.[17]

These were busy days of commuting between Padua and Venice, seeking the best the market had to offer, but they often proved to be fruitless, forcing Galileo to turn to Florence, as its glassworks produced lenses whose quality had rivaled Murano's for several decades.[18] His were not generic requests, as we can glean from the letter of September 19, 1609, from Enea Piccolomini, who sent him "crystals according to [Galileo's] instructions."[19]

• •

It was well into September by this time, and Galileo's research was turning into a determined quest to find increasingly better lenses. Yet his telescope was effectively at a standstill, stuck at the magnification of eight or nine he had achieved a month earlier. So when he went to Florence at the end of September, it was an instrument of this power that he showed to Cosimo II, as Galileo reminded Vinta on January 30, 1610: "That the Moon is a body very similar to the Earth has already been ascertained by me and shown to some extent to our Most Serene Lord, although imperfectly, because I did not yet have a spyglass of the excellence that I have now."[20] These words bear emphasizing, not only because they confirm that at the beginning of autumn 1609 Galileo's spyglass had not yet made any progress but above all because they tell us that, at the time, he had already turned it to the Moon, managing at least to discern that some areas were smooth and others rough. It must have been these initial results in astronomical observation that made the need for a more effective instrument so vital.

Improving the telescope was becoming nothing short of an obsession. To grasp its importance, we can cite an episode from Galileo's private life that has come to us through a rather singular document: a letter his mother, Giulia Ammannati, wrote on January 9, 1610, to Alessandro Piersanti, Galileo's servant in Padua. Giulia had returned to Florence at the beginning of the previous November, taking the scientist's eldest child, little Virginia, with her.[21] She complained that she had not heard from her son "for a month now" and thought his reticence was financially motivated. "I imagine he is not writing so he can avoid sending [Benedetto Landucci, Galileo's brother-in-law] the money [the latter] has had to spend on me and Virginia, or perhaps not to send him the two lenses he requested several times in vain." She immediately continued:

> However, dear Messer Alessandro, please arrange for [Landucci] to receive two or three [lenses], but not the type that is for short sight and is hollowed out, because he has the one Galileo left him, but the plane kind that go under the cannon; in other words, the ones that are at the bottom, so that when you look through that end, you can see things very far away. And since Galileo has many of those, it will be easy to take two or three or four of them . . . ; and I sincerely beg you to do so, because Galileo is so ungrateful to one who has done so much for him and continues to do so.[22]

This is an outright reprimand, but what we should glean here is Giulia's familiarity with a category of objects—from "lenses" that were "hollowed out" and "plane" to the "cannon"—and this tells us of their ubiquity in the Galilei household, both physically and as the topic of everyday conversation. Acrimony aside (and Giulia's letters were invariably rancorous), this letter gives us a snapshot of the condition in which Galileo's mother had left her son in Padua in October of 1609: a scientist surrounded by so many lenses he wouldn't even have noticed if a few went missing.[23] These lenses had been selected carefully and purchased from master spectacle makers in Venice, but they had ultimately proven inadequate to significantly improve the quality of his telescope.

• •

We do not know exactly when Galileo decided to get personally involved in making the lenses that allowed him to build the spyglass, the "excellence" which he boasted of to Vinta in January 1610.[24] The only indication we find in the *Sidereus* is that "no labor or expense" was spared. These words are too concise to give us insight into what type of work was involved or how long it took to make the instrument showing him objects "about a thousand times larger."[25] All these details instead emerge from the most unlikely place: a very ordinary envelope Galileo transformed into a long memo to himself on items to procure and things to do. The document was first reported by Antonio Favaro, editor of the monumental *Edizione Nazionale,* and was cited at the bottom of the letter the Veronese Ottavio Brenzoni sent Galileo on November 23, 1609 (see Figure 9).[26]

When we read the first set of items, it appears to be an ordinary shopping list prepared for an upcoming trip to Venice. There are things he needed to buy for his family, such as a pair of slippers *(scarfarotti),* a hat for his son Vincenzo, two ivory combs for his mistress Marina Gamba, and an array of supplies for the larder: lentils, chickpeas, rice, spelt, raisins, sugar, pepper, cloves, cinnamon, spices, preserves, oranges. Nevertheless, a rapid glance instantly tells us that we are looking at a rather special list, as it includes items entirely unrelated to the preceding ones: two artillery balls *(palle d'artiglieria),* a tin organ pipe *(canna d'organo di stagno),* polished German glass *(vetri todeschi spianati),* rock crystal *(cristallo di monte),* pieces of mirror *(pezzi di specchio),* iron bowls *(scodelle di ferro),* Tripoli powder *(tripolo),* an iron plane *(ferro da spianare),* Greek pitch *(pece greca),* felt, mirror to rub *(feltro, specchio per fregare),* and fulled wool *(follo).* Altogether, the latter items account for more than half of the entire list and tell us that toward the end of November 1609 Galileo was setting up his own optical workshop. After all, the practical side of science had always been part of his work as a mathematician, and he had set up a small workshop in the large house he had moved into in 1599 in the Borgo dei Vignali district in Padua, close to the

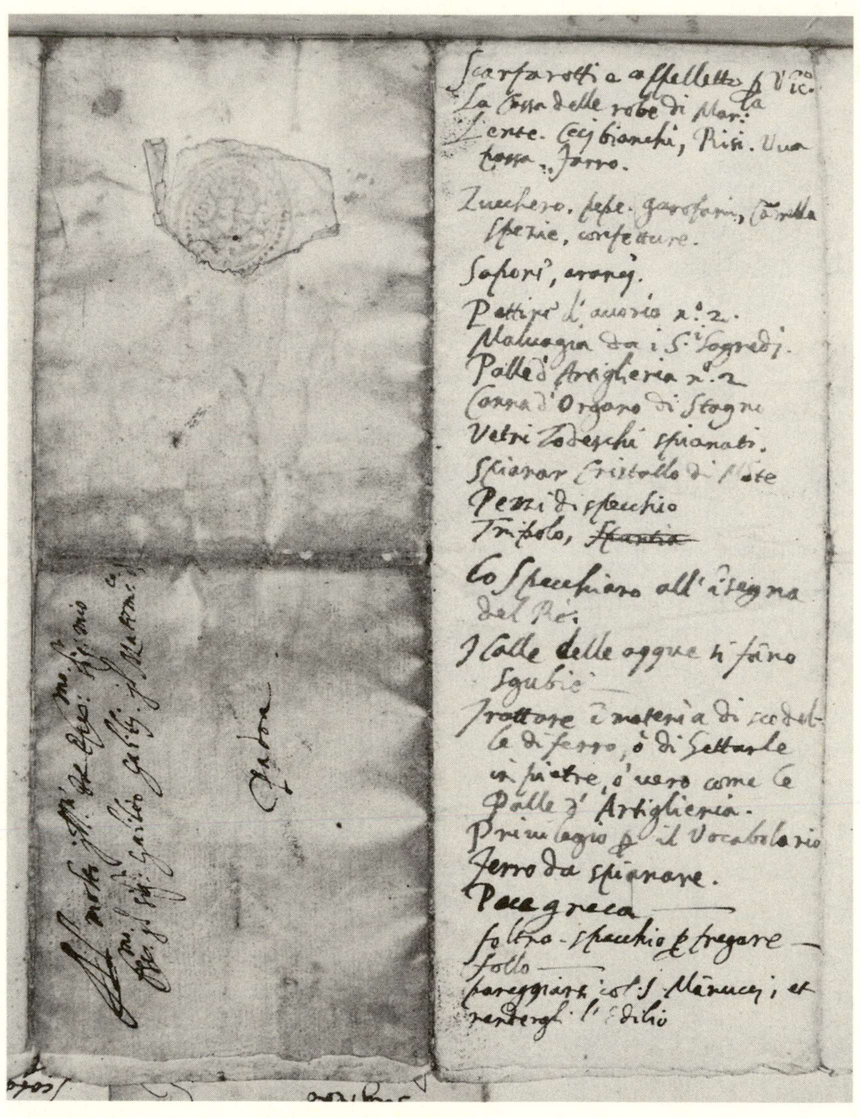

Scarfarotti a affibbiatura p. Vinc.
La cassa delle robe di Mar.a
Lente. Ceci bianchi, Risi. Uva passa, Farro.

Zucchero, pepe, garofani, cannella
spezie, confetture.

Sapori, aranci.
Pettini d'avorio n.° 2.
Malvagia da i S.ri Sagredi.
Palle d'artiglieria n.° 2
Canna d'organo di stagno
Vetri todeschi spianati.
Spianar cristallo di monte
Pezzi di specchio
Tripolo, [illegible]

Lo specchiaro all'insegna del Re.

[illegible]

Trattare in materia di [illegible] palle di ferro, ò di gettarle in pietre, ò vero come le palle d'artiglieria.
Privilegio p. il Vocabolario
Ferro da spianare.
Pece greca
Feltro-specchio p. fregare
Feltro
Pareggiare col S. Marucci, et rendergli l'Edilio

Padova

Figure 9. Galileo's shopping list; autograph note written on the back page of the letter sent by Ottavio Brenzoni on November 23, 1609

Basilica of St. Anthony. That year he had hired a craftsman, Marcantonio Mazzoleni, to make the instruments he designed, such as the magnificent geometric and military compass in wood and brass, but also squares, quadrants, and magnets he then intended to sell.[27] In short, this represented know-how he had amassed over the years and would need to draw on for the adventure of the telescope.

Naturally, since this was a note to himself, Galileo did not bother to explain the purpose of that array of items, but it is easy to understand. These were materials and tools used by spectacle makers, alluding to the practical knowledge that Galileo had acquired by frequenting their shops. On his list they were condensed into short, simple notes, but they are enough to allow us to picture the long procedure required to make lenses.

First of all, one had to procure the raw materials: polished German glass, which were sheets of glass from the furnaces of Murano and made of just one layer to minimize imperfections, using a method referred to as *alla tedesca,* or "in the German style"; prized and expensive rock crystal, which was far more perfect and clear than ordinary glass; and pieces of mirror, meaning well-polished sheets of glass that had not yet been coated with tin or silver to make mirrors. These raw materials were certainly not hard to find in a city such as Venice, where the production and trade of glass artifacts was a long-established industry. If anything, the difficulty lay in finding sheets of glass of high quality; those with air bubbles, slight scratches, or any other flaw would be unsuitable for lenses. This required an expert eye and, above all, constant contact with the numerous shops in the center of town, where the *occhialeri* and *specchieri*—spectacle and mirror makers—vied for customers with small retailers.[28]

Galileo had made note of two shops in particular, perhaps because he knew he was sure to find exactly what he needed. The first was that of a certain "mirror maker under the insignia of the king" *(uno specchiaro all'insegna del Re),* but Galileo does not list his name or location. For the second one, he gave a rather specific geographical location: in "Calle delle aqque [*sic*] they make cutting tools [*sgubie*]." However, he did not go to the alley close to St. Mark's Square solely for the tools used to score and cut glass. As it happened, one of Galileo's most esteemed spectacle makers,

Giacomo Bacci, also worked there. According to the Bologna astronomer Carlo Antonio Manzini, Bacci "had a shop in Calle dalle [*sic*] Acque . . . as reported by Mr. Francesco Ferroni, Venetian mirror maker, who has now lived in Bologna for many years and says that, at the time, he was the friend of Gio. Battista, son of said Giacomo."[29] Galileo would continue to patronize Giacomo even after he returned to Florence, commissioning lenses through his friend Sagredo.[30]

Finding the materials he needed was merely the first step, however. The real work began when a disk was cut from the selected sheet of glass and an iron plane was used like a file to remove the excess along its edges. At this point, to obtain a lens in the desired shape, the glass disk had to be ground on a suitable mold: artillery balls for concave lenses and iron bowls (though the bowls could also be made of stone) for convex ones. This was a very complicated step entailing the use of an adhesive, Greek pitch, to secure the lens on an implement fitted with a handle, and then rubbing it on the mold covered with Tripoli powder—an abrasive powder named after the Syrian city whence it came—to obtain a very smooth surface. But the process didn't end there, as the ground lens was still opaque. To make it clear, it had to be sprinkled with more abrasive powder—Tripoli again, but much finer for this phase—and polished until it shone, using a piece of felt or fulled wool and going to great lengths to avoid altering its curvature.[31]

It was through this painstaking process that Galileo managed to prepare his lenses to be mounted at the ends of a tin organ pipe—the "lead tube" described in his *Sidereus*—and make himself a more powerful telescope. In order to do this, he had turned himself into a skilled artisan, for he had personally ascertained how difficult it was achieve results that met his expectations. As he wrote to Vinta on March 19, 1610, "Very fine lenses that can show all observations are quite rare and, of the more than sixty I have made, with great effort and expense, I have only been able to retain a very small number."[32]

• •

If this is how things stood, however, and if Galileo limited himself to exploiting the techniques perfected by spectacle makers to manufacture glass and lenses, why did he immediately attempt to discredit the empirical background of his telescope? Why did he bother to specify that he had been guided by the "science of perspective" *(scientia di prospettiva)* and the "science of refraction" *(dottrina delle rifrazioni)*?[33] Naturally, as emerges from *Il Saggiatore,* he had every reason to emphasize that the device he had perfected arose from an exclusively theoretical study, "through reasoning," juxtaposing his work as a mathematician-philosopher with that of a "simple maker of ordinary spectacles."[34] Nevertheless, it would be misleading to view his distancing himself merely as an attempt to claim paternity for the "real" invention of the instrument.[35] Galileo may have represented himself this way in the *Sidereus,* but this does not mean that the emphasis on his optical knowledge was ungrounded.

Galileo's first encounter with the "science of perspective"—*perspectiva,* the term used in the Middle Ages for the study of optics—went as far back as the 1580s, when he participated in the classes given by the mathematician Ostilio Ricci at the workshop of the court artist and engineer Bernardo Buontalenti in Florence. In his lessons, also attended by the future painter Ludovico Cigoli and Don Giovanni de' Medici, Ricci taught not only perspective for artists but also the rudiments of optics, using a manual, *Della prospettiva,* attributed to Giovanni Fontana, a pupil of the medieval philosopher Blasius of Parma (Biagio Pelacani da Parma).[36]

Later, between 1592 and 1601, Galileo again took up optics in Padua, transcribing the *Theorica speculi concavi sphaerici* of the Venetian physician and mathematician Ettore Ausonio, famous for his "concave mirrors of appreciable size."[37] This short treatise on practical optics discussed image formation and perception in concave mirrors.[38] That Galileo read it suggests that he was familiar with the tradition of medieval optics, given that Ausonio paraphrased the analyses of reflection and refraction developed by the Polish monk Witelo in his *Perspectiva.*[39] Moreover, specific references to Witelo's work are found in one of the writings from 1606 attributed to Galileo.[40] But that is not all. In his library Galileo also had a copy of the *Opticae Thesaurus,* the book published in 1572 by Friedrich Risner with the Latin translation of Alhazen's treatise on optics along with Witelo's *Perspectiva:* two classics that had spawned the medieval tradition.[41] He also owned Della Porta's *De refractione* (1593), which explained image formation in convex and concave lenses, albeit with rather sketchy and arbitrary geometric demonstrations.[42]

It is difficult to assess the extent to which Galileo investigated this knowledge, but at the very least he was well acquainted with the theoretical explanations of the optical properties of lenses available at the time. Yet there was one significant treatise he was not familiar with: the work Kepler published in 1604 with the seemingly modest title *Ad Vitellionem paralipomena* (Supplement to Witelo), but which in reality struck a mortal blow to Witelo's *Perspectiva* and those who continued to embrace his doctrine. As opposed to the medieval tradition of optics, according to which visual perception occurred in the crystalline humor, Kepler maintained that the retina was the eye's sensitive organ and that the crystalline humor was merely a refractive lens through which light rays converged on the retina and produced a real image.[43] This was the first correct explanation of the mechanism of vision, stemming from his extensive study of refraction. Kepler had tried unsuccessfully to find an exact law of refraction, but his understanding of refraction at a plane surface sufficed for determining the optical properties of lenses, at least qualitatively.[44] Kepler, however, made no mention of the possibility of magnifying faraway objects using a combination of lenses of different shapes. In any case, the *Paralipomena* did not enjoy the success it deserved, not only because those who read the work found it overly complicated—bordering on the abstruse—but also because it was hard to obtain.[45] The latter difficulty was underscored by Galileo on October 1, 1610, when, after the *Dissertatio* had been published, he wrote from Florence to Giuliano de' Medici, who was in Prague, to request a copy: "I beseech your most illustrious lordship to kindly send me the *Optics* [the *Paralipomena*] by Kepler and his treatise *De stella nova*, because I have not been able to find them either here or in Venice."[46]

Kepler aside, Galileo's reference to the "science of perspective" and the "science of refraction" alludes to the very latest literature, in which he could find the concepts he needed to understand how the Dutch spyglass worked and improve on its performance. In other words, he was able to grasp that the ability of lenses to form images depended on the phenomenon of refraction and that to magnify objects he had to exploit the joint effects of concave and convex lenses. As we know, this knowledge alone could not suffice to determine which specific laws of optics were responsible, and things went no further until 1611, when Kepler published his *Dioptrice*, containing the first mathematical theory of the lens and the telescope.[47] Nevertheless, there is no question that the same set of concepts known to Galileo was generally considered perfectly capable of explaining the optics of his instrument.

This is demonstrated by a letter that the architect and mathematician Sergio Venturi sent to Naples on February 26, 1610, about two weeks before the *Sidereus* was published.[48] A native of Siena, Venturi lived and worked in Rome, where he had devised and executed various projects on behalf of the Borghese family. It was thanks to them that, as early as the spring or summer of 1609, he learned about one of the first Dutch spyglasses to reach Italy: the one that Guido Bentivoglio had gone to great lengths to deliver to Pope Paul V in April of that year.[49] Venturi may also have heard about Galileo by November, given that the news had already spread throughout Rome, as attested to by an unpublished letter from the physician Giulio Mancini to his brother in Siena: "[In Rome] one can see those mirrors that multiply sight, of the kind that Galilei is said to have invented."[50] Nevertheless, information may also have reached him through another channel: Bonifacio Vannozzi of Pistoia, a habitué of the Medici court and likewise part of the Borghese coterie.[51] Indeed, it is Vannozzi who provides valuable testimony to the fact that within a short time Venturi had started building spyglasses, gaining a reputation that—at least in Vannozzi's opinion—rivaled Galileo's. "Sergio Venturi of Siena, the great mathematician, has made one of those trumpet-shaped spyglasses to see far away, which in perfection far surpasses those that have arrived from outside: and also Galileo has made another one, which is quite good," Vannozzi wrote. "But at sea, where such instruments would be very beneficial, they cannot be used, as they must be steady and immobile, and there they can never be positioned properly due to constant motion. On high ground, they allow us to see enemies from afar and spy galleys that could attack us from the water."[52]

In any event, that Venturi had made a reputation for himself is confirmed by his letter dated February 26, 1610, a document that is something more than a mere letter of accompaniment for the "spyglass to see at a shorter distance" he had been commissioned to make. It is a short treatise in which the author listed the "principles" that, in his opinion, illustrated how it worked, and which took up ancient writers but also Alhazen, Witelo, and above all Della Porta. All of this led him to the following conclusion: "Having established these principles, I will apply them to the instrument and its parts, which are two spherical sections, one concave and the other convex of the kind of eyeglasses that are said to be for seeing sharp and for seeing large, that is, they represent the visible things smaller and larger than natural vision. Since Mr. Giovanni Battista Della Porta deals extensively with these things in his work on refraction, and your lordship has this very author with him, a man much esteemed in these matters, I refer in this part to this work."[53]

Nevertheless, Venturi's is not the only example we can mention. On April 5, after reading the *Sidereus*, Leschassier turned to Sarpi concerning doubts the book had raised in his mind, noting that in Paris a friend of his had managed to make a spyglass similar to the one "invented in Holland," thanks to the demonstrations provided by Della Porta in his "writings on optics."[54]

In short, before the *Dioptrice* was published, it was on the basis of this practical optical knowledge that Galileo and his contemporaries explained the effects of the Dutch spyglass. Nevertheless, this was a body of knowledge that Galileo had combined with that of the Venetian spectacle makers, from whom he had learned empirical methods in order to select the best lenses and determine their curvature so he could grind and polish them properly. Therein lay the secret to his success: the ability to combine rather elementary optical knowledge with the effective procedures employed by craftsmen to make lenses. This is why he became an impromptu spectacle maker, one so expert that by the end of August 1609 his "long-sighted cannons" were five times more powerful than any others available in Europe.[55]

Yet Galileo did not merely improve the performance of the Dutch spyglass. In just three months' time, he transformed a rudimentary optical device—one virtually everyone planned to exploit for military and naval purposes—into a refined and ingenious astronomical instrument. It was the beginning of a new era: the age of celestial discoveries, destined to change the image and perception of the universe forever.

• •

Galileo turned his telescope to the Moon around September or October 1609.[56] It was not until late November, however, that he embarked on systematic observation, when he finally managed to make a twenty-power instrument. Working extremely quickly, he finished it within a matter of days, which is the time it took to procure the items he had jotted down on his shopping list. This tells us that he knew exactly how and where to work on the lenses to achieve the optical properties he had sought so long. His demanding work as a craftsman and his roaming from spectacle shop to spectacle shop finally yielded results: a telescope allowing him to identify the Moon's previously inaccessible details. This marked promising progress on both the technological and astronomical levels, so much so, in fact, that he felt compelled to make circumspect mention of it just once during this period, in a letter dated December 4. It was addressed to Michelangelo Buonarroti, the master artist's grandnephew, informing him that he would be traveling to Florence the following summer, and then hastily adding: "I will have with me a few improvements to the spyglass, and perhaps some other invention."[57] There is no question, however, that he was alluding to the lunar discoveries he was making with his new telescope. On January 7, 1610, he sent another Florentine correspondent, Antonio de' Medici or perhaps Enea Piccolomini, a long and updated summary of what he had seen with "a spyglass," which represented the "face of the Moon" with "a diameter twenty times larger than what appears to the natural eye." He noted, "The Moon does not have the even, smooth, and clean surface that most people think it and other celestial bodies have but is instead rough and uneven, and, in short, by showing itself to be such, the only conclusion one can reach is that it is full of prominences and hollows, like the mountains and valleys scattered over the terrestrial surface, but much bigger."[58]

This letter is the first draft of what, in Latin and in greater detail, would form the third part of the *Sidereus*. It contains all the lunar observations

Galileo was able to make as soon as he had a powerful enough telescope. It was then that he decided to scrutinize it for an entire cycle, from waxing to waning, and to paint it to illustrate its morphological features. Exploiting the pictorial skills he had cultivated from an early age, he executed seven marvelous watercolors that are striking in their realism and the way they manage to render the plasticity of the surface of the Moon (Plate 6).[59] They are the only images we have of these observations, and four of them were reproduced as engravings in his *Sidereus*. Although they are undated, a comparison of the watercolors and the engravings suggests that the watercolors depict his observations on seven different nights, the first six between November 30 and December 18, 1609, and the last one on January 19, 1610.[60]

Furthermore, it seems that Galileo painted his watercolors "live"—as he was looking at the Moon with his telescope.[61] Before he could commit to paper any of what he saw, however, it was essential for him to use the telescope properly, as he indicated very clearly in his letter of January 7. "The instrument must be held firmly, and therefore, in order to escape the shaking of the hand that results from the motion of the arteries and by breathing itself, it is best to fix the cannon in some stable place," he wrote. "The lenses must be kept clean and clear of the fog or mist created on it by one's breath, humid or foggy air, and even the vapor evaporating from the eye, particularly when it is warm. And it is good if the cannon can be lengthened and shortened a bit, by about three or four digits, because I find that to see nearby objects distinctly the cannon must be longer, and shorter for things further away."[62]

Another adjustment was necessary: "It is best that the convex lens [*il vetro colmo*], which is the one further from the eye, be partly covered and that the opening that is left be oval in shape, since in this way objects are seen much more distinctly."[63] Reducing the aperture of the objective (the convex lens) with an oval diaphragm served the purpose of selecting the area of the lens that was optically best, neutralizing any distortions caused by imperfect grinding of its edges.[64]

The act of seeing, however, was merely the first step. To make what he saw understandable, Galileo also had to turn his observations into pictures. The way he rendered those images gave his observations substance and turned them into something tangible, so to speak. And since his twenty-power telescope had a very limited visual field, just 15 arc-minutes, Galileo's observations had to be reassembled carefully to form a final image.[65]

But let's try to imagine him at work as he turned his telescope to the Moon. He could observe one-fourth of the celestial body at a time, which meant that for complete scrutiny the instrument had to be shifted from one area of the lunar surface to another, requiring at least four explorations. In the meantime, he had to remember each individual portion accurately, try to set it down on paper, and then connect it with the next one to reproduce a final image that would reflect the real object as faithfully as possible. In short, finding the best glass and then grinding and manufacturing the lenses were not the only demanding tasks. Galileo also had to resolve one of the most delicate aspects of the complex operation of viewing: transposing what he saw into an image.

• •

The discovery of the rough, irregular moonscape, with mountains and valleys, unquestionably marked a watershed. It forced an overhaul of the traditional hierarchy of space, demolishing the entrenched conviction that there was a substantial difference between Earth and the celestial bodies. In January 1610, however, Galileo began to observe something that, as he wrote in the *Sidereus*, "greatly exceeds all admiration."[66] Jupiter, surprisingly, had three other bodies near it. Although the planet had been visible since the previous December, it was not until the following month that it seemed to attract his attention, particularly because of that odd configuration. And the fact that it caught him entirely unawares is witnessed by a unique document: another envelope (see Figure 10).

At first glance, it looks like a child's game of tic-tac-toe scribbled on a blank sheet of paper, or perhaps absent-minded doodling. This tiny sketch appears on the envelope containing the letter that Sagredo sent Galileo from Aleppo on October 28, 1609, and it is almost unquestionably Galileo's first diagram of Jupiter's satellites.

We can easily picture the scene. Galileo has been gazing at the sky, his eye glued to the telescope, when he sees something surprising. He lets go of his instrument and grabs the first piece of paper he finds lying on his

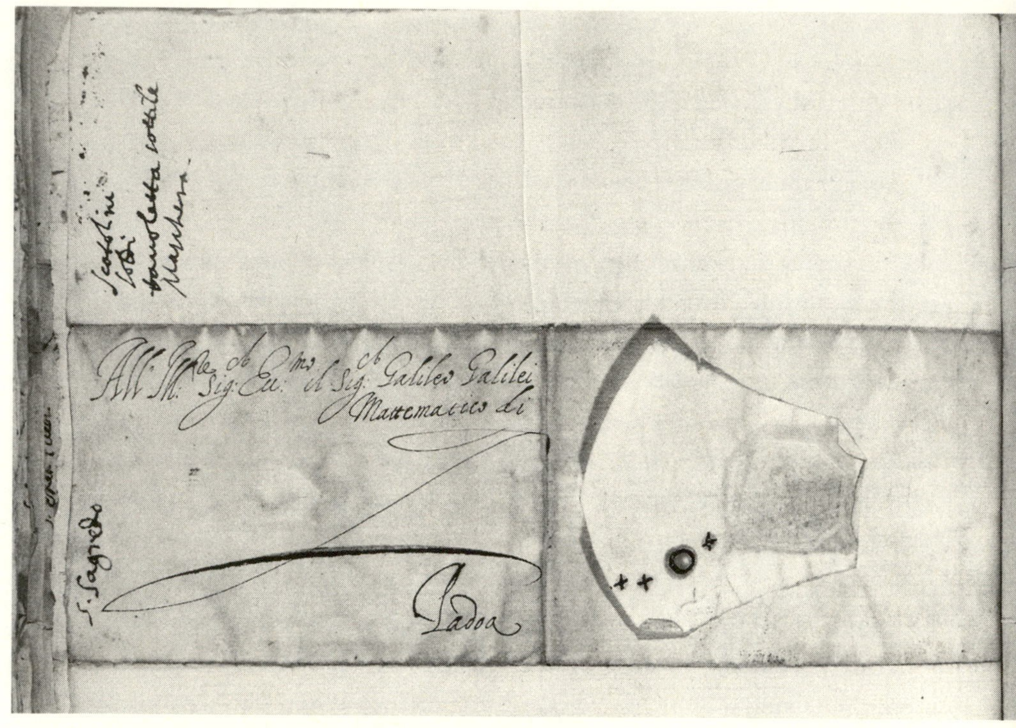

Figure 10. Galileo's first representation of Jupiter's satellites. Autograph note on the envelope of the letter sent by Giovan Francesco Sagredo on October 28, 1609

table: the envelope from a letter he has just received. He draws three small crosses with a little circle in the middle, recording the unexpected observation that he cannot yet explain.

For those in our line of work, discovering a document like this is one of those embarrassing, head-slapping moments: you think about all the times when you were poring over archives and, faced with immense and often excruciatingly boring folders of letters, decided to skip the stack of nearly blank papers with nothing but the name and address of the recipient. It's almost as if you mused, "This document has already been read, so I can skim those 'blank' pages and keep going." This time, however, the document we need is right there on the envelope, clearly evident. This is how Favaro described the information, citing it as a footnote to Sagredo's letter:

Alongside the address, we find this note written by Galileo:

"Small boxes.

Money.

Thin panel.

Mask."

Elsewhere on this paper is a diagram of the Medicean planets.[67]

That's it. And perhaps this is also why such a valuable document has remained unpublished until now.

But when did Galileo receive this letter? We know that Sagredo answered an earlier letter from Galileo, since lost, that was sent from Padua on April 4, 1609, and reached Aleppo "through Constantinople" nearly five and a half months later, on September 16.[68] If it took the same amount of time for Sagredo's reply to be delivered, it would have reached Galileo in early April 1610. However, a series of elements tells us that this was not the case and that Galileo's letter of April 4 was delivered extremely late, perhaps because it had not been entrusted to a professional service.

Starting in 1535, the dispatches and correspondence of the Venetian representatives in the territories of the Ottoman Empire were handled by couriers working exclusively for the Venetian Republic, and they followed a regulated itinerary for which the Dalmatian city of Cattaro (modern Kotor) served as a hub. Mail from Egypt, Syria, and Anatolia was sent to Constantinople, where couriers carried it overland to Cattaro; from there, special ships would transport it to Venice. The same thing happened in the opposite direction, the sole difference being that in Cattaro the letters were first sorted by destination. From Constantinople to Venice, correspondence took an average of thirty days (with a minimum of twenty and a maximum of forty), and this timing essentially remained unchanged between the late sixteenth century and the middle of the eighteenth century.[69]

From Aleppo, where Sagredo was serving as consul, the mail left at the end of every month and took fifteen to twenty days to reach Constantinople.[70] As a result, when there were no major obstacles (bad weather, raids by highwaymen), a letter tended to take between six and eight weeks to

Figure 11. Galileo's sketch of Jupiter's satellites. Letter to Antonio de' Medici (or Enea Piccolomini), January 7, 1610

travel from Aleppo to Venice. Even before mail delivery was regulated, one dispatch sent from Aleppo on December 20, 1507, reached Venice on January 31, 1508.[71] Nevertheless, another consideration forces us to conclude that Galileo must have received Sagredo's letter no later than early January 1610. If he had received it in April, it would not have made sense for him to jot down that configuration of Jupiter, not only because his *Sidereus* had already been published a month earlier but also because it was on January 15—as we are about to see—that he started to keep a log of his observations.

Therefore, there is little doubt that the envelope containing Sagredo's letter bears Galileo's first drawing of the moons of Jupiter. And we also know exactly when he made it: the evening of Thursday, January 7, 1610. That night, in closing the aforementioned letter he was writing to Antonio de' Medici or Enea Piccolomini on his lunar discoveries, he was so struck by what he had just seen that he added—practically in real time—the same diagram he had sketched a short time earlier on the envelope (Figure 11). He wrote, "And besides the observations of the Moon, I have observed the following in the other stars. First, that many fixed stars are seen with the spyglass that are not discerned without it; and only this evening I have seen Jupiter accompanied by three fixed stars totally invisible because of their smallness, and they were configured in this form."[72]

As astonished as he may have been, Galileo had not yet understood just how crucial his observation would prove to be, nor did it cross his mind that the "three fixed stars" next to Jupiter could be satellites. Things quickly precipitated, however, ushering in the most exciting week of his entire scientific career.

• •

Yet there are other surprises in store. Along with the envelope, there is another document that plays a key role in this story: a loose sheet on which, in August 1609, Galileo had begun drafting a letter in order to present his eight-power telescope to the Doge of Venice. In the remaining blank space on the page, he recorded his observations (see Figure 12).[73]

In the wee hours of Friday, January 8, 1610, spurred by the understandable desire to verify what his telescope had shown him the previous evening, Galileo took another look at the sky and again saw Jupiter, comparing it to the diagram of January 7, which he took pains to copy. This time, however, all three little stars were on the right. This was rather unusual, because it meant that, as opposed to what he had expected based on the ephemerides, the planet was not moving in the direction opposite its normal motion: "It was therefore direct and not retrograde." At this point, his sheet shows several gaps in the chronological sequence, as there are no notes for January 9, 10, or 11. They resume on the twelfth, and the two diagrams labeled simply "10" and "11" have nothing to do with his actual observations for those days.[74] But on the night of Wednesday, January 13, Galileo saw Jupiter with "four stars" around it, one to the east and three to the west. He sketched them twice, and in the square on the bottom right he made three small diagrams of the stars with arrows indicating their motion with respect to Jupiter along a straight line, as if they were oscillating back and forth. The following night he did not observe the sky because of the weather: "It is cloudy." On the fifteenth, however, all four were visible to the west, and the fourth, the one farthest from the planet, showed a slight deviation from the straight line. Galileo specified that "the space between the three western ones was not larger than the diameter of Jupiter and they were in a straight line."

Although in the *Sidereus* Galileo stated—clearly alluding to the Copernican system—that on January 11 he realized there were "three stars wandering around Jupiter like Venus and Mercury around the Sun," he reached

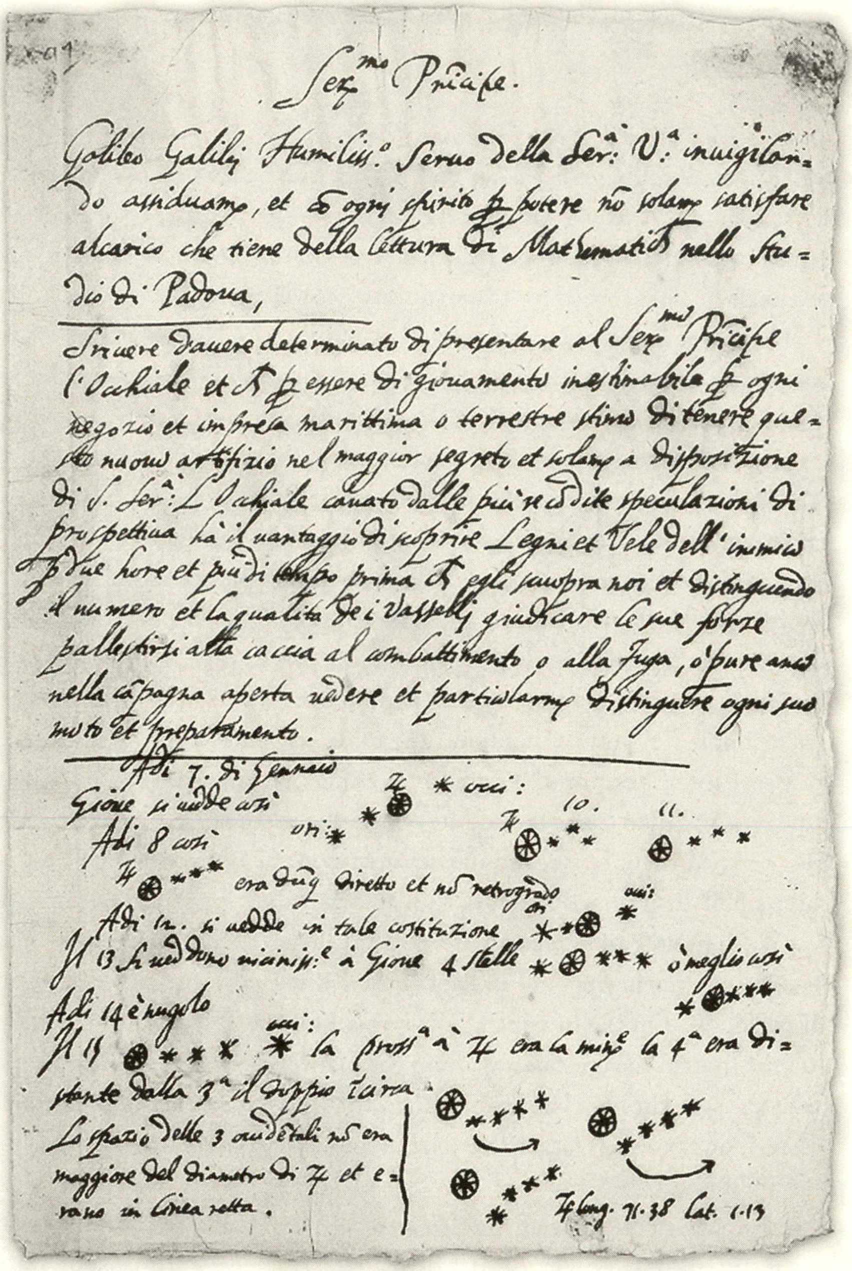

Ser.mo Principe.

Galileo Galilei Humiliss.o Servo della Ser.tà V.a invigilando assiduam.te, et con ogni spirito per potere non solam.te satisfare al carico che tiene della lettura di Matematiche nello Studio di Padova,

Scrivere d'havere determinato di presentare al Ser.mo Principe l'Occhiale et poter essere di giovamento inestimabile per ogni negozio et impresa marittima o terrestre stimo di tenere questo nuovo artifizio nel maggior segreto et solam.te a disposizione di S. Ser.tà L'Occhiale cavato dalle più recondite speculazioni di prospettiva ha il vantaggio di scoprire Legni et Vele dell'inimico per due hore et più di tempo prima che egli scuopra noi et distinguendo il numero et la qualità dei Vasselli giudicare le sue forze palle[s]tirsi alla caccia al combattimento o alla fuga, o pure ancora nella campagna aperta vedere et particolarm.te distinguere ogni suo moto et preparamento.

Adi 7. di Gennaio Giove si vedde così * * ♃ * così: 10. 11.
Adi 8 così ♃ * * * così: * ♃ * * ♃ * * *
era dunque diretto et non retrogrado
Adi 12. si vedde in tale costituzione * * ♃ * così: or.
Il 13 si veddono vicinissime à Giove 4 stelle * ♃ * * * o meglio così * ♃ * * *
Adi 14 è nugolo
Il 15 ♃ * * * * così: la pross.a à ♃ era la minore la 4.a era distante dalla 3.a il doppio in circa.
Lo spazio delle 3 occidentali non era maggiore del diametro di ♃ et erano in linea retta.
♃ long. 71.38 lat. 1.13

Figure 12. Galileo Galilei, report of observations of Jupiter's satellites, January 7–15, 1610; autograph note on a draft of the letter sent to Leonardo Donà on August 28, 1609

this conclusion on the fifteenth, when he decided to keep a regular log to record his observations.[75] In it he copied all the notes and drawings (see Figure 13) that had initially been entrusted to the loose piece of paper, filling in the gaps with data he may have jotted down on another sheet that has not been preserved.[76]

The evening of January 15 proved to be decisive. Galileo observed Jupiter until very late that night, and in his sketch of the stars accompanying it he changed his mind and emphasized that "they did not form an entirely straight line."[77] The seemingly odd motions of the stars he had followed for seven consecutive nights suddenly had a completely obvious explanation: they were satellites revolving around Jupiter.

The discovery was so sensational that it immediately had to be translated into illustrations. "They will be cut in wood, all on one block, the stars white and the rest black, and then they will be sawn into strips," he noted.[78] These are the last two lines written in his log in Italian, while all the subsequent notes, including the observations made later that night, are in Latin. There can be only one explanation for this change: his realization that the discovery could bring him great prestige and had to be circulated in a language accessible to an international audience.

By January 30 Galileo was already in Venice to stipulate an agreement for publishing his work.[79] Indeed, that day he wrote to Vinta: "I am currently in Venice to print certain observations I have made of the celestial bodies using my spyglass." He told him about the lunar discoveries, and then that he had identified "a multitude of fixed stars" and had established "the nature of the Milky Way." He closed his letter by saying:

> But the greatest marvel of all is that I have discovered four new planets and observed their proper and particular motions, which differ in relation to each other and from all the other motions of the other stars; and these new planets move around another very large star, not otherwise than Venus and Mercury, and perchance the other known planets as well, move around the Sun. Once this treatise is printed, which I will send all the philosophers and mathematicians as a message, I will send a copy to the Most Serene Grand Duke, along with an excellent spyglass, which he can use to verify all these truths.[80]

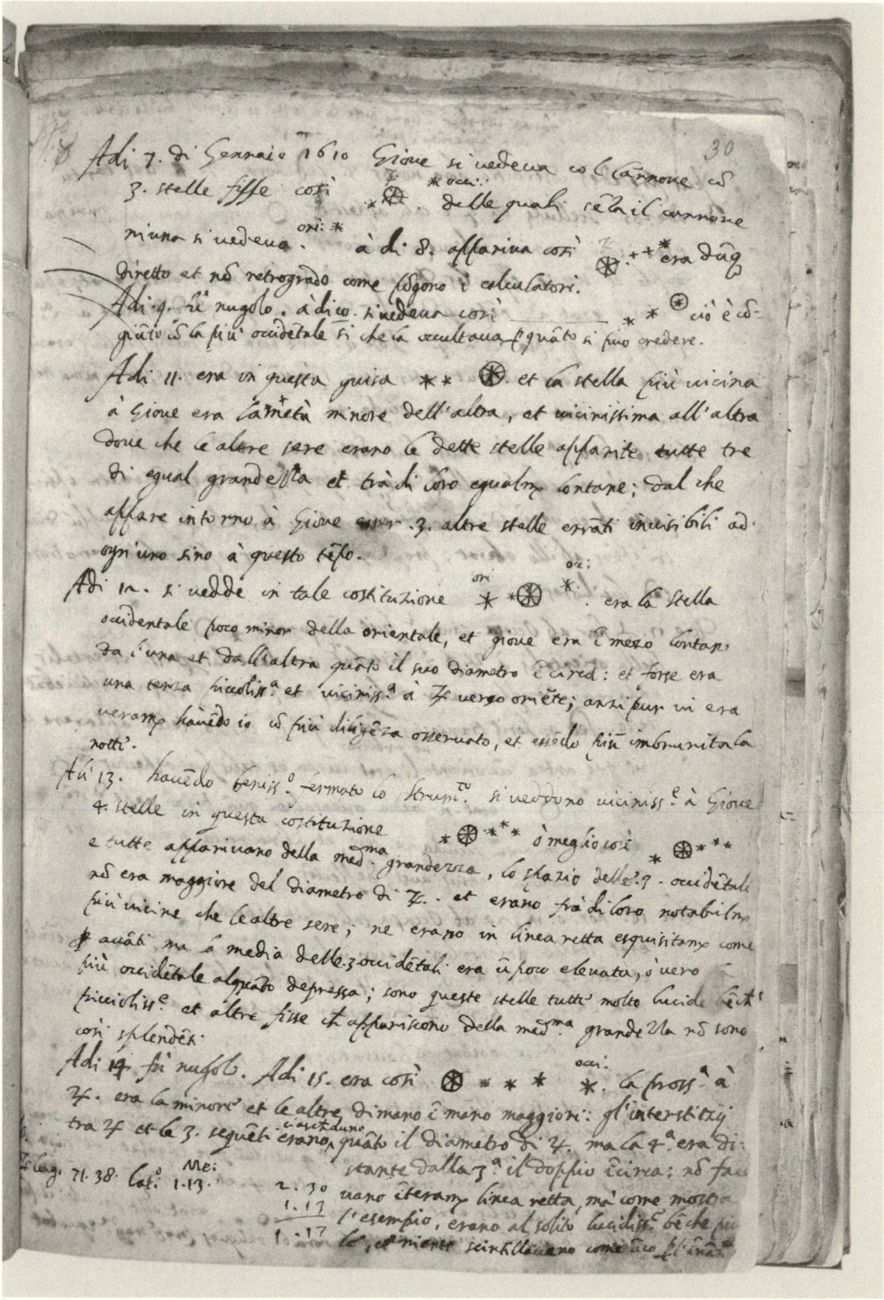

Adi 7. di Gennaio 1610 Giove si vedeva col cannone con 3. stelle fisse così * ⊛ * delle quali senza il cannone niuna si vedeva. a dì 8. appariva così ⊛ * * * era dunque diretto et non retrogrado come pongono i calculatori.
Adi 9. si nugolò. a dì 10. si vedeva così * * ⊛ cioè congiunto con la più occidentale sì che la occultava per quanto si può credere.
Adi 11. era in questa guisa * * ⊛ et la stella più vicina à Giove era la metà minore dell'altra, et vicinissima all'altra dove che le altre sere erano le dette stelle apparite tutte tre di egual grandezza et tra di loro egualmente lontane; dal che appare intorno à Giove esser 3. altre stelle erranti invisibili ad ogn'uno sino à questo tempo.
Adi 12. si vedde in tale costituzione * *⊛ *. era la stella occidentale poco minor della orientale, et Giove era in mezzo lontano da l'una et dall'altra quanto il suo diametro in circa: et forse era una terza piccolissima et vicinissima à ♃ verso oriente; anzi pur vi era veramente havendo io con più diligenza osservato, et essendo più imbrunita la notte.
Adi 13. havendo benissimo fermato lo strumento si veddono vicinissime à Giove 4. stelle in questa costituzione *⊛*** ò meglio così *⊛*** e tutte apparivano della medesima grandezza, lo spazio delle 3. occidentali non era maggiore del diametro di ♃. et erano fra di loro notabilmente più vicine che le altre sere; ne erano in linea retta esquisitamente come per avanti ma la media delle 3. occidentali era un poco elevata, overo la più occidentale alquanto depressa; sono queste stelle tutte molto lucide benchè piccolissime et altre fisse che appariscono della medesima grandezza non sono così splendenti.
Adi 14. fu nugolo. Adi 15. era così ⊛ * * * *. la prossima à ♃. era la minore et le altre di mano in mano maggiori: gl'interstizii tra ♃ et le 3. seguenti erano tutti uguali quanto il diametro di ♃. ma la 4a era distante dalla 3a il doppio in circa; non facevano esattamente linea retta, ma come mostra l'esempio, erano al solito lucidissime benchè piccole, et non scintillavano come anco per l'innanzi.

Long. 71.38. Lati. 1.13. Me: 2.30 1.13 1.17

Figure 13. Galileo Galilei, first page of the autograph record of observations of Jupiter's satellites, 1610

This was followed by an exchange of letters that continued almost until the eve of the publication of the *Sidereus nuncius.* On February 6 Vinta informed him that the Medici had expressed an "extraordinary desire to see these observations as soon as possible."[81] Galileo replied on the thirteenth, when he returned to Padua, and he told Vinta that he intended to dedicate the new celestial bodies to the grand duke. However, he needed some advice: "I am unsure about one thing: whether I should dedicate all four to the Grand Duke alone, calling them Cosmici after his name, or if, given that they are exactly four in number, I should dedicate them to all the brothers with the name Medicea Sydera."

He thus entrusted Vinta with the task of establishing which name showed "more decorum," but he also had two other requests: "One is the secrecy that has always accompanied your other most serious dealings; and the other is an immediate reply, because for this alone I am holding up the printing, as all I have left to establish is this matter of the title and dedication. Tomorrow I am returning to Venice, where I will await your response, which, if you please, you might entrust to the postmaster there, so that it will not fall into someone else's hands and be sent to Padua."[82]

On February 20 Vinta sent his opinion (the new planets would be called the Medicea Sydera), which arrived just in time to be incorporated into the dedication and the title page. During these negotiations, in addition to following the various printing phases closely, Galileo continued to stargaze; on January 31 he saw the Pleiades and on February 7 the constellation of Orion's belt and sword.[83] The last observation of Jupiter's satellites included in the published text was that of March 2.

After two months of feverish work, during which he also oversaw the engravings that were to accompany the publication as well as the final administrative formalities for its imprimatur, Galileo completed the dedicatory letter to Cosimo II, dated March 12. His *Sidereus nuncius* finally came off the press of Tommaso Baglioni on the night of Saturday, March 13.[84]

FOUR
In a Flash

Sarpi and Galileo fell out with each other once and for all that day. According to the Venetian theologian, the *Sidereus* bore an unequivocal unwritten message that had all the air of betrayal. Its author had eliminated any mention of one of the parts—by no means negligible—that had made it possible, and what could have become an extraordinary opportunity for the international success and prestige of the Venetian Republic had been transformed into unexpected and bitter disappointment. The book said nothing about where the telescopic observations had been made, and above all it expressed no sign of gratitude toward Venice and the Venetians who had helped make the spyglass. This was enough to make Sarpi comment that he hadn't even read the book. After all, shouldn't the man who had instigated the whole affair be paid back in the same coin?

Viewed from a perspective outside the palaces of Venetian politics, however, those spare pages of the *Sidereus nuncius* were so powerful that no interpretation based on the present could belittle them. The images of the lunar mountains and valleys, the clear outlines of the stars composing the Praesepe and the Orion Nebula and the constellation of the Pleiades, and the discovery of the four celestial bodies around the planet Jupiter, with the day-by-day depiction of their motion and exact position, made this text unique among the books of the era. It contained a different way of seeing not only the Earth but also humans and their relationship with the world and nature. In the intellectual history of the modern age, perhaps no other book circulated as quickly and widely as the *Sidereus nuncius,* and only maps can instantly reveal the full extent of its success. Only the publication of Charles Darwin's *Origin of the Species* (1859) would achieve the same international acclaim.

The celestial message and its unknown herald seemed to have come from another world. One merely had to turn those few pages to grasp that it would very quickly become an exceptional and prolific generator not only of books but also of pieces of history and real life capable of triggering a rapid chain of events, each of which would become an integral part of this boundless adventure.

• •

On the evening of Saturday, March 13, an unbound copy—its pages still wet—was promptly sent to Belisario Vinta in Florence. The grand duke's beautifully bound book would be shipped the following week, the time it took for Galileo to return to Padua and send the spyglass along with it.[1]

The grand duke and the friends at court who supported Galileo's plans to move to Florence were thus foremost in the scientist's mind. He wrote to Vinta, in particular, to warn him that using the spyglass might be disappointing, "because those unaccustomed to it need great patience at first, as they do not have someone to adjust the instrument and stabilize it properly," and he alerted him that in summer the four satellites could not be observed because of their proximity to the Sun.[2]

As this initial information was traveling toward Florence, it crossed paths with other news. While in the fall of 1608, following the invention of the Dutch spyglass, The Hague had been the center of constant correspondence among the principal European countries, at this point it was Venice that would play a pivotal role. One of the first to communicate this was Mark Welser, a Bavarian banker very well connected to the Jesuits. Just one day before the *Sidereus* was published, Welser sent a letter from Augusta to Rome, to Christoph Clavius, the prince of mathematicians of the Society of Jesus, saying, "From Padua I have received trustworthy written reports that Galileo Galilei, mathematician at that university, using the new instrument many have named *visorio*, of which he establishes himself as the author, has found four planets new to us, as they have never been seen by any mortal person as far as we know, and also many more fixed stars, never known or seen before, and wonderful things [*mirabilia*] in the Milky Way. I am well aware that *tarde credere est nervus sapientiae* [slowness in believing is the sinew of wisdom], so I cannot commit myself to anything, but I beg Your Reverence to tell me your opinion on this freely and confidentially."[3]

Although the source (probably the Paduan Lorenzo Pignoria) was exceedingly reliable, the news was too surprising to be accepted without being investigated further. Before believing it, Welser needed an authoritative opinion that could confirm what he had learned, and there was no one in the world more expert and up to date than Clavius.

The English ambassador to Venice, Sir Henry Wotton, also talked about *mirabilia*. This was too sensational to wait another moment to inform England. It was "the strangest piece of news (as I may justly call it) . . . ever yet received from any part of the world," he confessed to Sir Robert Cecil, Earl of Salisbury, on March 13, the day the work was published. "Come abroad [published] this very day of the Mathematical Professor at Padua who by the help of an optical instrument (which both enlargeth and approximateth the object) invented first in Flanders, and bettered by himself, hath discovered four new planets rolling around the sphere of Jupiter, besides many other unknown fixed stars."[4]

He sent this immediately, saying that by "the next ship your Lordship shall receive from me one of the above-named instruments, as it is bettered by this man." Unlike Welser, however, Wotton also provided a summary of the real innovations: the four new satellites, the discovery of the "true cause" of the Milky Way, the multiple prominences on the Moon, and—what he considered the oddest thing of all—that, as opposed to what was believed and taught, the Moon not only was illuminated directly by the Sun but also received reflected light from Earth.

He was so well informed, in fact, that he did not hesitate to provide a general opinion, immediately underscoring the debates this work was bound to spark—and not only from an astronomical standpoint. "So as upon the whole subject he [the author] hath first overthrown all former astronomy for we must have a new sphere to save the appearances and next all astrology. For the virtue of these new planets must needs vary the judicial part, and why may there not yet be more? These things I have been bold thus to discourse unto your Lordship, whereof here all corners are full. And the

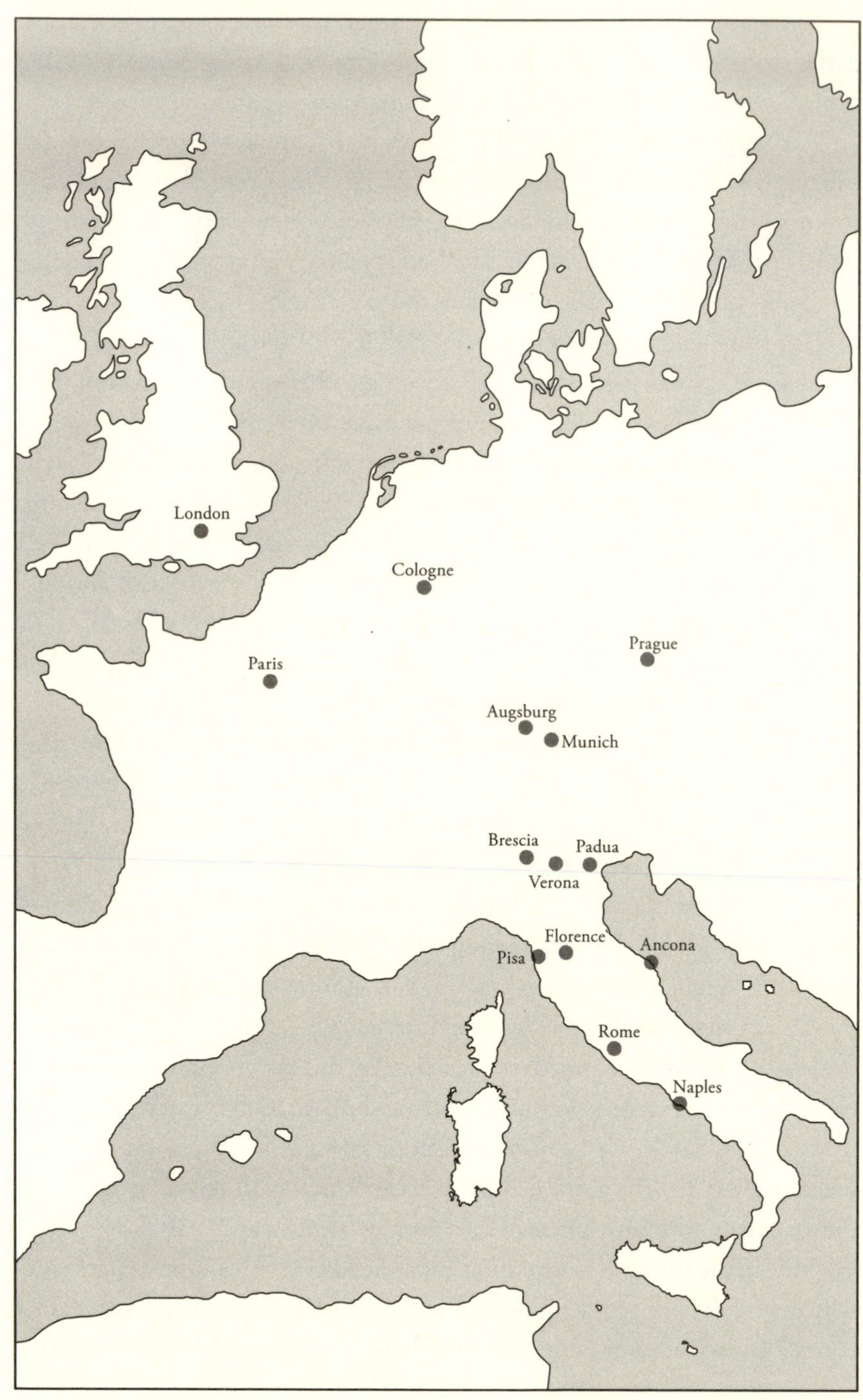

Map 2. The publication of *Sidereus nuncius:* spread of the news of the publication and circulation of the copies, from March 13 to April 30, 1610

author runneth a fortune to be either exceeding famous or exceeding ridiculous."[5]

This was, to say the least, stunning news that had everyone in Venice abuzz; if proved true, it would have incalculable consequences far beyond the peaceful and narrow territorial waters in which the books of mathematicians and astronomers tended to navigate. This is why he immediately wrote to the court of King James: because it was a topic that could not be limited to the specialists and professionals of culture, but instead involved the broader sphere of politics.

Although the *Sidereus* was addressed to philosophers and mathematicians, as specified on the title page, the radical changes it portended were universal in scope, and this is precisely why it would also be read by people outside these fields. It was a book *for everyone* that had nothing in common with the philosophy and astronomy books of the day. One did not even need to know Latin to understand it, as, paradoxically, it didn't matter if one actually read it or not. One merely had to leaf through it and look at the five plates illustrating the Moon, the four of the Milky Way, and the "little sketches" of the Jovian satellites, arranged chronologically, to realize that this little book was nothing like the others describing the heavens and measuring the distances between the planets. It was the images that made it uncommon, and the secret lay not in the written text, nor in the astronomy books of the past, but *outside* the realm of words and in the existence of a new instrument. The infinite distance separating this book from all the previous works *de coelo* lay precisely in the predominance of images over text. This is what made it so innovative and original: words accompanied images, rather than the other way around.

• •

The news swiftly reached Rome. It was Giovanni Battista Amadori, one of Galileo's Florentine friends, who informed Cigoli that the book had been published. On March 18, just five days after it had been printed, Cigoli urged Galileo to send him a copy.[6] That same day in Naples, Giovan Battista Manso, the poet friend of Torquato Tasso and Giambattista Marino and the future author of the celebrated biography of Tasso, praised the man akin to a "new Columbus" for his courage in embarking on "paths no longer taken by the human intellect."[7] Even though the lenses he had were mediocre and did not allow him to discern "those valleys and those mountains, and that rough [lunar] surface," the news Paolo Beni had sent him from Venice left no room for doubt: these were not deceptions produced by the lenses, but real things. In spite of this, the information did not fully satisfy him. Manso felt that these discoveries had to be circulated, starting with a philosophy capable of supporting them, yet no such philosophy existed. Consequently, he felt that he did not know how to categorize a Moon as mountainous as the Earth "in philosophy, neither solely as the fifth element imagined by Aristotle, but neither according to Plato's principles." And then there were the four new planets and their movements, "which do not seem to coincide with any opinion of philosophers nor astrologers, and even less so with the observations and demonstrations made so far by Ptolemy or Copernicus or Fracastoro."[8] Moreover, like Wotton, Manso also noted the disquietude and protests of astrologers and physicians who, faced with the circulation of the first but unverified information regarding this telescopic news, already feared for their own professions. "They think that [due to the discovery of new celestial bodies] astrology will perforce be ruined and most of medicine demolished, so that the distribution of the houses of the zodiac, the essential dignities of the signs, the qualities of the natures of the fixed stars, the order of chroniclers, the government of the age and of men, the months of the formation of the embryo, the reasons of the critical days, and thousands of other things depending on the sevenfold number of the planets would essentially be destroyed from their very foundations."[9]

Interest in the *Sidereus* far exceeded the boundaries of astronomy (the book was immediately listed in the catalogue at the fair in Frankfurt).[10] This is unsurprising, given the multiple meanings the people of the day attributed to interpreting the heavens. By March 19—just six days after it was published—the 550 copies had already sold out. Of the thirty the author himself was supposed to receive from the printer, Galileo managed to obtain only six.[11]

Galileo decided to print another edition forthwith, this time in the vernacular and enriched with new observations and, above all, images of complete lunation.[12] That day, he mailed two letters to Florence. The first, to the grand duke, accompanied the book and a "rather good" spyglass.[13] The second, however, of which there is an unfinished autograph draft of the original, was addressed to Vinta and contained a detailed list of the personages to whom the book and lenses should be sent as soon as possible. This was a task Galileo could not handle personally, as it required the full efforts of diplomacy and was thus an official deed of the grand ducal state. Yet there was a limitation that was by no means negligible, for of the more than sixty spyglasses he had made, few were truly excellent. "These few I planned to send to great princes, and in particular the relatives of the Most Serene Grand Duke; and I had already received requests from the Most Serene Dukes of Bavaria and the Elector of Cologne, and also the Most Illustrious and Reverend Cardinal Dal Monte; to whom I shall send it as soon as possible, along with the treatise. My wish would be to send some also to France, Spain, Poland, Austria, Mantua, Modena, Urbino, and wherever Your Most Serene Highness may wish; but without some support and coverage of expenses, I would not know how to ship them."[14]

These were places "friendly" to Florence; more important, they were all strictly Catholic. It would be up to the grand ducal machine to get things moving to circulate the book and lenses as widely as possible, and this had to be done not so much to promote the work of a "servant" and "vassal" but because it would celebrate the prestige and power of the Medici family, whose name—before that of any other sovereign—had been carved in the heavenly book of God.

Book and spyglass had thus become an affair of state. Galileo declared this to his main political liaison in no uncertain terms: "My sole aim is to continue this great undertaking *involving* Our Most Serene Lord."[15] The undertaking would be even greater if "as many people as possible" could "see and recognize the truth."[16] To achieve this goal, however, the collaboration of Tuscan diplomacy was indispensable, and it was promptly provided, as can be noted based on these specific arrangements that Vinta communicated to Asdrubale Barbolani:

> If Galileo Galilei wants to send a few of his books printed about the Medicean stars to Court of Caesar [the imperial court], kindly accept them

> and place them in the parcel of the Most Serene Master, which will please His Highness, and if he wishes also to send his telescopes in order to observe these stars, accept them as well. But you must think carefully on how to reduce them to a size that can be packaged so as to take up little space, because if they are as long as trumpets we must think of some other way, either through merchants or couriers, and if he should like to send such things to London, to Secretary Lotto, accept them as well, and help him, and facilitate their proper delivery, thereby pleasing the Most Serene Master.[17]

The *Sidereus nuncius* thus became the *Medicea sidera*, the "book printed about the Medicean stars," to be sent to the ambassadors in Prague and London via diplomatic courier, as was done for the dispatches on military, religious, and political matters that were sent to every corner of Europe nearly every day. Therefore, it comes as no surprise that the message and telescope were delivered to princes and cardinals rather than mathematicians and astronomers, "thereby pleasing the Most Serene Master," as Vinta had observed, and despite the less-than-flattering opinions about Galileo himself that members of the Medici entourage continued to send to Florence.[18] Among them, the views that stood out once again were those of Giovanni Bartoli, ever ready to recall the close friendship between Galileo and Sarpi and to emphasize the rumors going around about him in a number of milieus, painting him as shrewd and unscrupulous. "He is a very capable man and is a very good friend of Fra Paolo," Bartoli repeated on several occasions. He could not help but add that in Venice his book was "read by everyone . . . [and] in [it] he shows that, with his spyglass, he has found four other planets and seen another world on the Moon, and similar things that provide pleasant food for thought to the professors of those sciences, above all because of the title *Sidera Medicea*."[19]

• •

It was March 27, and two weeks had elapsed since the printing of the *Sidereus*. That day, writing respectively from Pisa and Florence, Enea Piccolomini and Alessandro Sertini sent Galileo a letter full of admiration, informing him of the great interest his instrument had aroused.

The former told him that the grand duke was eager to try the spyglass, "hoping to see a new miracle," unlike those who had already shown themselves to be determined to reject "those things you affirm you have seen and wish to show everyone."[20] Sertini instead informed him that "Florence is full of spyglasses brought in from Venice at the request of a number of people, which are quite reasonable." The city—and not only the court—was ready to give a worthy welcome to those celestial novelties and their discoverer, and with delightful irony Sertini recounted that the eagerness to see and try the spyglass was growing daily. "Yesterday morning, upon arriving at the Mercato Nuovo, Filippo Mannelli came up to me, telling me that Piero, his brother, had written to him that the courier from Venice had a little box for me from you," he wrote. "This news spread and I could not defend myself from those who wanted to know what was in it, thinking it might be a spyglass; and when they learned that it was the book, their curiosity did not abate, above all among men of letters, and I believe that Don Antonio [de' Medici] will be busy illustrating it. Last night at Messer [Francesco] Nori's house, we read a passage, the part that deals with new planets; and in the end it was considered a great and marvelous thing."[21]

What is striking is the growing number of people in Florence who learned about the discoveries in such a short time, to the point that poor Sertini found himself in a bind once word got out that he had received a "little box" from Galileo. These were the people who, a short time later, would shock the Dominican friars of San Marco, as they were open to the "thousand impertinencies being sown throughout our city, maintained very Catholic by her good nature and by the watchfulness of our Most Serene Princes."[22]

It took less than two weeks for the news about the spyglass and the starry messenger to spread across Europe. London, Augsburg, Rome, and Naples were the first cities to hear the name of an unknown mathematician who gained consensus but also sparked heated debates. Other cities would soon follow, starting with Prague, the residence of Emperor Rudolf II, and Bologna, where Giovanni Antonio Magini taught. And Bologna would turn out to be one of the first cities Galileo visited to "christen" his spyglass.

FIVE
Peregrinations

Bologna, April 25, 1610: a Sunday. It was the feast of Mark the Evangelist, and on the city's packed calendar of rituals, the day was celebrated with particular devotion. Ceremonies started early in the morning with the pilgrimage of the Compagnia dell'Aurora to the sanctuary of Madonna del Sasso and continued until evening, culminating with the crowded procession of the Greater Litanies and a solemn Mass sung before the archbishop.[1]

St. Mark was the patron of notaries, scribes, glassmakers, glass painters, and opticians, so he was clearly the ideal patron for what Galileo was doing in town that day. Not far from the church dedicated to the saint in Piazza di Porta Ravegnana, the scientist was trying to convince a select audience of local doctors and scholars that Jupiter's satellites unquestionably existed.

The meeting took place at the residence of the nobleman Massimo Caprara, now known as the Palazzo Caprara-Montpensier.[2] Galileo arrived there after having participated in a similar encounter the previous evening at the home of Giovanni Antonio Magini, with an audience of about twenty people.[3] Those in attendance included the Bohemian Martin Horky, who lived at Magini's house, serving as copyist and secretary; the indefatigable compiler of astrological yearbooks Giovanni Antonio Roffeni; the musicologist and mathematician Ercole Bottrigari; and the theologian Paolo Maria Cittadini. It is likely that this coterie included the philosopher Flaminio Papazzoni, said to be intent on refuting celestial novelties "explicitly" *(ex professo)* at public schools.[4] Everyone was keenly interested in personally verifying the most incredible of the discoveries announced in the *Sidereus nuncius:* the Medicean planets. The curiosity of many of these figures also concealed a spiteful hope for failure, as they assumed these astonishing revelations would prove to be exactly what they suspected: a trick—clever and perhaps even brilliant, but a trick nonetheless.

Magini was indubitably the one best able to assess the matter. He had already expressed huge reservations about the reliability of the telescope and was thus quite skeptical about the existence of Jupiter's satellites. An excellent astronomer famous for his skill in astrology and calculating planetary motion, he enjoyed a solid international reputation; only a few days earlier, in fact, Kepler had invited him to Prague to help develop new ephemerides.[5] So it makes sense that he had a hard time swallowing Galileo's success. Moreover, he had yet to forgive the scientist for occupying the university chair in his own birthplace, Padua. As the very well-informed Martin Hasdale wrote from Prague, "Magini is displeased that others have moved forward, particularly in his homeland, and if it were elsewhere, it would sting less." This triggered Magini's determination to "efface the merits" of his rival in that "profession in which he would merely like to be a Phoenix."[6] Despite their antagonism, Magini welcomed Galileo to his home and, in his guest's honor, even laid out "a sumptuous and delicate banquet."[7] Magini did not own a telescope and perhaps had never even had the chance to use one. Nevertheless, he opined that the spyglass was an insidious generator of optical illusions. "As to Galileo's book and instrument, I believe it is a trick," he wrote, "because when I used the colored spectacles I made to observe the solar eclipse, I saw three Suns; so I think this is what has also happened to Galileo, who must have been deceived by the reflections of the Moon."[8]

His assistant, Martin Horky, shared this view and was determined to join the fray against these celestial discoveries. At the end of March Horky had already asked Kepler for an opinion on the Jovian satellites, stating, "Est res miranda, est res stupenda: vera an falsa ignoro" (it is a wonderful thing, an astonishing one: true or false, I don't know).[9] On April 6 he revealed that he wanted to refute the *Sidereus nuncius* and its "quatuor ficti planetae." Ten days later he denounced the pernicious astronomical consequences of the Galilean "fabula," if anyone were to believe them: the ephemerides "secundum fundamenta Tychonica" that Magini planned to publish would have to contemplate no less than eleven planets, rather than the usual seven.[10]

But Magini and Horky were not the only skeptics in Bologna. Others were of the same opinion, as can be gleaned from Roffeni's little-known testimony, and they were prepared to give these astounding discoveries the welcome they felt they deserved.

> One cannot deny that as soon as the *Sidereus nuncius* arrived in Bologna, many minds were astonished by the news it contained. In fact, some found it incredible that, after so many centuries, the four planets encircling Jupiter had never appeared to the many excellent observers of celestial phenomena. Others stated that those new marvels in the heavens should be rejected as a visual hallucination caused by the refraction of the concave and convex lenses set at the ends of the telescope. . . . Nevertheless, each one wanted to try [Galileo's] spyglass, so that by using it, once they rejected this opinion after careful experience, they could acquire some sort of certainty.[11]

The expectation was palpable, and Galileo's arrival in the city had the air of a big event. For anyone who read the *Sidereus*, all that was left to do was to ascertain the reliability of that information—using his very own spyglass.

• •

According to Horky, the observations conducted at the houses of both Magini and Caprara failed miserably. On April 27, he wrote to Kepler: "I never slept the day and night of April 24 and 25, as I was testing Galileo's instrument in thousands and thousands of ways, both on land and in the sky. On land it does wondrous things, but it is deceptive in the sky, as it shows the fixed stars as if they were double."[12]

In the *Brevissima peregrinatio*, the brief essay he composed a short time later, he repeated: "That night, at the house of the most illustrious Messer Massimo Caprara and in the presence of numerous very noble lords, with your [Galileo's] telescope we saw the Spica Virginis doubled. The doubling of this star was first shown to you by Dr. Antonio Roffeni; you instead denied seeing it double, as admitting mistakes is something harder than diamonds."[13]

As was the case with Magini, Horky too found that the reflection of the light of the stars was deceptive. Yet the telescopic experience was not new to him, as he had had the chance to test the spyglass in Padua the previous March during public demonstrations at Galileo's house.[14] It was perhaps because he was familiar with use of the telescope that he did not deny having spied the satellites at the Bologna encounter, noting that "on both April 24 and the following night I saw only two small globes or, rather, two very small spots." However, "on April 25, since God answered my prayers and the sky was as clear as possible, Jupiter rose, showing itself in the West and appearing with all four new companions [*novis quatuor famulis*] over our horizon in Bologna."[15]

Yet Horky continued to maintain that Galileo's discovery had been refuted and that those "very small globes or spots" could not be acknowledged as an astronomical reality. They were not stars that had never been seen before but hallucinations, optical illusions, a result of the instrument's unreliability. To back his stance, Horky cited the testimony of "excellent men and illustrious doctors" who "made various observations of the sky on the night of April 25, and everyone acknowledged that the instrument was deceptive."[16] Demoralized and humiliated, Horky wrote, Galileo was devastated by the debacle, physically ailing, and on the verge of a nervous breakdown.

> Throughout Bologna he has a terrible reputation: he is losing his hair; all of his skin is devastated by the French disease; his skull is ruined and his mind delirious; his optic nerves have been destroyed because he has observed the minutes and seconds around Jupiter with too much curiosity and presumptuousness: he has lost his sight, hearing, taste, and touch; his hands suffer from chiragra as he has illicitly pilfered the treasure of philosophers and mathematicians; his heart suffers from palpitations because he has passed his celestial fable off to everyone; since he can no longer convince scholars and illustrious persons, his intestines have issued an unnatural tumor; and since he has wandered hither and yon, his feet show signs of gout. Blessed the physician who can restore the sick *Nuncius* to health.[17]

This description is so bleak that even his worst enemy could not have been more ruthless. There was not a speck of truth in all this, but as we all

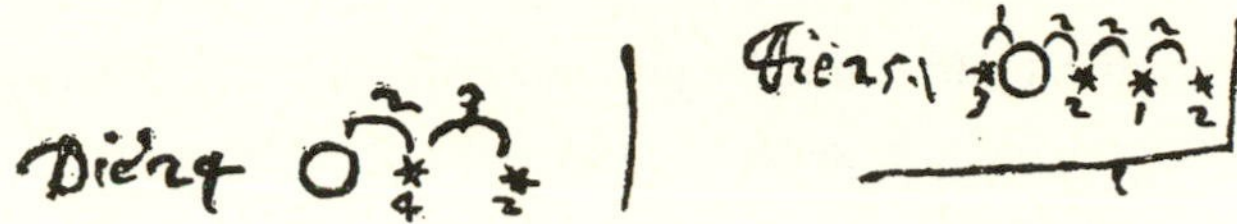

Figure 14. Galileo Galilei, diagram of the positions of Jupiter's satellites, April 24 and 25, 1610

know, slander and disparaging remarks can wreak great havoc and write far more pages of history, then and now, than we can possibly imagine. Naturally, there is little point in saying that Galileo—not Horky or Magini—was right. Nor that his observations of April 24 and 25 (Figure 14) attested to the presence of two satellites the first day and all four the next night.[18]

The fact is, however, that in the short term Magini and Horky's words prevailed over Galileo's. Because of their obsessive campaign, news of the Bologna event reached European courts in a matter of weeks. In three letters sent to Prague, no less than twenty-four experts "in the profession"—eyewitnesses to Galileo's attempt to "demonstrate his book with his instrument"—declared that "not one would confess he had seen something, and all stated that they could not see any of what [Galileo] said he could see."[19] On May 26, even the most authoritative among them, Magini, unhesitatingly confirmed to Kepler that the telescope was a flop: "I find it hard to believe that Galileo will emerge victorious in this. On April 24 and 25 he stayed at my house with his telescope, as he wanted to show us Jupiter's new satellites. He was unable to do so. More than twenty of the most educated men were there, yet none of them could see the new planets distinctly."[20]

In the end—again according to Horky's recount—early on the morning of Monday, April 26, a glum Galileo, bitter over the objections and criticisms that had been raised, had no choice but to make his sad way back to Padua, without even bothering to thank his host, Magini, for the favor or for all the food for thought he had received in Bologna.[21]

• •

Brevissima peregrinatio contra Nuncium sidereum: this is the title of Martin Horky's booklet, the first printed work against telescopic discoveries.

The *peregrinatio* was a fortunate literary topos exalting the concept of travel as an ethical and cognitive journey of perfection, comparing it to a path of truth in which the sequence of situations imparts useful teachings to the world and ourselves. As the Spanish poet and playwright Bartolomé Cairasco de Figueroa wrote:

> Peregrination is not that disorderly,
> restless, and anxious wandering
> of one who walks out of curiosity.
> Nor of he who, because of melancholy fate
> or need or vainglory
> or idle intent, is peregrine.
> The peregrination that on earth and in heaven,
> is worthy of praise and memory
> and is praised in this story
> is one that, with pious zeal,
> out of desire or obligation,
> visits the places the sky shows us here.[22]

Visiting "the places the sky shows us" with a spirit of seeking the truth: this is precisely what Horky set out to do. The sky was now Galileo's new sky, and the place to explore naturally had to be Jupiter, with four enigmatic "new planets" revolving around it. As Horky wrote, "I have wandered the Earth enough; now the *Sidereus nuncius,* which reveals great and extraordinary marvels to all mortals, pushes me to another most noble celestial peregrination around the star of Jupiter. I have certainly not commenced this perilous journey of my own free will, as I was led to the sky of Jupiter by the father of this *Nuncius*."[23]

This was about restoring truth, the same truth that we seek both in ourselves and the world on journeys from place to place and city to city. Horky viewed the celestial *peregrinatio* as the continuation of his uninterrupted roaming on Earth. His ceaseless search for unknown places and new experiences brought him from his native Lochowicz, in Bohemia, to France and then Italy to study philosophy and medicine.[24] A letter sent to Kepler in January 1610 gives us a more accurate idea of his European itinerary:

> I visited Tübingen in the year 1608; I was in Strasbourg, Heidelberg, Altdorf, Basel, Freiburg; then I left for Paris, in the company of the noble Silesian Valentin Zeidlic, to see the beautiful French capital. After a year there, I went to Venice, lady of the sea, or, rather, the den of Venus and (according to Lipsius) enviably beautiful for everything concerning the pleasure of Venus, opulent and extraordinarily pleasant: I spent three months visiting her. But since she was not wholly to my liking—in fact (again according to Lipsius), she is more a friend of commerce than of knowledge—I went to Padua, which I also did not like. I wished instead to enjoy the vision of the mother of studies and scholars: Bologna. I shall stay here for a few years to study medicine and mathematics.[25]

Despite his plans, Horky's Bolognese sojourn was very short and he was soon on the road once more. We know little about him, but we do know that after a quick stop in Prague, he returned at the end of 1610 to his native city, where he entered the medical profession. Between 1616 and 1618 he was in Constantinople, holding positions at the imperial embassy. By the time he returned to his homeland in 1619, the Thirty Years' War had already broken out; expelled from Bohemia, Horky began to roam across Germany.[26] In 1624 he was probably in Rostock, where he printed a pamphlet with instructions on how to avoid getting the plague.[27] In 1632 he settled in Hamburg, working as a physician and writing forecasts. That same year, he published in Leipzig a satirical pamphlet of astrological prophesies, a genre in which he later made numerous contributions (up to his death, seemingly in the early 1670s).[28]

His wanderlust seems to have subsided (albeit only briefly) when he arrived in Bologna in November 1609 and was a guest at the home of Magini, who gave him work as a copyist.[29] Horky fell in love with both the city and the university, and greatly admired them: "Bona respublica, bona Academia. In hac boni ductores, in illa boni Doctores."[30] He especially appreciated the *libertas academica* of the Bologna university and even praised it in a rather paradoxical comment, considering the polemical figure to whom it was addressed: "The *Nuncius* [Galileo] must then remember that at universities the authority of the *ipse dixit* of the Peripatetics does not apply—so that one must rely solely on the authority of Galileo and the telescope, and discuss things as if they were part of a family—but everyone loves to philosophize and no one is obliged to swear on the master's words."[31]

As early as the end of May 1610, Horky announced to Kepler that he had "rather harshly" criticized the discovery of the Jovian satellites. Nevertheless, he added that he did not want to release his *Brevissima peregrinatio* until he returned to his homeland, because in Italy "nothing can be published until it has been seen and approved by the Inquisitor, appointed by Pope Paul V."[32]

In addition to the difficulties involved in obtaining an imprimatur, publication was also hindered by Magini's opposition, as the latter seemed to show fearful respect for Galileo, whereby—in Horky's words—"a fox does not bite another fox nor a dog bark at another dog."[33]

Keen to carve out a prime role for himself, Horky decided to throw himself headlong into overcoming all these obstacles and unswervingly proceeded to print his slim volume. In perfect keeping with the title of the work and its author's lifestyle, the operation entailed yet another *peregrinatio*, but this time truly *brevissima*, as it merely involved going from Bologna to Modena, a journey of around 25 miles. To escape Magini's ban on its publication, in the middle of June Horky secretly traveled to Modena, where he printed five hundred copies of the book at his own expense.[34]

News of its publication immediately reached an irate Magini. Furious over the "most serious indignity" of printing such "slanderous writing," he harshly rebuked his copyist and promptly kicked him out. Consequently, on the evening of Sunday, June 20, poor Horky was seen "on the road to Modena, wretched and in a sorry state."[35]

Publication of the *Peregrinatio* created serious problems for Magini, who was afraid he would be considered an accomplice (or even the instigator) in his assistant's onslaught. To avoid any possible suspicion, he swiftly announced that he had had nothing to do with the "cock-up of that German of mine."[36] In particular, he shrewdly decided that he would immediately alert Antonio Santini, his go-between in communications with Galileo, to assure him that Horky had printed the little book entirely behind his back.[37] Other trusted friends also stepped in to guarantee that the *Peregrinatio* was Horky's initiative alone and that the Bolognese professor was extremely vexed by the situation. For example, Paolo Maria Cittadini commented, "I am grieved by the great anguish of Magini's spirit. He is suffering because his servant, whom he supported with meals and science, is now trying, if not with the sword then with slander and invectives, to damage before all humankind the one who did him so many favors."[38]

To corroborate his opposition to the publication, Magini went so far as to say that he threatened Horky in order to wrest the work from him, but the latter, frightened, fled to Pavia (to Galileo's old enemy, Baldassarre Capra) and then found refuge in Milan.[39]

Despite these reactions, rumors that Magini had been "aware of and agreeable to everything" continued to circulate.[40] His ties to his assistant were too close and their common hostility toward celestial discoveries too open. As Isabelle Pantin noted, the pronouncements defending the astronomer "prove that Horky's project had already been known in Bologna for some time." Yet the apologies of Magini and his friends "were not entirely in bad faith."[41] In reality, Horky made his decision to publish the *Peregrinatio* completely on his own, and it was largely inspired by his ambition to make a name for himself and be hailed as the one who had exposed the "tricks" divulged in the *Sidereus nuncius*. In his quest for visibility, he established ties with other figures farther away geographically, but indubitably quite close to him in both sentiment and aspiration.

At the end of May 1610, Horky wrote to Florence to inform the Vallombrosan monk Orazio Morandi that he needed to find allies in his battle against Galileo.[42] Although they had never met, Horky stated that he knew about the monk's anti-Galilean stance and had heard that four "erudites" were planning to weigh in against the *Sidereus*. Therefore, he asked if he could consult with the other Florentine opponents, mentioning an otherwise anonymous "Secretary of Lady Moon" ("Segretario di Madonna Luna") as well as Francesco Sizzi.[43] The latter was in the midst of preparing a text against the telescopic discoveries, the *Dianoia astronomica, optica, physica,* which would be published in February 1611.[44] In fact, it is thanks to the *Dianoia* that we know he was in touch with Horky. Sizzi describes his correspondence with Magini's "servant" (this is how Horky is described) even before the *Peregrinatio* was published, confirming that during the writing process the two kept each other apprised of progress on their respective texts.[45]

This means that Horky developed his critique in direct contact with Galileo's Florentine adversaries, a circumstance that Roffeni specifically mentioned as proof that Magini had had nothing to do with the initiative.[46] In reality, however, all the figures involved in the anti-Galilean network maintained close ties with Magini, and Horky himself must have discovered Morandi's opinion (and thus felt encouraged to write to him) through the monk's correspondence with his employer.[47] In short, even if he did not participate personally in his assistant's writing efforts, Magini had some responsibility for the drafting of the *Peregrinatio,* not only because the work took up what he had maintained regarding the celestial discoveries but also because he played a major role in facilitating Horky's contacts with other opponents of Galileo.[48]

The "Florentine" group was certainly an odd one: a monk from the Vallombrosan Order (Morandi) and the young scion of a noble family (Sizzi), plus a tenaciously peripatetic littérateur and philosopher (Ludovico Delle Colombe), who was also writing against Galileo. They were all tied to Giovanni de' Medici, the illegitimate son of Cosimo I and Eleonora degli Albizi, and shared his aversion to celestial discoveries along with his keen interest in astrological studies.[49]

Influences and prognostications were Horky's daily fare, and this certainly helped him feel at ease among his new friends. To demonstrate his divinatory talent, in his letter to Morandi he employed an astrological clue to prophesy the impending doom of the "father of the *Nuncius.*" To wit: "Galileo is under the threat of Jupiter itself, because when he was born, the planet was in conjunction with the evil Saturn in the Twelfth House (the one astronomers call the house 'of calamity'). I think that, for Galileo, this conjunction of Jupiter and Saturn portends many ordeals from astronomers (who know Jupiter and the Moon well, and can distinguish the stars), due to the four mysterious new planets."[50]

In other words, even the heavens portended Galileo's imminent demise. In Horky's mind, of course, this well-deserved punishment for the imaginary revelation of the chimerical "companions of Jupiter" would come from none other than his own work, the *Brevissima peregrinatio.*

⬩ ⬩

As soon as it was published, Horky sent copies "everywhere, so the truth may be judged."[51] Initially, he even planned to deliver a copy to Galileo personally.[52] As he waited for an opportunity (which never arose), he mailed the book to various people. Matthäus Welser received one in Prague and, in turn, had Kepler read it (a copy had also been sent to the latter, but perhaps never arrived), while one was delivered to Paolo Sarpi in Venice.[53] In the wake of the lively debate on celestial discoveries, the work was also discussed in Padua; in fact, an Englishman who was passing through town, Sir Thomas Berkeley, mentioned it in his diary and briefly recapped its theories.[54] In Tübingen, Kepler's teacher, Michael Maestlin, was delighted to read a book that by revealing the visual deception of the telescope would, as he put it, "fell Galileo with his own sword."[55]

The *Peregrinatio* also reached Florence immediately. Sizzi received it in June, along with a cordial letter in which Horky said he was sure that, following his confutation, Galileo would nevermore *(in aeternum)* be able to demonstrate the four Jovian satellites in the sky.[56] Horky was convinced that his initiative would be embraced not only by the Italian enemies of the "father of the *Nuncius*" but also by Kepler himself. He felt that Kepler's *Dissertatio cum Nuncio sidereo* was fully aligned with his *Peregrinatio* and that, in fact, it was the theoretical premise that had allowed him to add the necessary empirical corroboration: "I know the source of the deception. You [Kepler] discovered it with your last argument on page 34 of the *Dissertatio,* while, on the contrary, using Galileo's own telescope, I found and discovered the error in the sky."[57]

In short, Horky acknowledged Kepler as the main inspiration for and tutelary god of his work. According to Horky, the *Dissertatio* not only questioned the priority of Galileo's discovery of the mountains on the Moon and of the Milky Way but also dramatically repudiated the most astounding of the celestial discoveries: Jupiter's "companions." Unfortunately for him, however, the passage quoted to support his interpretation had absolutely nothing to do with the satellites, as it merely suggested a possible cause for their difference in size.[58] As Galileo himself noted, of all the "immense nonsense" in the *Peregrinatio,* what stood out most was the fact that Horky, "understanding nothing about the reasoning imagined by Mr. Kepler . . . regarding the appearance of the Medicean planets, sometimes larger and sometimes smaller, says that this is what chiefly destroys me."[59]

This was a gross misunderstanding, just as the objections raised against Galileo were equally gross and shoddy, despite the fact that they were passed off as fully corroborated by Kepler's authoritative support. As fate would have it, the German astronomer would be the one to administer the *coup de grâce* to the *Peregrinatio*'s delusions of grandeur, rejecting its rambling theories outright.

In Kepler's letter to Galileo, dated August 9, he described the work as "disgraceful pages that are simply a waste of time."[60] That same day, he also wrote to Horky to say he could no longer consider him a friend and to alert him that he had also written to Galileo. Kepler then told him that the secretary of the Spanish ambassador in Milan—the city to which the young Bohemian had fled was under Spain's jurisdiction at the time—was aware that Horky's father was a Protestant preacher and had informed numerous Italians in Prague, so Kepler advised Horky to leave Italy as soon as possible.[61]

The letter never arrived, so Horky never knew the great risk he had run.[62] In this case, his vagabond soul proved providential. Animated by his usual errant spirit, he soon left Italy to embark on his incessant peregrination yet again.

By October he was in Prague, where he met Kepler. According to the latter's account, Horky—unaware of the reactions of the preceding months—acted like a man who thought he had won the battle, a conqueror utterly convinced he was talking to an ally. He maintained that, in the face of disputes and harsh rebukes, he had expounded an opinion widespread in the academic world of Bologna and shared by many erudite Italians. In the end, however, pressed by Kepler's compelling arguments, he admitted his mistake and apologized.[63]

This humiliating self-critique put an end to Horky's attempt to nail the "father of the *Nuncius*" through his lies. A precipitous and insolent quest for glory ended in humiliating defeat.

• •

Seemingly heedless of this criticism, Galileo never even rebutted the attack against him. The insults in the *Peregrinatio* did not bother him in the least, and he even punned that he had no intention of "descending from heaven to Orcus" *(ex caelo denique descendis ad Orcum)*.[64]

Others took on the burden of responding in his stead. First came the Scotsman John Wedderburn, who wrote a *Confutatio* of Horky's work that was published in October 1610.[65] Another rebuttal came from Roffeni, who the following year had his *Epistola apologetica contra peregrinationem Martini Horkii* printed in Bologna.[66] It is likely that the text was written with Magini, just as Roffeni himself had promised at the end of June.[67] This is also attested to by a work now at the Biblioteca Universitaria in Bologna, the last page of which bears the following gloss by an anonymous hand of the era noting that Magini was the real author: "Sed auctor verus fuit Maginus."[68]

The repeated professions of innocence, along with Roffeni's assurances, may have convinced Galileo that Magini had nothing to do with Horky's initiative. On his way back to live in Tuscany for good, in September 1610 Galileo stopped in Bologna once more, and many things had changed since April. For some time, Magini had been using the telescope to verify the validity of the theses propounded by the *Sidereus*.[69] Nevertheless, the "confusion" and the debates raging over the Jovian satellites as well as the other discoveries were as animated as ever. In August, in a letter to Kepler, Galileo sketched out the various reactions:

> In Pisa, Florence, Bologna, Venice, and Padua, many have seen, my dear Kepler, but everyone remains silent and hesitates. . . . What am I to do? Laugh like Democritus or cry like Heraclitus? I am willing, dear Kepler, to laugh at the extraordinary stupidity of the masses. But what about the leading philosophers of this university who, with the stubbornness of a snake, were never, ever, willing to observe the planets, the Moon, and the

> spyglass, though I offered it to them thousands of times? . . . This ilk thinks that philosophy is a book like the *Aeneid* and the *Odyssey*, and that the truth must be sought not in the real world or nature, but (using their words) by comparing texts.[70]

There was so much resistance to overcome, and so much depended on the opinions and stances of the most authoritative astronomers of the era. Among them, Kepler, the emperor's mathematician, played a prime role. The debate under way in Prague proved to be crucial, as it was there that the fate of the "sidereal messenger" would be decided.

SIX
The Battle of Prague

Prague, April 1610, Kepler to Georg Fugger, imperial ambassador to Venice:

> The *Sidereus nuncius* is much talked about at court, as by now I suppose it is everywhere. (Would that the *Astronomia nova* had attracted such attention!) The Emperor graciously let me glance through his copy, but otherwise I had to contain myself as best I could until a week ago, when Galileo himself sent me the book, along with a request for my opinion on it, which I suppose he wishes to publish. The courier returns to Italy on the 19th, which leaves me just four days in which to complete my reply. Therefore I must close now, in the hope that you will forgive my haste—and also that you will not take amiss my response above to your touching & much appreciated gestures of support for me. In these matters of science, it is a question, you see, not of the individual, but of the work. I do not like Galileo, but I must admire him.

This letter doesn't exist. It's a fake. Or, rather, it's fiction—magnificent fiction—that is the brainchild of a writer of the caliber of John Banville, and it is in the fourth chapter of his *Kepler: A Novel.*[1] The chapter, entitled "Harmonices Mundi," is composed solely of letters written by Banville/Kepler between 1605 and 1612. There are twenty in all, twenty replies to twenty letters Kepler actually received.

The notion is so enticing it's almost annoying. These letters dovetail so well with the original documents, we find it hard to believe that all this merely reflects the inventiveness of an author who lived four centuries later. If we set the imaginary Banville/Kepler reply alongside the actual letter Fugger sent Kepler on April 16, the former is so realistic in its philological rigor and the accuracy of its historical and scientific arguments that we are astonished not to find it in the *Gesammelte Werke* containing the German astronomer's entire correspondence.

> Regarding Galileo's *Nuncius aethereus,* I received it recently. But since many who are well versed in the study of mathematics considered that arid discourse, so far removed from philosophical presuppositions, to be a well-dissimulated hypocrisy, I did not dare send it to his Holy Majesty the Emperor. This fellow loves to collect the feathers of others everywhere to dress himself up like Aesop's crow, and in this case as well he wants to be considered the inventor of that ingenious spyglass, whereas it was a Belgian who first brought it, a traveler who came here through France, who showed it to me and others. When Galileo saw it, he made others resembling it and perhaps added something of his own to the invention, which would also be easy to do.[2]

It was with this ponderous dispatch that the battle of Prague began. A little over a month had passed since the *Sidereus* had been printed, and the discussion on celestial discoveries grew more and more intense, transforming itself into a question of prime importance that immediately exploded outside strictly scientific milieus. Fugger reported opinions that must have been quite widespread in Venice and asked for confirmation from one who, better than anyone else (thanks to his position) and with minimal effort, could shelve the ambitions of this Italian parvenu and put an end to something that had become far too sensational. Yet some went even further, considering the *Nuncius aethereus*—as Fugger referred to it—a sham filled with things sought after and imagined but never actually seen. The information the physician and astrologer Ottavio Brenzoni had collected in Verona, his city, and among his correspondents asserted just that. "They say that the spyglass is the reason for all those apparitions on the Moon and of those stars and planets that have not been seen again," he wrote, "first, because of some spot or unevenness on the glass; then, because looking through a shiny glass, a tired observer can easily see some great vapor as a shiny body."[3]

That sky wasn't there. It didn't exist. It was the spyglass that created lunar mountains and other hitherto unseen phenomena, ordaining the fictional as real, because nothing similar could be found in nature. Thus, those "dots" Galileo called bodies revolving around Jupiter, and the interplay of lights and shadows on the Moon that he considered mountains, were merely optical illusions produced by flaws and irregularities in the lenses. In short, his announcement contained a series of mistakes and had no grounds. This criticism was far more radical than Fugger's but essentially not as dangerous, and it would shortly be swept aside by the first of many confirmations regarding the reliability of the spyglass.

Fugger conveyed other kinds of information to Kepler, however, and asked for his opinion. His words did not question so much the value of the instrument itself as Galileo's claim that he had invented it and his overweening ambition to claim priority for that invention. Truth be told, this was a far more serious and malicious criticism than what had been leveled against him by those who, a priori, refused to ascribe any scientific merit to the telescopic observations. It was saying that Galileo was the worst of plagiarists, an unprincipled opportunist, the author of an inelegant booklet and a doctrine that did not even deserve to be presented to the emperor as a gift.

In all likelihood, Fugger had not yet read a single page when he wrote to Kepler. After all, he let it slip that he had just received it. Yet the ambassador was not incompetent in scientific matters. By this time, he had lived in Venice for a number of years and was a well-known figure there—and not only in the palaces of politics. For example, we know that he corresponded with Magini and dabbled in mathematics, even taking private lessons from a still-unknown Giovanni Camillo Gloriosi.[4]

Galileo's successor for the chair of mathematics at the University of Padua, the Neapolitan Gloriosi had been in the city on the lagoon since at least 1606, and although he had not yet published anything, he knew Galileo and had been befriended by Magini. He was keenly interested in algebra and admired Viète, so Sarpi must have known him as well. We also know that he frequented Agostino Da Mula and two other mathematicians who were in Venice during those years, Antonio Santini and Marino Ghetaldi.[5] Furthermore, he corresponded with foreign scientists and naturalists based in Rome, such as the Jesuit Christoph Clavius, as well as the future Lyncean—member of the Accademia dei Lincei—and missionary

Johannes Schreck (Terrentius). It was in response to Schreck, who had asked Gloriosi what he thought about the *Sidereus nuncius,* that on May 29, 1610, he wrote a long and detailed letter maligning the book and its author. An examination of the letter quickly reveals that we are looking at considerations taking up what his "pupil" Fugger had written to Kepler just a month earlier. In effect, the two testimonies overlap on several points, so it would not be far-fetched to think that the ambassador's letter merely recaps Gloriosi's opinion.

Like Fugger, Gloriosi observed that before building his own spyglass Galileo had seen an example (offered to the Republic of Venice by a Belgian) and that, unlike what he had asserted in the *Sidereus,* it was untrue that his telescope had been built based on the principle of refraction. But there is more. The accusation of plagiarism was leveled not only against the telescope but also against the geometric and military compass that, in Gloriosi's mind, Galileo had contemptibly claimed as his own invention, whereas credit should instead go to yet another Belgian, Michel Coignet, mathematician to Archduke Albert of Austria.[6]

The feathers of others—to take up Fugger's fabulistic motif—were thus far more numerous than one might originally have imagined, starting with the lunar feathers, given that Pythagoras and Plutarch in the past, followed by Maestlin and then Kepler in his *Ad Vitellionem paralipomena,* had already said everything there was to say. Galileo's extraordinary discoveries about Jupiter also were trounced, as allegedly they were not entirely his work either.

> Word has it that others already discovered two of them by means of the spyglass. It is said publicly that Agostino Da Mula, a Venetian aristocrat, was the first to observe these stars and that he informed Galileo, who knew nothing about them. His Excellency Messer Fugger told me he learned that the satellites of Jupiter were also observed by the Dutch, who invented the telescope. Perhaps Galileo was inspired by them and, to gain glory and money, even though he was not the first observer, he nevertheless wanted to be the first to write about them. You should be aware that the Florentines are shrewd and industrious: hence, grasping this propitious opportunity, he proclaimed himself author and inventor of the telescope and of

the new planets, and received many honors and advantages from both the Republic of Venice and the Grand Duke of Tuscany.[7]

There is no question about it: Gloriosi and Fugger were doing excellent teamwork, matched by Magini and Horky in Bologna at the very same time. The aim was to discredit and expose the "shrewd" Tuscan who was only interested in reaping "glory and money." Just a month after the *Sidereus* had been printed, instead of gaining approval and tokens of respect, Galileo was being hit from every side. This was a well-organized barrage, marked by the emergence of an alignment of interests that just slightly preceded the Florentine one led by Giovanni de' Medici and Archbishop Alessandro Marzimedici.[8]

• •

It would take far more, however, to transform this gossip from insinuation into something solid that might gain widespread acceptance. This is why the likes of Magini, Horky, Fugger, and Gloriosi looked toward Prague. They were waiting for *him*, the successor of the incomparable Tycho Brahe, to step in, as they harbored the ill-concealed hope that the man who had adorned himself with the feathers of others would be shamed in front of the whole world.

In Venice, Gloriosi had procured a good telescope. Training it skyward, he had found two bodies near Jupiter, but was unable to say if they were fixed stars or planets. What he wanted most of all was to clarify things, and this is also why he focused on Prague, writing, "A more precise instrument, more time, and more observations are needed if we want to obtain better results. Nevertheless, I will not avoid saying that everything we are examining lies in the realm of probability, and perhaps it will only be confirmed with time. And there will be no shortage of people involved, and above all Kepler, who constantly observes the stars, will not fail in this."[9]

First the Low Countries, then Venice, and now Prague: the venue of our story shifts yet again, this time to the Bohemian capital, a great cultural crossroads, and a free and tolerant city—though not for long. Prague became a place to conquer, a battleground of discoveries, real or alleged, whence the response of the imperial astronomer was eagerly awaited. Yet

Prague would almost have remained silent if the letters of a largely unknown figure had not been saved: a man by the name of Martin Hasdale or Hastal. We do not know his nationality (possibly German, perhaps Walloon), but between April and December 1610 he informed Galileo of the discussions and raging debates sparked by the arrival of the first copies of the *Sidereus* at court and at the main embassies. Hasdale is the linchpin, the key connection between Galileo and Prague, and he supplied the Italian scientist with valuable information regarding the sides that were squaring off on issues that directly affected him.[10]

There are ten letters in all, but not one of Galileo's replies has survived. Nevertheless, reading the first missive, dated April 15, allows us to grasp how decisive the Hasdale piece is in reconstructing our puzzle and better understanding the texts being penned about the celestial affair. "I have been planning to return to Italy for quite some time," he wrote, "and particularly to Padua and Venice, more to enjoy that most kind conversation with you than anything else; and this desire grows in me even more when I see new parts of your most felicitous and divine intelligence; of which the latter, titled *Nuntius sydereus*, has recently captivated this entire court with admiration and wonder, and all these ambassadors and barons are hastening to call these mathematicians here to see if they can refute your demonstrations."[11]

The tone is friendly. Even though we do not know how many months or years Hasdale spent in Venice, there is no question that he met Galileo several times and talked to him at length. His closing also tells us that he was accustomed to doing so with "Father Master Paolo and Father Master Fulgentio," that is, Sarpi and Micanzio. Indirect confirmation comes from Micanzio, who in his *Vita di Paolo Sarpi* paints a vivid picture of Hasdale, describing him as an untrustworthy man, obstinate Calvinist, and unscrupulous wheeler-dealer. "One Martino Asdrale, Walloon, an excellent man for spying, came to Venice since he had had enough and could no longer stand the court; he has long apprenticed at the shop of Secchini [a Calvinist merchant]. . . . No one more than he had better information about what was going on in Rome regarding that controversy [the Interdict], no one more openly condemned the rage of the pontiffs."[12]

The comment is not entirely unfounded, given that on July 5, 1607, a "Martino Asdrale della città di Liege"—Martin Hasdale of the city of Lieges—was called before the Heads of the Council of Ten and, "in a grave and severe manner," was asked to leave Venice immediately: "Within three days you must leave this city, and within another four our entire State, to which you shall not return without the permission of this Council, under penalty of death, adding that your pensions shall be paid to your legitimate intervenors."[13]

So he was a man in his forties (this is about how old he must have been when he arrived in Prague) who was the perfect court animal, well versed in palace affairs and intrigue and its endless gossip. When he was forced to leave Venice in 1607, Hasdale arrived in Prague after wandering around various German universities, but he quickly became a right-hand man to Rudolf II.[14] As the ruler's librarian and chamberlain, he was part of the small group of collaborators and advisors who regularly had access to him. Among them were the likes of Caspar Rutzky, Daniel Frosch, Cristoph Kuhbach, Christian Haiden, Hans Marcker, and Cornelius Drebbel, all of whom were arrested and imprisoned in the days following the emperor's death. Rutzky hanged himself; the others, including Hasdale, were tried but then released.[15]

• •

The *Sidereus nuncius* had "recently captivated this entire Court with admiration." Hasdale's letter to Galileo begins with these words and focuses entirely on the slender volume that was creating such an uproar at court, where not one ambassador or baron had failed to query "these mathematicians here to see if they can refute your demonstrations."[16] Little does it matter here how much Hasdale may have exaggerated this early information in a move to curry the favor of his illustrious correspondent. What counts is that the news contained in it and the letters that followed are extremely detailed, giving Galileo a virtually complete map—in real time—of the positions for and against him. This sweeping information allowed him to appreciate the caution of those waiting for further proof before asserting their stance, but it also let him expose the specious reasons of others and discover testimony that would otherwise have remained unknown. One example is the long conversation that, on the morning of April 15, Hasdale had with Kepler when they were lunching with the ambassador of Saxony.

> I asked him what he thought about that book and about you. He responded that for many years he has been in touch with you by letter and that, in reality, he knows of no greater man in this profession than you, nor has he ever known one; and that as great as Tycho was reputed to be, you yourself are far ahead of him. Insofar as the book is concerned, he says that truly you have demonstrated [in the *Sidereus nuncius*] the divinity of your talent; but that you have given some cause for complaint not only to the German nation but also your own, as you have not mentioned any of the writers who have hinted at and given you the chance to investigate what you have discovered now, citing among them Giordano Bruno as an Italian, and Copernicus and himself, professing that he has mentioned similar things (but without evidence, unlike you, and without demonstrations): and he brought with him his book, to show the place to the Saxon ambassador.[17]

Proof that these were not merely vivid impressions is demonstrated by the fact that these selfsame considerations can be found in the *Dissertatio cum Nuncio sidereo* Kepler was drawing up at the time, a manuscript copy of which he sent to Galileo on April 19. (The work was published in Prague in early May.) In all likelihood, the book Kepler brought with him and planned to show the ambassador, along with the *Sidereus,* was the *Paralipomena,* discussing the various properties of lenses, providing the first correct explanation of the mechanism of vision, and, from an optical standpoint, clarifying several celestial phenomena, including the spots on the Moon. The aim was clear: showing his interlocutors the theoretical weakness of Galileo's book and the inadequacy of its explanation of the working of the telescope. But he also wanted to illustrate how much had been overlooked, how many works and authors had not even been mentioned, in order to make the importance and sheer innovation of those discoveries shine more brightly. At the same time, there was yet another aim: demonstrating the "divinity of [Galileo's] talent," as Hasdale reported and as Kepler wrote in the *Dissertatio* when he expressed his unconditional support for Galileo and his discoveries "against the obstinate critics of innovation, for whom anything unfamiliar is unbelievable, for whom anything outside the traditional boundaries of Aristotelian narrow-mindedness is wicked and abominable."[18]

Even before producing any observational confirmation (and Kepler would not do so until October 1610, when he printed his *Narratio*), the German Copernican was already allied with the Italian one. "I may perhaps seem rash in accepting your claims so readily with no support from my own experience," he wrote. "But why should I not believe a most learned mathematician, whose very style attests the soundness of his judgment? . . . Does he not make his writings public, and could he possibly hide any villainy that might be perpetrated? . . . Shall I not have confidence in him, when he invites everybody to see the same sights and, what is of supreme importance, even offers his own instrument in order to gain support on the strength of observations?"[19]

Despite its lack of essential references to other works and the results of Galileo's predecessors, the *Sidereus* was an exceptional book that perfected the work undertaken by Brahe. Just as the calculations and observations of the latter had been helpful for studies about Mars (which had led Kepler to discover the elliptic orbits), with the aid of the spyglass and Galileo's valuable collaboration Kepler now hoped to perfect his *Hipparchus,* a most ambitious work he had been toiling over for some time. In the *Dissertatio,* he wrote, "I want your instrument for the study of lunar eclipses, in the hope that it may furnish the most extraordinary aid in improving, and where necessary in recasting, the whole of my 'Hipparchus' or demonstration of the sizes and distances of the three bodies, sun, moon, and earth."[20]

It was early May. Kepler was perfectly aware that people were expecting his opinion, and not only in Prague. Horky and Magini had already urged him a number of times to step in, and Fugger had done the same. So did Mark Welser, whom Hasdale had met in early April at the residence of the Spanish ambassador, as he "wanted to hear Kepler's opinion about the book."[21] Then there was Baron Johann Matthaeus Wacker von Wackenfels, his protector but also one of the emperor's most trusted advisors and a staunch supporter of Bruno's thesis of the infinite universe. Wacker was the first to inform Kepler orally of the discovery of four new planets (later proven false), plunging him into the darkest despair.

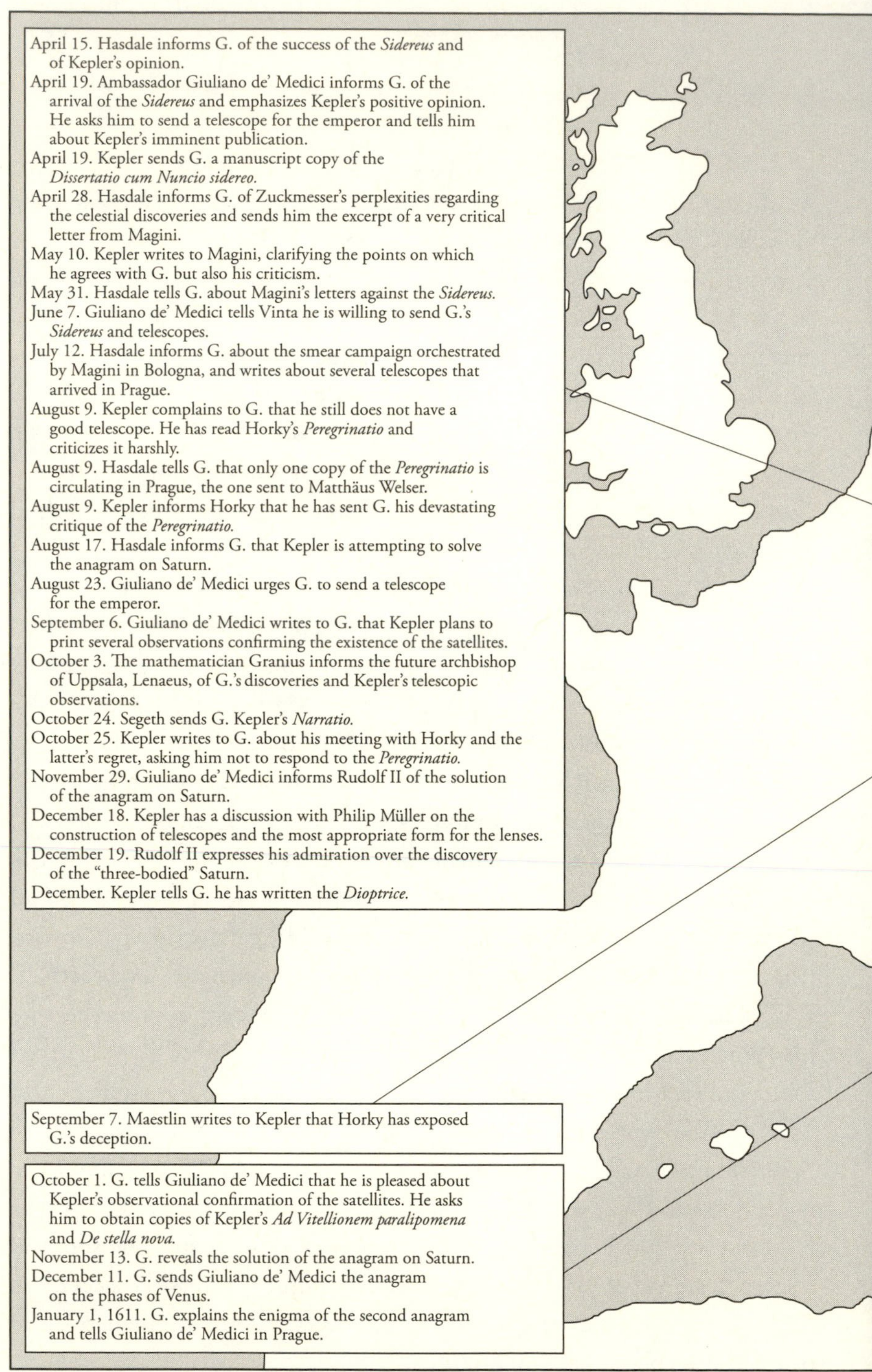

April 15. Hasdale informs G. of the success of the *Sidereus* and of Kepler's opinion.
April 19. Ambassador Giuliano de' Medici informs G. of the arrival of the *Sidereus* and emphasizes Kepler's positive opinion. He asks him to send a telescope for the emperor and tells him about Kepler's imminent publication.
April 19. Kepler sends G. a manuscript copy of the *Dissertatio cum Nuncio sidereo.*
April 28. Hasdale informs G. of Zuckmesser's perplexities regarding the celestial discoveries and sends him the excerpt of a very critical letter from Magini.
May 10. Kepler writes to Magini, clarifying the points on which he agrees with G. but also his criticism.
May 31. Hasdale tells G. about Magini's letters against the *Sidereus.*
June 7. Giuliano de' Medici tells Vinta he is willing to send G.'s *Sidereus* and telescopes.
July 12. Hasdale informs G. about the smear campaign orchestrated by Magini in Bologna, and writes about several telescopes that arrived in Prague.
August 9. Kepler complains to G. that he still does not have a good telescope. He has read Horky's *Peregrinatio* and criticizes it harshly.
August 9. Hasdale tells G. that only one copy of the *Peregrinatio* is circulating in Prague, the one sent to Matthäus Welser.
August 9. Kepler informs Horky that he has sent G. his devastating critique of the *Peregrinatio.*
August 17. Hasdale informs G. that Kepler is attempting to solve the anagram on Saturn.
August 23. Giuliano de' Medici urges G. to send a telescope for the emperor.
September 6. Giuliano de' Medici writes to G. that Kepler plans to print several observations confirming the existence of the satellites.
October 3. The mathematician Granius informs the future archbishop of Uppsala, Lenaeus, of G.'s discoveries and Kepler's telescopic observations.
October 24. Segeth sends G. Kepler's *Narratio.*
October 25. Kepler writes to G. about his meeting with Horky and the latter's regret, asking him not to respond to the *Peregrinatio.*
November 29. Giuliano de' Medici informs Rudolf II of the solution of the anagram on Saturn.
December 18. Kepler has a discussion with Philip Müller on the construction of telescopes and the most appropriate form for the lenses.
December 19. Rudolf II expresses his admiration over the discovery of the "three-bodied" Saturn.
December. Kepler tells G. he has written the *Dioptrice.*
September 7. Maestlin writes to Kepler that Horky has exposed G.'s deception.
October 1. G. tells Giuliano de' Medici that he is pleased about Kepler's observational confirmation of the satellites. He asks him to obtain copies of Kepler's *Ad Vitellionem paralipomena* and *De stella nova.*
November 13. G. reveals the solution of the anagram on Saturn.
December 11. G. sends Giuliano de' Medici the anagram on the phases of Venus.
January 1, 1611. G. explains the enigma of the second anagram and tells Giuliano de' Medici in Prague.

Map 3. Prague, March 31, 1610–January 1, 1611: circulation of Galileo's *Sidereus nuncius* and telescope

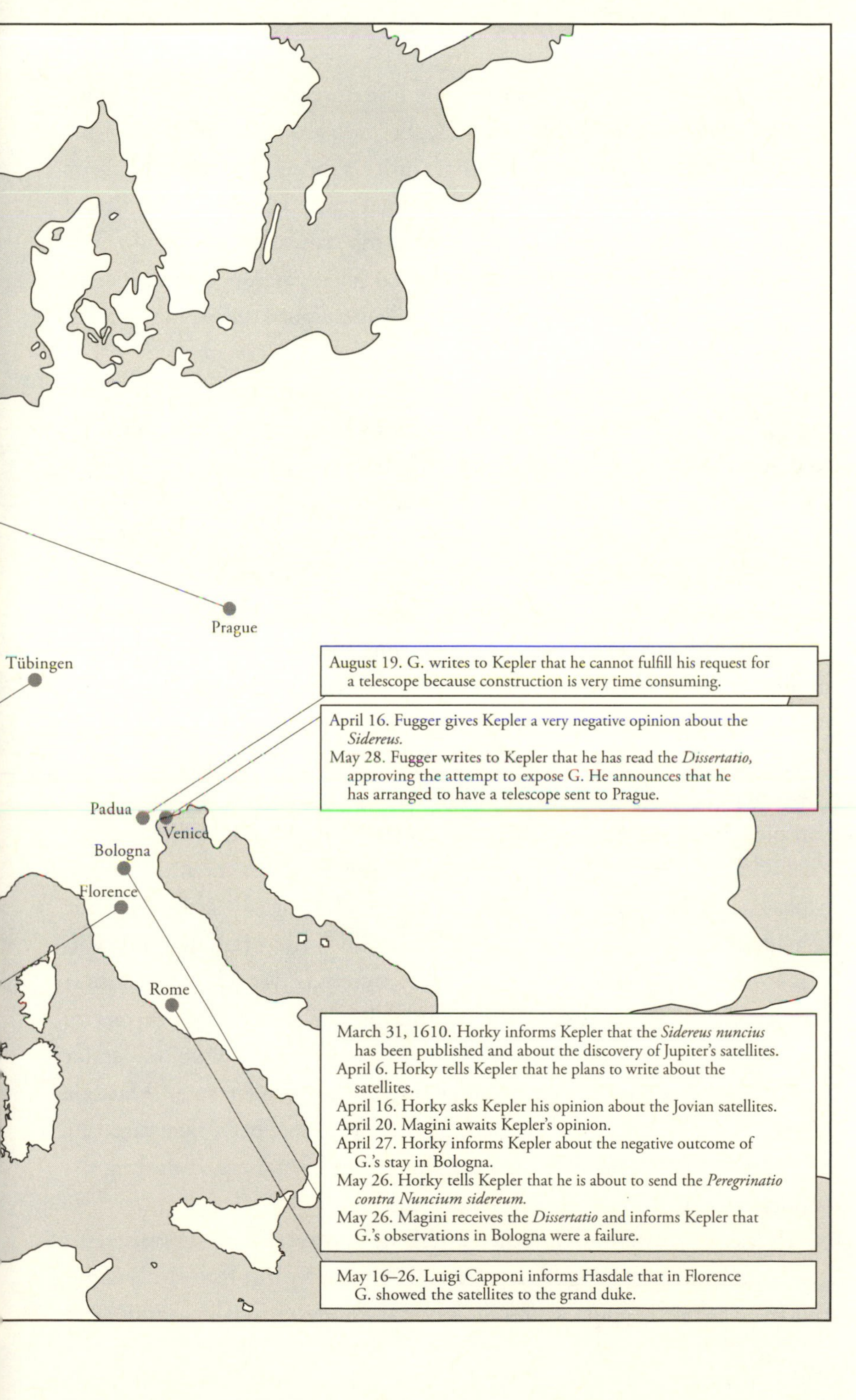
Prague
Tübingen
August 19. G. writes to Kepler that he cannot fulfill his request for a telescope because construction is very time consuming.
April 16. Fugger gives Kepler a very negative opinion about the *Sidereus.*
May 28. Fugger writes to Kepler that he has read the *Dissertatio,* approving the attempt to expose G. He announces that he has arranged to have a telescope sent to Prague.
Padua
Venice
Bologna
Florence
Rome
March 31, 1610. Horky informs Kepler that the *Sidereus nuncius* has been published and about the discovery of Jupiter's satellites.
April 6. Horky tells Kepler that he plans to write about the satellites.
April 16. Horky asks Kepler his opinion about the Jovian satellites.
April 20. Magini awaits Kepler's opinion.
April 27. Horky informs Kepler about the negative outcome of G.'s stay in Bologna.
May 26. Horky tells Kepler that he is about to send the *Peregrinatio contra Nuncium sidereum.*
May 26. Magini receives the *Dissertatio* and informs Kepler that G.'s observations in Bologna were a failure.
May 16–26. Luigi Capponi informs Hasdale that in Florence G. showed the satellites to the grand duke.

As Isabelle Pantin has observed, the *Dissertatio* "was not only the outcome of reflection conducted in solitude, because it was also affected by discussions at court."[22] We must add that this text failed to convince Galileo's countless opponents. Nevertheless, it helped introduce the discoveries of the *Sidereus* to the German world. In fact, in August 1610, in his almanac *(Schreibkalender)* for the following year, Paul Nagel, astronomer and rector at the Lutheran school in Torgau, specifically mentioned the "four Medicean planets."[23]

Naturally, Kepler was well aware of the information in Hasdale's hands. First of all, there was the news about Magini's virulent campaign to disparage the Tuscan, a plan ardently backed by Johann Eutel Zuckmesser, mathematician to Duke Ernest of Bavaria and elector of Cologne. In Prague, word had it that Magini had written to the mathematician to have him side against Galileo, and that he was doing the same thing "with all the mathematicians of Germany, France, Flanders, Poland, England, and so on; [Hasdale] learned this not from just one person, but from a number of different countries, all persons who represent persons of princes."[24] In some cases, it was Kepler who gave Hasdale firsthand information. For example, toward the end of April he showed him one of the letters Magini had sent the emperor saying that the spyglass was a complete hoax, terming the discovery of the four celestial bodies "ridiculous."[25] In many cases, this uncontrolled welter of rumors and insinuations naturally concealed political motives, shoving all other scientific considerations into the background. Yet again—and Hasdale is our witness—the *Sidereus* ended up being interpreted as dangerous because it was strongly anti-Spanish. "The Spaniards think that, for reasons of state, your book should be suppressed as harmful to religion, under which cloak it becomes licit do anything to foster their rule," Hasdale wrote. "This league, which is here against you, has been fabricated by none other than them, and their employees and followers, including the resident of Lucca."[26]

Particularly after the Interdict, anything and everything represented a good opportunity to go against "heretical" Venice, Spain's historical enemy. But from a strictly political and diplomatic standpoint, the reasoning was flawless: the idea that those discoveries by a professor from the prestigious Paduan university might enhance the cultural fame of La Serenissima could be transformed into an undisputed political success.

• •

Throughout 1610, Kepler, Hasdale, and Giuliano de' Medici, ambassador for the Grand Duke of Tuscany, were in very close contact, but the one who pestered Kepler the most with questions about the spyglass and the *Sidereus* was the emperor himself. Rudolf's thirst for information was boundless. As soon as the book arrived in Prague, the emperor showed it to Kepler, asking him to study it carefully and offer an opinion about the observations. He was also the one who, just three months earlier, had peppered him with questions about the nature of lunar spots, "as [the emperor] was convinced that the images of countries and continents are reflected in the Moon as though in a mirror."[27] He even offered one of his spyglasses so Kepler could confirm or refute Galileo's discoveries.

Rudolf's passion for "secrets, charms, sorcery, sleight of hand, alchemy, painting, and sculpture"—as the ambassador Roderigo Alidosi mentioned to the Tuscan grand duke—is too well known to dwell on here.[28] The same can be said for his predilection for anything to do with astrology, "and perhaps even necromancy," as another Tuscan ambassador, Cosimo Concini, was embarrassed to report.[29] Then there was his obsessive attraction for any type of human invention, any kind of arcana and automata, which he would have people seek around the world to add to his already rich *Kunst-Wunderkammer*, the most famous and monumental collection of artwork and curiosities in Europe, situated in the sprawling west wing of the Hradschin Palace and shown to only a fortunate few.

Naturally, many anecdotes circulated about his quirks and melancholy spirit. Moreover, a number of people doubted he was truly Catholic. "Not only did his Majesty not confess, he did even not display any sign of contrition," commented Cardinal Scipione Borghese following the emperor's death.[30] In a dispatch to the grand duke, Concini relayed that the emperor "asked a minister if he would be loyal to him and when he said yes on the communion he had taken, [the emperor] said, 'So you have had communion,' and took a number of steps back."[31]

"[His Majesty] delights in hearing about things both natural and artificial, and whoever is able to deal in such matters will always find the emperor's ear ready," stated the Venetian ambassador Tommaso Contarini.[32] The archdukes of Austria wrote, "His Majesty is interested only in wizards, alchymists, kabbalists and the like, sparing no expense to find all kinds of treasures, learn secrets and use scandalous ways of harming his enemies."[33] In this scenario, the *Kunst-Wunderkammer* played a leading role, symbolically representing the emperor as the omniscient dominator and sovereign. Therefore, it comes as no surprise that the "long-sighted spectacle," the latest and most fashionable *secretum*, would fascinate him to the point of obsession, above all after the announcement of the celestial discoveries. But it is through Hasdale's letters that we find firsthand insight into just how much the new instrument revolutionizing all of astronomy held sway over Rudolf, representing the almost miraculous symbol of the union between the celestial and terrestrial worlds. This is why he had to have Galileo's spyglass, the best around at the time, in his microcosm of *naturalia* and *artificialia*.

In July 1610, the emperor and Kepler were personally engaged in astronomical observations, but with poor results. The spyglasses Rudolf had received and continued to receive above all from Venice—through his ambassador Fugger, Ottavio Panfili, and Ferdinando Taxis, Grand Master of the Imperial Post in Venice[34]—were indubitably better than the first Dutch spyglasses that, being an admirer of Della Porta and enthralled with anything optical, he had promptly procured, but they were not as good as he had hoped.[35] He eagerly expected to receive another spyglass as a gift from the Tuscan ambassador at any moment, one built by Galileo himself. It was never delivered, because the one intended for him was "purloined" at the last minute by Cardinal Scipione Borghese. Rudolf was outraged, and it seems that the "theft" almost sparked a diplomatic incident. Wrote Hasdale to Galileo,

> I whetted the appetite of C. [Caesar, i.e., the emperor] but also drew his acrimony by reminding him that Cardinal Borghese had taken from his hands the spyglass you had made. His Majesty burst out with these

> words: "So these priests want everything." I was ordered to write to you on his behalf, but I got out of it by saying that you had written to the Lord Ambassador of Tuscany, who would surely send one twice as good as the one Borghese took. Upon seeing that His Majesty was not placated, I finally stopped him by saying that the Grand Duke specially convened you to Florence to make a few to send the various princes.[36]

But things didn't end there, and a week later the affront still rankled. "Then, regarding the glasses, His Majesty had me tell His Excellency of Tuscany that I should write to you to lend greater consideration to the instruments you make with your own hand, as His Majesty has been most displeased that the priests took from his hand what was intended for him," Hasdale wrote to Galileo. "In short, he expects the most perfect ones as soon as possible, at least one."[37]

He demanded "at least one," a "most perfect" spyglass. Rudolf's insistence is almost endearing. It seems that during those months there was nothing he desired more. Not even the spherical machine that, according to its inventor, the Dutchman Cornelius Drebbel, produced perpetual motion could rival those fragile but immensely powerful lenses. The great esteem and credibility Galileo had acquired in Rudolf's eyes is also demonstrated by the fact that, again through the ubiquitous Hasdale, the emperor turned to the Tuscan to find out if he was familiar with the workings of one of the most sublime inventions of the human mind, alongside the legendary dove of Archytas: the burning glass, the "secret of the parabolic mirror of Archimedes that burns from afar."[38]

◆ ◆

It was the twenty-fourth of August. So far, no one in Prague had observed the Medicean planets, and the battle surrounding the *Sidereus* still raged. Despite the emperor's attraction to the new instrument, it alone was not enough to change current stances, nor had the publication of the *Dissertatio* managed to convince the opponents of Galileo's work. In fact, it was thought to confute rather than confirm the Tuscan's discovery.

Anyone who read the text from a very skeptical stance could find other excellent reasons for opposing the *Sidereus*. In effect, by detailing the names of predecessors whose works had come close to the true explanation for the celestial phenomena (with the exception of the Jovian satellites) through mere speculation, wasn't Kepler condemning many of the innovations present in the *Sidereus* and asking Galileo to return to his principles? Furthermore, which authoritative witnesses had been with Galileo to confirm those discoveries?

The game was still wide open. If Galileo had known about the correspondence between Maestlin and Kepler, he would have had even more reason to worry about the outcome of the battles still raging in Prague. This is what the famous mathematician from Tübingen, Kepler's teacher and a Copernican himself, wrote on September 7, 1610:

> In your work, you have done very well to pluck the feathers that were not Galileo's in the first place [*Galilaeum deplumasti*]; because he was not the first inventor of this new spyglass; he was not the first to understand the roughness of the Moon's surface; nor was he the first to demonstrate to the world that, in the sky, there are many more stars than those mentioned so far in the writings of the ancients. . . . Until now, no astronomer has recognized these four planets. . . . Now Martin Horky has rid us of this concern. He has understood the deception inherent in this vision, and not through a similar spyglass but Galileo's instrument, and he has felled the author with his own sword. . . . We'll see what [Galileo] has to say. To be honest, I hope that you too will not remain silent against him.[39]

It almost seems as if the wrong letter has fallen into our hands and we're reading Fugger's words. Maestlin firmly held that Galileo could claim no credit. The only remaining doubt involved the four planets. On this point as well, however, he relied on the truthfulness of Horky's arguments—he had read and liked his *Peregrinatio*—but, above all, he hoped his favorite pupil would again step in to "defeather" the unknown but highly ambitious professor from Padua once and for all. Paradoxically (and this is a key element), just as he was receiving Maestlin's opinion and advice, Kepler was also finishing his *Narratio,* his second text devoted to Galileo and his discoveries. It was with the *Narratio* that the battle of Prague would reach its watershed, shifting the tide to the Italian scientist, despite anything Maestlin had to say.

But let us go back to August 24, when Hasdale told Galileo that the emperor was impatiently awaiting his spyglass. That same day, Rudolf had turned to Hasdale for another reason: to see Galileo's latest letter to the ambassador Giuliano de' Medici. "When I gave His Majesty a summary of your most recent letter to His Excellency the Ambassador, he wanted to see the original, which I procured and gave to him and then also retrieved," wrote Hasdale. "In short, His Majesty is anxious to know the meaning of that code of shifted letters, which conceal what you have most recently found."[40]

Knowing the content of the letter wasn't enough, Rudolf wanted to see the original, announcing Galileo's most recent great discovery: his observation of "three-bodied" Saturn. Such a request might have seemed quite strange, but there was a reason the announcement had been made in such an unusual way. This was the first time an astronomer announced the outcome of his work in code, through an anagram addressed to the emperor and his mathematician. In all, there were thirty-seven letters arranged in an arcane sequence, with the exception of the word "poeta" between the thirteenth and seventeenth letters, and the word "tauri" between the thirty-first and thirty-fifth: *smaismrmilmepoetaleumibunenugttaurias.*

After the *Sidereus,* Galileo stepped in again, but this time he overcame his adversaries' final doubts. The astronomical part suddenly became an utterly sophisticated and mysterious game, the kind enjoyed by Rudolf and the alchemists, astrologers, wizards, and Kabbalists who populated the Bohemian capital during this period and had a passion for encryption. It seemed magical and surreal—and perfectly in keeping with the atmosphere at court. At the same time, however, it was also a way to open up another front, yet another challenge to arrive at the truth: a gauntlet Galileo threw down to the emperor's mathematician, the only one who could rival him, and the only one—among mathematicians and astronomers alike—to whom he sent the anagram for Saturn, soon followed by one for Venus.

Why did Galileo decide to launch a new offensive at this point, and with such a mocking air, turning himself into a scientist who played with words?

The answer may lie in the *Dissertatio* and Galileo's interpretation of the work, diametrically opposed to that of Maestlin and Magini. The text indubitably gave rise to contrasting interpretations, and it is equally true that it would take the *Narratio* to confirm the celestial discoveries and end the Prague challenge once and for all. But there is no question that, for Galileo, the *Dissertatio* marked a turning point entirely in his favor. Or at least that's what he wanted people to think. Proof comes from his letter of May 24, 1610, to the physician and astrologer Matteo Carosio in Paris: "Here there were still twenty-five who wanted to write against me; but so far, finally, all that has been seen is a text by Kepler, the emperor's mathematician, confirming everything I have written, without rejecting one iota."[41]

In his letter of thanks to Kepler on August 19, he wrote, "You were the first and perhaps only one who, without making any observations . . . unconditionally trusted my statements."[42] Kepler's words reassured him, but they also galvanized him, making him more determined than ever to rebut his increasingly confused adversaries and then doggedly continue the work he had undertaken. The news he received in Rome, again through Hasdale, was encouraging. "With the most recent information from His Excellency Cardinal Capponi," Hasdale informed Galileo, "I have heard that the mathematicians of Rome and Tuscany grasp your invention, so I wanted to show it Kepler to comfort him and Zugmesser to confound him."[43]

As the able defender of Galileo's interests, Hasdale knew precisely what to do to back the scientist's credibility, sending "the imperial mathematician's booklet," the *Dissertatio,* to Luigi Capponi in Rome.[44] Then there is the fact that news of Kepler's telescopic observations would soon circle the globe, as evidenced by the letter of October 3, 1610, that the Swedish mathematician Nicolaus Andreae Granius sent from Prague to the future archbishop of Uppsala, Johannes Canuti Lenaeus: "I do not know if you have heard the news of the recent observation of four new planets orbiting around Jupiter, first discovered between the 8th of January and 4th March this year by Galileo, the Paduan professor, with his most wondrous telescope. It is a MARVELLOUS, MARVELLOUS thing. . . . Kepler, the Mathematician of the Emperor, showed me the telescope, by benefit of which any object appears a thousand times greater than when seen without using the telescope."[45]

After Kepler's decision to back Galileo, the scientist could finally take center stage once more. Naturally, he did this in his own decidedly anti-academic way (in keeping with his style since his youthful *Contro il portar la toga*), applying playfulness to astronomy.

Hasdale had written, "His Majesty is anxious to know the meaning of that code of shifted letters."[46] On October 24 that sentiment was echoed by Thomas Segeth (whom Galileo knew well because the Scotsman had spent a number of years in Padua) with words that, if anything, were far too restrained, considering the court's unbounded curiosity: "Mr. Kepler and I and all the finest spirits are most anxiously awaiting the disclosure of your new observation. Please, if this is something that can be known without detriment to you, kindly include me."[47] This was no mere divertissement. The encoded game had been transformed into a way of grasping reality.

• •

By the end of October, more than two months had elapsed since the *aenigma astronomicus* had been sent, but no one had managed to unravel it. During that same period, along with Segeth's letter Galileo received a copy of the *Narratio*, the first printed report confirming the existence of the Jovian satellites and the mountainous moonscape. These few pages were published hastily, like a latter-day *Sidereus nuncius*, in which Kepler listed the observations made between August 30 and September 9 with other collaborators (Segeth, Kepler's pupil and assistant Benjamin Ursinus, Tycho's former assistant and son-in-law Franz Tengnagel, and the imperial councilor Tobias Schultetus) and in which he also attempted to decipher Galileo's anagram.

Kepler immediately thought about Mars, *his* planet. The best he could do was a verse *semibarbarus*, as he defined it (*Salve umbistineum geminatum Martia proles*, "Hail, furious twins, sons of Mars"), based on the observation of moons around Mars that he then ascertained were fixed stars.[48] It didn't even occur to him that those mysterious letters could refer to Saturn. (Though who could have figured it out without a spyglass as powerful as Galileo's?) *Altissimum planetam tergeminum observavi*, "I have observed the highest triple-bodied planet": this was his solution to the anagram. Kepler even tried to observe Saturn on the evening of September 5, but he noted

Figure 15. Galileo Galilei, autograph transcription of Kepler's solution of the anagram of Saturn

nothing unusual: "I saw Saturn and there was no star in its immediate vicinity."[49]

Meanwhile—and we can be sure of this—this game amused Galileo immensely. Needless to say, he provided no clues; in fact, at a certain point in this debate, letter by letter he marked off the mistakes in Kepler's solution (Figure 15).

He left his "companion in the quest for truth"—as he had called him fifteen years earlier—to sort it all out on his own.[50] After all, he was an expert in mysteries and secrets, given that he had entitled his first work *Mysterium cosmographicum*. As if that were not enough, in December he decided to double the ante by sending another anagram, this time devoted to the discovery of the phases of Venus and, here as well, addressed to the emperor through a letter to Giuliano de' Medici. "In the meantime, I am sending you the cipher of another detail I observed again, which underlies the decision of great controversies in astronomy, and in particular it encompasses a bold argument for the Pythagorean and Copernican constitution," Galileo wrote, "and, in time, I will publish its decryption."[51]

The same story and the same incredulity—and it drove Kepler mad yet again. *Haec immatura a me iam frustra leguntur oy*, rearranged, yields *Cynthiae figuras aemulatur mater amorum*, "The mother of love [Venus] imitates the forms of Cynthia [the Moon]": this was the meaning of the anagram.

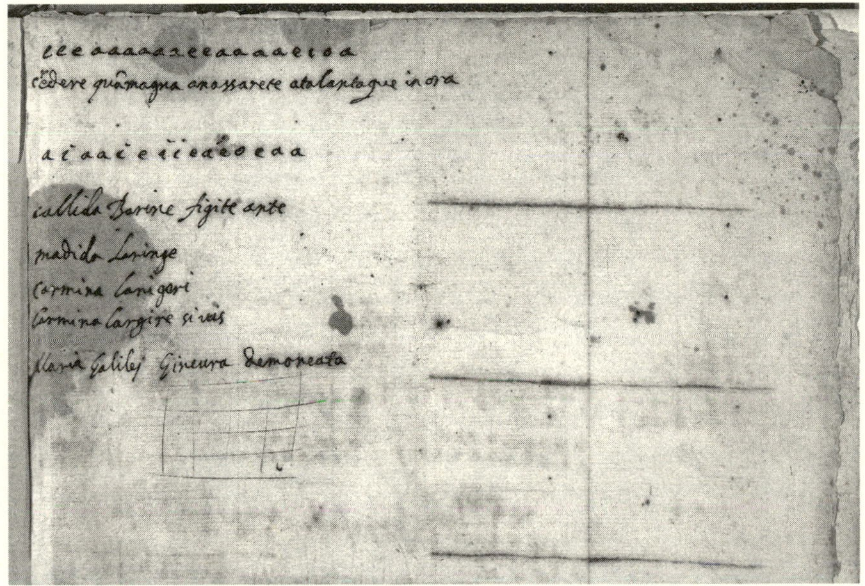

Figure 16. Galileo Galilei, autograph experiments with anagrams

Here again, however, Kepler's idea was brilliant, but miles away from the solution: *Macula rufa in Iove est gyratur mathem etc.* ("There is a red spot on Jupiter that revolves mathematically, etc.").

As far as we know, the anagrams of Saturn and Venus were the first cases of word puzzles applied to astronomy. Others, starting with Huygens, would follow this example. That said, however, Galileo loved wordplay in the first place, as confirmed by his notes on the *Sidereus,* which contained riddles like the one in Figure 16.[52]

Galileo abandoned these experiments for some unknown reason, but he was unable to let go of this form of communication. It is interesting to note that he decided to send *per aenigmata* the odd configuration of Saturn and the observation of the phases of Venus to the court of Prague alone. No other prince received them. In fact, to take precautions against anyone who wanted to claim discovery, he wrote to Vinta that "the star of Saturn is not just one but is composed of three, which touch each other, and among them, they neither move nor change."[53] He defined it a "most extravagant marvel," but at the Medici residence it had to be concealed until the *Sidereus nuncius*—by then impossible to procure—could be reprinted in the vernacular and add to its reputation through this and other discoveries.

* *

The solutions were communicated to Giuliano de' Medici, but Kepler disseminated them, publishing Galileo's letters in the introduction to the *Dioptrice* and even translating them into Latin.[54] We know that Kepler was unable to see either of the two sensational discoveries because his telescopes were not powerful enough. Moreover, his first two observations concentrated on Jupiter and, if we follow the thread of his report, they were shaky and uncertain.

In the first ten days of September the sky over Prague was almost always cloudy and the presence of the Moon complicated things. Kepler's telescope was not of the best quality and was anchored to the floor, hindering observation. Lastly, his lack of specific measuring devices meant that estimating the distance between the satellites and Jupiter was rough at best. But let's not overlook another aspect. Galileo had built that spyglass, but it had not been sent directly to Prague. Kepler had borrowed it from the elector of Bavaria, and it was the best available to him. "In August, while returning from Vienna in Austria, Ernest, the Most Reverend and Serene Archbishop of Cologne, Elector and Duke of Bavaria, lent me an instrument he said had been sent to him by Galileo," Kepler noted, "and he said that, regarding the quality of images he could obtain from it, it was not as good as that of others he had with him, complaining that the stars reproduced with it were quadrangular."[55]

We can couch this in other terms. Despite all his promises, Galileo never sent a spyglass to Prague. The few truly excellent ones had already been sent as gifts—along with the book—in the weeks following publication of the *Sidereus nuncius* (and the elector of Bavaria was one of these recipients). So why not send one to the emperor and thus his mathematician, the only one who had publicly defended him so far?

We know Galileo's response, which is reticent at best, and if we compare it with other documents, it lacks credibility. On August 19 he directly informed Kepler that the spyglass he had in Padua at the time could not observe the Medicean planets. The best one had been appropriated by the grand duke "to place it in his Tribune and safeguard it among his most distinguished objects," and he had been unable to build others of the same quality. But Kepler didn't have to worry. As soon as Galileo got to Florence he would make other machines to build and clean lenses and would send one as soon as possible.[56]

This was yet another delay that kept Kepler from obtaining what he considered the most desirable thing in the world, as he had written just ten days earlier.[57] The imperial mathematician had little choice but to make the best of things.

> The spyglasses we have here, and I am talking about the best, enlarge the diameter ten times; the others barely three. Only one of mine can multiply twenty times, but the visibility is weak and not very good. I do not know the reason and will try to improve this. None of those I have been able to see so far allow me to distinguish the small stars, save one I built myself. Although it does not multiply the diameter more than three or four times, it can distinctly show the numerous stars of the Milky Way. This is truly surprising, especially if we consider that it was built to create optical illusions. The reason is its luminosity, in that it lets in much light. In fact, as opposed to other lenses, the edge of the convex lens is not covered, so its entire surface can be used. And in this way one's gaze can cover a broader region, and I can easily find what I seek.[58]

Galileo was thus aware of Kepler's efforts to build the high-quality spyglass essential to continuing his observations. Moreover, there could be no doubt regarding his intentions, as the observations of the numerous stars in the Milky Way sounded like yet another confirmation of the *Sidereus nuncius*. Yet just months earlier a few spyglasses had been sent from his Paduan workshop to the Roman cardinals Borghese, Farnese, Montalto, and Dal Monte.[59] They might not have been excellent and were probably not as powerful as the one that he had sent the ultra-Catholic Maximilian, Duke of Bavaria, in late May, but they were certainly better than the one Kepler had in Prague.[60] That said, even the French regent, Maria de' Medici, had to wait until the end of September to receive one, and perhaps others had to queue as well.[61] Regardless of how things went, what is striking is that until the end of his days the emperor insistently demanded "those lenses to make Galileo's spyglasses."

> One night there came here a Flemish alchemist, one greatly in His Majesty's favor, to ask me if, on his behalf, I could write to our Lord the Grand Duke to ask in his name for two of Galileo's lenses to make spyglasses, along with some glass he will then have ground here, like the two ground glasses he desires, which is something His Majesty wants more than anything else.
>
> His Majesty has reminded me again of Galileo's spyglasses and lenses, which I wrote to you about last week; and to fulfill His Majesty's desire even more, if you think fit, they can be sent by post, as done for crates of oil. And so you can judge the Emperor's mood as he awaits these things, and his wanting to hold up marriages, as I wrote to you last week. And, again, an alchemist arrived in the empire, with whom he is spending the entire day when he is not assailed by the fear of a successor.[62]

Giuliano de' Medici sent these dispatches to Vinta on November 14 and 21, 1611, just one year after the *Narratio* was printed. Rudolf's "desire" had not waned and, if anything, had become even more obsessive, typical of his melancholy moodiness. Bordering on tediousness, the Tuscan ambassador harped on the perfection of the spyglass, "which, as you already know, is greatly coveted, and I, in particular, would be most grateful for one."[63] First from Padua and then from Florence, Galileo tried to appease him about the delay, noting, "I must accommodate the equipment I use to work my spyglasses, part of which must be mounted on walls and cannot be transported: so Your Most Illustrious Highness should not be surprised if I tarry again in sending yours; but I shall ensure that the delay will be compensated by the excellence of the instrument."[64]

It must be said, however, that although Galileo never sent Kepler a telescope, he did ship him something between August and December 1610: the calculation of his Jovian observations conducted between March 9 and May 21 (when the planet and its satellites were visible because they were not too close to the Sun yet) and two puzzles with transposed letters, the anagrams on Saturn and Venus. As the Tuscan ambassador reported, Kepler "exhausted himself trying to discover what it was, and he is imagining a thousand things, and says he cannot still his soul."[65]

Galileo must have been overjoyed to receive this news. There is no denying that his challenge against the imperial astronomer was unquestionably one of the reasons behind the puzzles, so hearing about all of Kepler's difficulties and anxiety would rightly have been a point of pride for him. Both were die-hard Copernicans, but even from their first correspondence nearly fifteen years earlier, they were clearly competing against each other, as their projects and approaches were too different to allow them to collaborate. So far, the game had been led by Galileo, and he certainly had no intention of passing it on to one who had proven to be an ally but was also a formidable competitor, as the *Dissertatio* had amply demonstrated.

But what about Kepler? Why had he fought this battle? For the truth, as he stated time and again; in this case, battling for the truth meant standing up for Galileo but also for himself. If Galileo had proven to be the winner, it was thanks to him and the publication of the *Dissertatio* and, above all, the *Narratio*. Nevertheless, this was a victory that had to be consolidated in order to become a true conquest. Those discoveries were too scant to suffice. It would take much more to get a new and revolutionary *systema mundi* accepted. Galileo was a very capable hunter of stars, the author of a brilliant observatory report, but not the builder of an *astronomia nova*. We need merely take a close look at the *Dissertatio* to grasp this, and there is no doubt that Galileo must have realized it too. Consequently, when on August 19 he leaked to Kepler that he had just been appointed "Phylosophus et Mathematicus" of the Grand Duke of Tuscany, the announcement acquired a very polemical air of revenge against those who, both then and later, refused to acknowledge the title he had so long desired. Wrote Galileo:

> You me ask for other witnesses, my dear Kepler. And I produce the Grand Duke of Tuscany who, having often observed the Medicean planets in Pisa over the past few months, gave me a gift of more than a thousand gold pieces upon my departure, and now he has called me back to the homeland with an annual stipend of a thousand gold pieces and the title of His Majesty's Philosopher and Mathematician, and moreover with no other duties, so that I may have all the time I need to perfect my writings on mechanics, the constitution of the universe, and local motion, both natural and violent,

> of which I have geometrically demonstrated many unsuspected and marvelous properties.[66]

In short, in response to Kepler's request for further proof and cautions about his numerous dogged opponents ("I do not want to conceal from you that here in Prague there have appeared letters from countless Italians who vehemently deny having observed those planets with your spyglass"), Galileo declared that his appointment as the grand duke's mathematician and philosopher was a tangible sign of success.[67] He had willingly accepted Kepler's most prestigious assent in a moment of great difficulty, but now he neither requested nor wanted anything else. There were other projects afoot, and they did not mesh with those of the German—and Lutheran—astronomer who, to boot, was the pupil of another Lutheran, the heretic Maestlin, whose works had helped plump up the already long lists of the *Index Librorum Prohibitorum*.[68] Could it be mere happenstance that the best spyglasses Galileo built following the publication of the *Sidereus* were sent first of all to Roman cardinals and prelates, while not even one found its way to Prague or London?

SEVEN

Across the English Channel: Poets, Philosophers, and Astronomers

As the battle commenced at the court of Rudolf II, a copy of the *Sidereus nuncius* arrived at another European court, that of James I in London. The English ambassador Sir Henry Wotton sent it from Venice on March 13, 1610, the very day it was published, and he attached his own report on the Galilean discoveries. However, to understand how the first news of the *Sidereus* spread throughout England, another document is even more significant than Wotton's report: a long, detailed letter from Sir William Lower replying to one from Thomas Harriot that has since been lost. Lower's immediate reaction was enthusiastic, to say the least: "Me thinkes my diligent Galileus hath done more in his threefold discoverie [the mountainous surface of the Moon, the real cause of the Milky Way, and the four satellites around Jupiter] than Magellane in opening the streights to the South Sea or the dutchmen that weare eaten by beares in Nova Zembla. I am sure with more ease and saftie to himself & more pleasure to mee. I am so affected with this newes as I wish sommer were past that I mighte observe these phenomenes also."[1]

The date was June 11, 1610, and Lower was quite struck by the singular coincidence that Harriot's letter had been delivered to him just when "we Traventane [Trefenty] philosophers were a consideringe of Kepler's reasons [in the *De stella nova in pede Serpentarii*, 1606] by which he indeavers to overthrow Nolanus & Gilberts opinions concerninge the immensitie of the spheare of the starres." While agreeing with many aspects of Kepler's astronomical theory, Lower did not reject the idea that Bruno and Gilbert were right, because between the distant stars Kepler supposed to be fixed on the sphere enclosing the world and the orbit of Saturn there could be a considerable number of stars "which by reason of their lesser magnitudes doe flie our sighte." Recalling that he had often heard Harriot say precisely the same thing, he pointed out the possibility that "aboute Saturn, Jupiter, Mars, etc. ther move other planets also which appeare not." But then came the discovery of the Jovian satellites. At this point, he mused that "probablie experience hath made good" regarding the infinite nature of the universe.[2] Encouraged by these surprising results, which seemed to corroborate their hypothesis, the Traventane philosophers were feverishly excited. "We are here so on fire with thes things that I must renew my request & your promise to send mee of all sortes of thes Cylinders," he wrote. "My man shal deliver you monie for anie charge requisite, and contente your man for his paines & skill. Send me also one of Galileus bookes if anie yet be come over and you can get them."[3]

With his words, Lower not only draws a map of the earliest circulation of the new discoveries in England but also gives us the context in which they were received: the Northumberland Circle of enterprising "virtuosi" associated with Henry Percy, ninth Earl of Northumberland, and known in his day as the "Wizard Earl." So we must focus on this milieu, but first we should try to clarify a number of misrepresentations and attempt to identify the specific role of the various players.

Let's start by saying that the image of the Northumberland Circle as a cohesive and organized group of intellectuals is essentially a historiographical myth. In fact, a "philosophy" of the Northumberland Circle never existed, at least not as an articulated and coherent doctrinal corpus to which its alleged disciples adhered, nor were its meetings characterized by research and scientific experiments conducted by members working closely together.[4] The only certainty is that between 1590 and 1620 an informal network of literati, mathematicians, astronomers, natural philosophers, physicians, and alchemists gravitated around the Wizard Earl, striving to radically update current knowledge. In addition to Harriot and Lower, the group included Walter Warner, Nathaniel Torpoley, and Robert Hues, who were later joined by Nicholas Hill and John Protheroe. All of them rejected Aristotelianism and shared an atomistic or corpuscular view of the world, taking up the Renaissance naturalism of Telesio, Patrizi, and Bruno.[5] Moreover, they also shared another interest: Copernicanism.[6] The Northumberland

Circle was thus known for promoting the new philosophical and scientific culture, countering the Aristotelianism that reigned supreme at all universities. This was no trifling undertaking, given the difficult climate created in England when James I became king, as personal rivalries had become entwined with the political and religious tensions that had arisen during the extremely long Elizabethan period.[7]

Being linked with Percy was quite risky in those years, especially after the failed Gunpowder Plot of 1605.[8] Politically unpopular with the court, Percy was initially suspected to have participated in the scheme, in which a distant cousin, Thomas Percy, had played a leading role; the earl was charged with high treason and imprisoned in the Tower of London for sixteen years. The leading figures in his entourage—including Harriot, Torpoley, and Lower—were also arrested and underwent grueling interrogations but were finally released.[9] A few years earlier, in 1603, Sir Walter Ralegh, a close friend of Percy's, and Harriot's first patron, had been locked away in the Tower, where he was executed in 1618 for allegedly conspiring with the Spanish against James I.[10] Moreover, we must not forget that in 1592 Ralegh's circle was publicly denounced for atheism.[11] While nothing ever came of this, Harriot was branded for life as an "epicurean atheist," and this weighed heavily on him when attempts were made to tie him to the Gunpowder Plot.[12]

It was in this context that the scholars close to the Earl of Northumberland conducted their intellectual activities. As Harriot complained in a letter to Kepler dated July 13, 1608, the situation in England was so dire that it was impossible to "philosophize freely."[13] Therefore, it is unsurprising that Harriot and his friends were careful to avoid printing their ideas, opting instead to exchange thoughts and communicate almost in secret.[14]

But let us try to understand how Harriot found the information on Galileo's discoveries that he then sent to Lower. By mid-April 1610, news about those discoveries as well as a copy of the *Sidereus nuncius* had already reached London, but it seems unlikely that, unpopular as he was at court, Harriot would have had access to it. Naturally, this "strangest piece of news"—as the ambassador Wotton called it—leaked outside the walls of Whitehall, so Harriot may well have heard about it somehow. After all, this wasn't about delicate political or military matters calling for the utmost discretion, but concerned events significant only to those with an interest in science and astronomy.

Nevertheless, a handwritten note suggests a source less random than undocumentable hearsay: Harriot's reading of the *Dissertatio cum Nuncio sidereo*, published in early May 1610, in which, as we have seen, Kepler fully supported Galileo's celestial discoveries.[15] It is a very short comment that Harriot jotted down during his careful study of the *Astronomia nova*, regarding the point at which Kepler criticized the inaccuracy of Erasmus Reinhold's *Tabulae prutenicae* (1551), commonly used at the time.[16] Since calculations based on the *Tabulae* predicted that around October 16, 1610 (new style), Mars would be in opposition to the Sun, Harriot noted: "1610. Oct. 6. differt a calculo pruten 3° gr. In Dissertatione cum nuncio sidereo, p. 5."[17] This memorandum stemmed from his cross-referencing of the *Astronomia nova* and the *Dissertatio*, as in the wake of Galileo's success astronomers were also stimulated to hunt for satellites around Mars. According to Kepler—on the very page Harriot had marked—"the most propitious time will be next October, which will show Mars in opposition to the Sun and . . . nearest to the Earth, with the error in the predicted position [according to the *Tabulae prutenicae*] exceeding 3°."[18]

It was thus through the *Dissertatio* that Harriot learned about the recent astronomical discoveries. Lower's letter, quoted at the beginning of this chapter, also tells us that this must have occurred no later than the beginning of June 1610, when there were still very few copies of the *Sidereus* circulating in London.

The fame of the celestial discoveries reached England well before the book. What is even more intriguing, however, is the fact that the news did not initially reach the prestigious universities of Oxford and Cambridge, but promptly arrived at a remote estate in Carmarthenshire, in southwest Wales, called Trefenty or Tra'venti, more than 200 miles from London. From London to Trefenty: it was along this uncommon geographic line that the "diligent Galileus" found his first English advocates.

• •

The scion of an ancient and respectable family from Cornwall, Lower had moved to Trefenty in 1606, just a year after marrying Percy's stepdaughter,

Penelope Perrot. The estate at Trefenty was part of his wife's dowry, which went beyond her material assets, as Lower's marriage to Penelope also introduced him to the family and intellectual network of the Earl of Northumberland. In fact, when he went to London for sessions of Parliament, of which he had become a member in 1601, he stayed at Syon House, where Harriot had taken up residence several years earlier.[19] There he had the chance to meet various scholars close to Percy, such as the mathematician Walter Warner and Nathaniel Torpoley, also a mathematician and the only one in the group to declare himself an anti-atomist.[20]

On a scientific level, however, it was with Harriot that Lower established his closest and most important ties. It was a collaboration punctuated by their unfortunately patchy correspondence, as none of Harriot's letters have survived.[21] In this mutilated dialogue, we can nevertheless easily recognize the men's intellectual hierarchy: Harriot as the revered and undisputed teacher who guided and oversaw research, Lower as the disciple who scrupulously followed instructions. It is nevertheless a valuable dialogue that sheds light on the activity of a small and peripheral scientific community that had been established around Lower. We know neither the formation of the group nor the names of its scholars, with the exception of John Protheroe, who lived in Hawksbrook, a few miles from Trefenty, and was specifically mentioned in Lower's letter to Harriot of February 6, 1610.[22] In any case, this was by no means an unsophisticated coterie. Its members vaunted great philosophical, mathematical, and astronomical knowledge, as demonstrated by their discussion of Kepler's works. In fact, Lower loved to define his group as "the Trefenty philosophers," ever open to the challenges that continued to emerge on the horizon and thus eager to embrace Galileo's recent discoveries. After all, any interest in astronomy went back to before those discoveries had been announced in Europe and monopolized debates.

Lower had come up with a careful plan to track the motions of the bright comet that appeared in September 1607—later dubbed Halley's Comet—and kept a log of its position. As he wrote to Harriot on September 30, he had observed the comet regularly, both with the naked eye and using a "cross-staffe" to measure its astronomical distance from the various "fixed stars."

He first saw it around midnight on September 17 in Ursa Major, while he was crossing the Bristol Channel on a boat from Cornwall to Wales.[23] Weather permitting, he continued to observe it from Trefenty until October 6 and sent his findings to Harriot, who analyzed them attentively and compared them with his own.[24]

It was indubitably the comet of 1607 that seriously motivated Harriot to consider applying his knowledge of optics to study the heavens. But it is very likely that his decision was also influenced by Kepler, who had heard about Harriot's exceptional skills "in all the mysteries of nature" *(omnia naturae arcana)*, and "especially in optics," from Johann Eriksen, one of Tycho Brahe's first assistants. This was firsthand information, because Ericksen was in London and often saw Harriot.[25]

In his *Ad Vitellionem paralipomena* a few years earlier, in addition to setting out his revolutionary theory of vision, Kepler also sought to demonstrate how essential it was for astronomers to be well versed in optics. Therefore, as soon as Ericksen told him that Harriot shared his interests, he promptly contacted the Englishman. It was October 2, 1606, and Kepler asked Harriot for his opinion on various optical questions, such as the origin of colors, rainbows, and the haloes around the Sun. Above all, however, he wanted to receive experimental data that would allow him to pinpoint "the cause of refraction": "Send me the measurements of all the refractions you have found in your experiments and, at that point, all the rest will be done."[26]

Thus commenced a brief exchange of letters that was marked by mutual cordiality but was probably not very satisfactory, at least in terms of what Kepler had hoped to gain from it.[27] Harriot was reluctant to reveal the true nature of his experiments and merely provided a series of tables and an explanation of his own theory of refraction in terms too vague to allow Kepler to follow his line of reasoning. As the Englishman wrote at the end of his reply, Kepler had no choice but to wait for Harriot to publish his books:

> I have now conducted you to the doors of nature's house, where its mysteries lie hidden. If you cannot enter, because the doors are too narrow,

then abstract and contract yourself mathematically into an atom and you will easily enter, and when you have come out again, tell me what miraculous things you saw. Concerning the colors, there are great mysteries, not here to unfold. But when I have written on the rainbow, you will see their immediate causes and how they are mutually related. And likewise about many other things in natural philosophy, which I will describe, if God accord me spare time and health. In the meantime wait patiently.[28]

Harriot never kept his promise, either to Kepler or to those who, time and again, urged him to announce the results of his studies; he never published anything during his lifetime.[29] Nevertheless, the things he did report cannot be chalked up to pretentiousness, as his expertise in optics was extremely advanced.[30] Therefore, these are the skills we must examine in order to grasp their intrinsic merits and close ties with astronomical observation.

• •

Optics and astronomy: it was from this fortunate crossroads that Harriot hailed the discoveries described in the *Sidereus*. Harriot had been studying optics—and particularly the refraction of light in various media—since the 1590s, and would continue to do so for the rest of his life. This is documented by the thousands of manuscripts he has left to posterity, containing an astonishing number of observations, experiments, diagrams, tables, and calculations on the refraction of light on plane and curved surfaces, in prisms, and in lenses.[31] As early as 1602, this intense ongoing work allowed him to arrive at a correct formulation of the sine law of refraction.[32]

Harriot carried out most of his research at the Earl of Northumberland's magnificent Syon House, near the village of Isleworth, just a few miles from the center of London and set on the banks of the Thames opposite Kew Gardens.[33] There Harriot set up a scientific laboratory, where he performed alchemical experiments but also calculated the specific gravities of an incredible array of substances, the flow of water through pipes, projectile trajectories, and the motion of free-falling bodies.[34] That residence became the perfect place to cultivate his mathematical studies, make instruments, and, as he wrote to Kepler in December 1606, plumb the mysteries of "nature's house."[35] Even the garret was transformed into a kind of astronomical observatory so he could follow the motion of heavenly bodies and measure their positions.[36]

In the peaceful atmosphere of Syon House, Harriot was able to indulge in his vocation as a mathematical practitioner, taking up an approach that had become popular in Elizabethan England and consisted of applying the abstract methods of mathematics to real problems and needs.[37] It was from this immediate background, where measuring instruments played a central role, that he developed his interest in optics, attempting to solve questions that had to do with cartography, the science of navigation, and, above all, determining latitude at sea. His awareness of errors due to the effect of refraction in the atmosphere—when the cross-staff or the mariner's astrolabe was used to measure the altitude of a celestial body above the horizon—led him toward the systematic analysis of refraction, a proper understanding of which would help correct or at least mitigate those errors.

Harriot's extensive series of experiments on refraction had a specific aim: developing new techniques to improve the performance of navigational instruments. This was a complicated investigation demanding both mathematical and astronomical expertise, but also the ability to handle prisms and lenses of different shapes with which to verify the behavior of refracted light. In fact, he turned to a talented craftsman, Christopher Tooke, a lens grinder he hired between 1604 and 1605 who would serve as his main assistant at Syon House for the rest of Harriot's life.[38]

It was during these studies, entwining optical experiments with astronomical observations and measurements, that in July 1609 he managed to make a spyglass. We do not know the circumstances surrounding its construction, or whether he had heard about Lipperhey's "new invention." We know that he had not seen one himself, as extant documents state that the first spyglasses were sent from the Spanish Netherlands to England between November 1609 and February 1610, several months after Harriot had already built his own.[39] What is certain, however, is that news of the invention swiftly spread throughout Europe and at least one copy of the *Ambassades du Roy de Siam* had reached London by the end of 1608.[40] In any event, the instrument was so rudimentary that numerous craftsmen immediately reproduced it with very little trouble. So for someone like Harriot, who had long studied

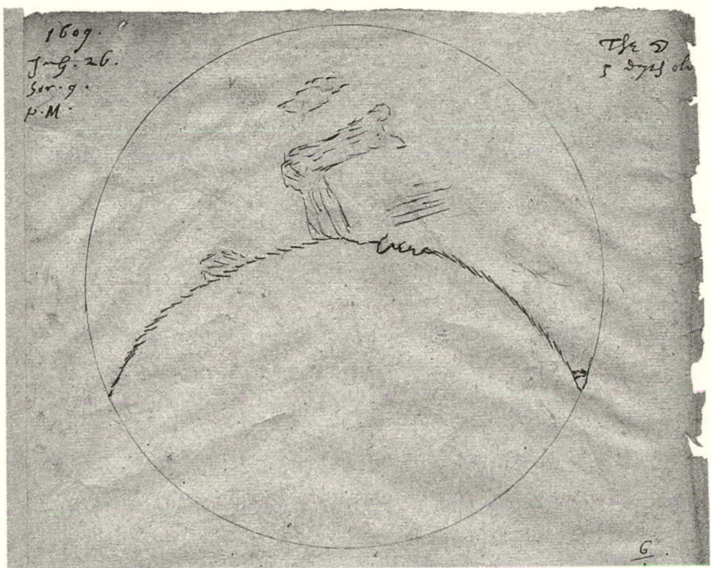

Figure 17. Thomas Harriot, drawing of the Moon, July 26, 1609

the properties of lenses, it would have been quite simple to implement the idea of inserting two differently shaped lenses at either end of a tube to make an optical device that could magnify objects. It is unsurprising that he would then promptly use it for astronomical purposes.

• •

The time was 9:00 P.M. on July 26, 1609. From the dormer window of his residence at Syon House, Harriot turned his six-power spyglass on the Moon, on the fifth day of its waxing phase. But he did not merely observe it. He decided to portray its surface in order to capture the image on paper exactly as it appeared to him through the instrument (Figure 17).

This was the first drawing of the Moon seen through a spyglass.[41] Four months before Galileo, and entirely independently, Harriot examined the lunar surface and made this drawing.[42] But what did he observe? Let's take a closer look at this curious depiction. The boundary separating the illuminated area from the dark one—the line referred to as the terminator—is clearly wrong, because the horns of the crescent extend well beyond the central axis of the lunar sphere. In fact, this configuration could occur only in the presence of an eclipse, but an eclipse is out of the question for a five-day-old Moon.[43] Furthermore, its topographic features are so vague that any attempt to identify them is futile. Harriot's silence certainly doesn't help, as his drawing bears no comment whatsoever that might help us understand if the bizarre patches that stand out on the illuminated part of the Moon are shadows cast by craters or mountains.

Various explanations have been advanced to explain the undeniable approximation of Harriot's sketch of the lunar surface on July 26, 1609.[44] It has recently been suggested that it is due to a particular type of cartography he developed when, at Ralegh's service, he mapped out the shores of the new lands discovered in America. In other words, in depicting the Moon, Harriot was allegedly inspired by a method in which the focus was not topographic features but, rather, the outlines of coasts and the boundaries between lands and oceans drawn with the relative spatial relationships in two dimensions. Seeing the patches of light and dark on the Moon as lands and oceans, Harriot simply reproduced their outlines, just as he had done years earlier with the painter John White when he charted the coasts of Virginia.[45]

Perhaps that's exactly how it went: that his mind-set as a cartographer led him to overlook or underestimate the topographic features of the Moon. A small clue that seems to confirm this possibility can be found in Lower's letter of February 6, 1610. Lower had started to explore the Moon with a "perspective Cylinder" Harriot had specially sent him—perhaps along with the drawing—in order to get his confirmation. From Trefenty, Lower reported back to him with the results of his observations in the company of his young assistant, John Protheroe:

> According as you wished I have observed the moone in all his changes. In the new I discover manifestlie the earthshine, a little before the dichotomie that spot which reprefents unto me the man in the moone (but without a head) is first to be seene. A little after neare the brimme of the gibbous parts towards the upper corner appeare luminous parts like stares, much brighter then the rest and the whole brimme along, lookes like unto *the description of coasts in the Dutch bookes of voyages*. In the full she appeares like a tarte that my cooke made me the last weeke. Here a vaine of

> bright stuffe, and there of darke, and so confusedlie al over. I must confesse I can see none of this without my cylinder. Yet an ingenious younge man that accompanies me here often, and loves you, and these studies much, sees manie of these things even without the helpe of the instrument, but with it sees them most plainielie. I meane the younge Mr. Protheroe.[46]

Lower's colorful description came with no illustration, but it almost seems to be the pendant to Harriot's silent drawing. It is as if the Traventane philosopher wanted to provide the caption missing from that drawing and, through a comparison with the "coasts" described in the "Dutch bookes of voyages," wanted to reveal how Harriot had probably sketched the outlines of the lunar surface. But Lower added nothing else, and on a verbal level, his report is just as vague as Harriot's drawing.

There is another important element. It seems that after July 26, 1609, Harriot did not systematically observe the Moon, nor did he wonder why its surface was spotted. In extant manuscripts, there is no trace of such observations until the summer of 1610. Therefore, it seems evident that if he decided to resume his telescopic observations, this must be related to an unexpected but also sensational event: publication of the *Sidereus nuncius*, of which, as we know, he was well informed by the beginning of June 1610. At that point, the things he had seen earlier through his spyglass—ahead of Galileo—acquired a significance he had not even suspected until then. Their importance is conveyed by the aforesaid letter from Lower to Harriot dated June 11, 1610, in which he unhesitatingly acknowledged, "In the moone I had formerlie observed *a strange spottednesse* al over, but had no conceite that anie parte therof mighte be shadowes."[47]

Lower's concise, sincere statement, which is often overlooked, tells us something quite simple: that merely observing a phenomenon was not enough to grasp its true nature, as it also had to be interpreted.[48] It was only this gnoseological endeavor, in which things represented signs to be interpreted and set in an order of constant relations, that kept the new things revealed by the spyglass from remaining an inaccessible realm. In the same letter, in fact, Lower showed his frustration at distinguishing what he saw through his "perspective Cylinder" and described how Galileo's discoveries "now" allowed him to dispel his doubts about the "7 starres" of the Pleiades: "[I had observed] the 7 starres also in Taurus, which before I alwayes rather beleved to be 7 then ever could number them. Through my Cylinder I saw thes also plainelie and far asunder, and more then 7 to, but because I was prejudgd with that number, I beleved not myne eyes nor was carefull to observe hou manie; the next winter now that you have opened mine eyes [in other words, now that Harriot had informed him that Galileo had identified them] you shall heare much from me of this argument."[49]

It was June of 1610, and neither Harriot nor Lower had read the *Sidereus* yet. Nevertheless, indirect knowledge of its contents had already marked a watershed in their ability to perceive celestial phenomena. Within just a month, the book became more readily accessible in England, as proven by the letter of July 6, 1610, that Sir Christopher Heydon sent from Baconsthorpe, near Cambridge, to his friend William Camden in London, where the latter had headed the prestigious Westminster School. Although he had published a work defending judicial astrology in 1603, Heydon shared Camden's unabiding passion for astronomical observation, which the *Sidereus* had obviously helped revive.[50] The tone of the letter leaves little room for doubt: "I have read Galileus, and, to be short, do concur with his opinion; for his reasons are demonstrative: and of my own experience, with one of our ordinary trunks, I have told eleven stars of the Pleiades, whereas no age ever remembers above seven."[51]

Harriot must unquestionably have obtained a copy of the *Sidereus* in early July, and we can readily imagine his excitement as he—the very one who, just a year earlier, had first observed and sketched the Moon—turned the pages of Galileo's evocative description of that satellite's mountainous surface, comparing it to Earth and illustrating it with marvelous engravings. Just a month had elapsed since Galileo's discoveries had arrived, filtered by Kepler's *Dissertatio*. But now Harriot could finally see with his own eyes just how impressive Galileo's images were, representing the constellations of Orion and the Pleiades, and, above all, the discovery of Jupiter's satellites, entrusted to a series of visual sequences, suggesting to the reader the importance of the time factor in observing their periodic motions.

Harriot's reading of the *Sidereus* injected new impetus into his astronomical investigations, but it also gave him a theoretical framework within which to interpret them. After a hiatus of nearly a year, he resumed his lunar observations on July 17, 1610, launching a very intricate series of studies that continued over the next two years, albeit not regularly.[52] Despite the fact that his first comments did not emerge until August, the second drawing—the one dated July 17—shows that Galileo's engravings had significantly oriented his drawings. In fact, a comparison between Harriot's schematic image (Figure 18) and Galileo's more accurate rendering (Figure 19), both of which illustrated the configuration of the Moon in the first quarter, reveals astonishing and undeniable similarities, such as the large round crater close to the center of the terminator.[53] This resemblance can be seen in the notes Harriot started to add to his illustrations, in which the descriptions of the lunar surface in terms of terrestrial geographic features clearly evoke various passages in the *Sidereus*.

On August 26 he commented that he had observed "greater eminences & some valleyes with shadowes," and he defined "about 1/3" of that as "the Caspian," establishing a comparison with Earth's well-known sea; on September 11, along the ragged line of the terminator, he noted "some ilands and pronomontoryes"; and on October 23 he wrote of a mountainous landscape "with an opening in the middest."[54]

• •

The astronomical program Harriot launched in July 1610 takes up the concepts outlined by the *Sidereus*. Well aware of the significance of Galileo's epoch-making book, he was so impressed by it that he made the celestial discoveries the key points of his work agenda. After all, in the face of phenomena that upended the traditional image of the celestial space, no astronomer or philosopher could have been untouched by this: certainly not a Copernican such as Harriot, who fully understood the cosmological consequences of Galileo's discoveries, starting with Jupiter's satellites.

This was the discovery that so excited Harriot and his disciples. In this case, however, it was not possible to observe them immediately. As Lower had noted in his letter of June 1610, one had to wait until summer was over.[55]

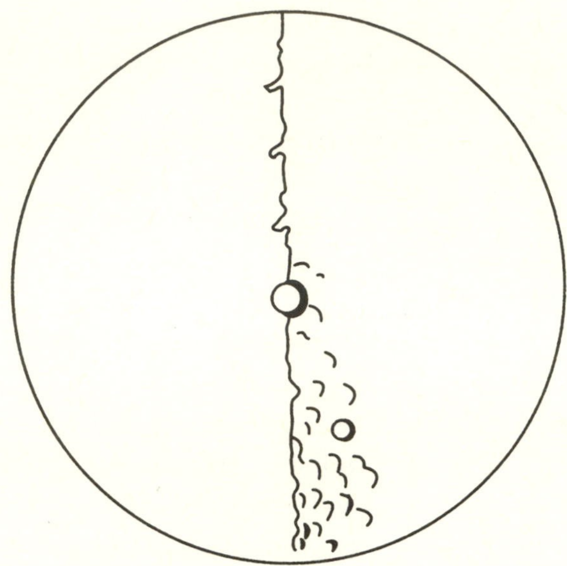

Figure 18. Thomas Harriot, drawing of the Moon, July 17, 1610

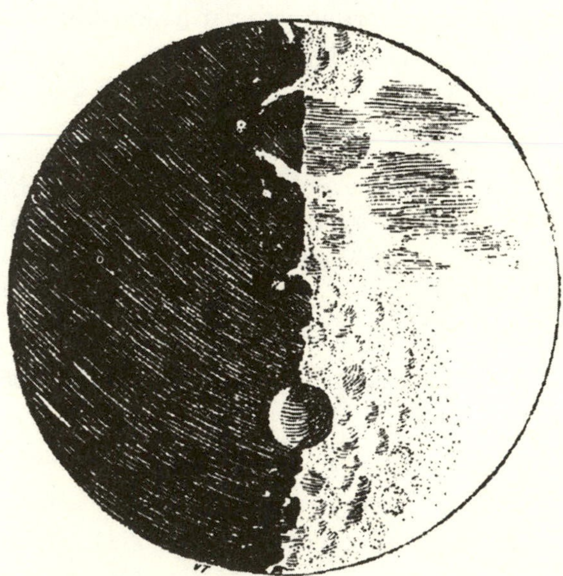

Figure 19. Galileo Galilei, drawing of the Moon, 1610

The satellites can be seen only when Jupiter is far away enough from the Sun. The best time for observation is when Jupiter and the Sun are on opposite sides of the celestial sphere—that is, when they are in geocentric opposition. In June 1610, however, Jupiter and the Sun were in geocentric conjunction, meaning that they were still very close; to see them, observers had to wait until autumn.[56]

Harriot made his "first observation of the new planets" on October 17, the first of a long series that would keep him busy until the spring of 1612.[57] His hopes were high, but he was disappointed. On October 17 he could see "but one [of these satellites] and that above."[58] Little changed in the months that followed, with poor weather conditions often keeping him from attaining Galileo's results. Nevertheless, he embarked on an intense work plan conducted chiefly at Syon House, but he also went to London on several occasions: on November 16, 1610, he observed from the area of Blackfriars; on February 16, 1611, he worked both from Greenwich and also "at D. Turners house in little St Ellens" (at the house of physician Peter Turner, who lived near the church of St. Helen, in the heart of the city).[59]

Every time he went to London to visit the Earl of Northumberland in the Tower or for personal reasons, Harriot brought a spyglass with him and took advantage of the trip to make his observations. Indubitably, he kept both the earl and his old patron Sir Walter Ralegh, likewise imprisoned in the Tower, informed of his activities. But he also kept them apprised of recent celestial discoveries, as attested to by Ralegh's *History of the World*, published in 1614 but commenced between 1610 and 1611, and in which we find praise of "Galilaeus, a worthy astrologer now living, who, by the help of perspective glasses, hath found in the stars many things unknown to the ancients."[60]

During this demanding activity, Lower arrived at Syon House on December 7, 1610. However, he too would be disappointed. Based on Harriot's notes, we know that at 9:00 P.M. they saw only one satellite, whereas at 5:00 A.M. they managed to see two.[61] Their difficulties were not solely the result of Jupiter's position or the weather, for it was by no means easy to see the four satellites. As Lower wrote to Harriot on March 4, 1611, "Concerninge the Joviall stares I writte nothinge of them last, because I had notinge to write. For indeede although both I and the young Philosopher at Hawksbrooke have often and in verie cleare nights . . . when wee I say have often diligentlie observed Jupiter wee could never see anie thinge. I impute it to the dullnesse of my sighte for onlie with your greate glasse I could see them in London."[62]

Therefore, nearly a year after the *Sidereus* was published, Galileo's advice to anyone who wanted to observe the Medicean planets—namely, to procure a spyglass with a suitable magnifying power—was more valuable than ever. That Harriot bore this in mind is evident from the zeal with which, thanks to the help of his trusted collaborator Tooke, he did his utmost to improve his "perspective trunckes" starting in July 1610. This intense work began to yield results just a month later, when he was able to make one with a magnification of twenty, and culminated in April 1611 with another one that had a magnification of thirty-two.[63] Despite these efforts, however, Harriot's spyglasses never achieved the resolution of Galileo's. Consequently, even when he finally managed to see all four satellites together on December 14, 1610, the quality of his observations—as he himself complained—still left much to be desired.[64] Naturally, this had a substantial effect on the development of his astronomical projects. Taking up the invitation that, in the *Sidereus*, was addressed to "all astronomers," in the spring of 1611 Harriot decided to make his own contribution to determining the periods of the new satellites.[65] This entailed long, complex work to which he devoted himself until 1614, and it is discussed in more than two-thirds of all his manuscripts on Jupiter.

However, it is not easy to find our way through the *mare magnum* of his calculations on the periods of the satellites.[66] The fact that they are "in dreadful disorder"—to quote John J. Roche, the scholar who most closely examined them—makes a reliable chronological reconstruction virtually impossible.[67] Obviously, this situation, which affects the entire collection of Harriot's manuscripts, is due above all to the fact that they are worksheets often rewritten and scribbled over, and not intended for publication.[68] Despite these difficulties, however, several elements emerge quite clearly. First of all, we can see that in order to determine the orbits and periods of the satellites, Harriot relied on data in the *Sidereus*. When he determined the period of the third satellite, the one we now call Ganymede, he used three series of Galilean observations to arrive at a greater degree of approximation.[69] Harriot naturally used his own calculations, but he always compared them against Galileo's, which he knew were the result of observations far more accurate than his own. It was thanks solely to this method that, in some cases, he managed to arrive at better values.[70]

Nevertheless, the Jovian satellites were not Harriot's only interest. At the same time, he frequently studied the Moon and began to analyze sunspots, which he first saw on the morning of December 8, 1610, after he and Lower had spent the previous evening gazing at the Medicean planets.[71] He did this a number of times, such as the night of January 11, 1611, when he observed the satellites and then the Moon in its last quarter.[72]

Harriot flanked this intense activity with his equally intense study of Kepler's astronomical works, which he usually managed to obtain immediately.[73] As a result, in May and June 1610 it was the *Dissertatio* that promptly informed him of the most recent discoveries, while at the end of the year he learned about Saturn's tricorporal appearance thanks to the *Narratio*. Galileo had announced the discovery through an anagram Kepler published in the latter work, unsuccessfully trying to solve it. Harriot likewise made an attempt, filling an entire sheet of paper with ingenious combinations of words, but without finding the answer.[74] He would discover the mystery concealed in that obscure anagram the following year when he read the *Dioptrice*, as in the introduction Kepler disclosed the letter from Galileo to Giuliano de' Medici containing the solution.[75]

◆ ◆

Harriot and his followers viewed the celestial discoveries as an important opportunity—not to be missed—to revive and fuel a debate on the reasons underlying heliocentrism, of which they had been the earliest proponents in England at the end of the sixteenth century. They were so enthusiastic that some of them even interpreted the Medicean planets as visible proof supporting Bruno's doctrine of the infinity of the universe and the multiplicity of worlds.[76] Those discoveries were gratifying because they confirmed their Copernican convictions, yet at the same time sparked the understandable desire to venture down the paths Galileo had opened.

However, the local political and intellectual milieu had forced them into what we can term preventive self-censorship. As a result, their impassioned correspondence regarding the *Sidereus* remained a private affair, and Harriot's observations, drawings of the Moon, and calculations on the periods of the Jovian satellites were never published. It seems unlikely that anyone outside the small circle of the Earl of Northumberland knew about Harriot's shrewd use of Galileo's data.

This attitude was rife with consequences because, at least initially, it allowed astronomical discoveries to be received and presented in England on an ideological level. In fact, when they began to be discussed publicly in 1611, every technical consideration, every assessment of their reliability, fell by the wayside and finally disappeared entirely. What prevailed above all was a reflection on the destiny of man and society. Francis Bacon, one of the first to have access to the copy of the *Sidereus* that Wotton had sent James I due to the favor he enjoyed at court, chose to examine the inescapable cosmological questions the Galilean discoveries had raised, but the terms of the debate did not budge. This was not because Bacon rejected Galileo's interpretation, despite acknowledging the authenticity of his discoveries, but for the simple reason that the works in which they were discussed, the *Descriptio globi intellectualis* and the *Thema coeli*, written around 1612, went unpublished until 1653.[77]

Surprisingly, the person who made those discoveries known in England was neither an astronomer nor a philosopher but a poet: John Donne, in the now-famous verses of his *Anatomy of the World*, composed in the first half of 1611 and published a short time later.[78]

Donne's reading of the *Sidereus* elicited a reaction entirely different from Harriot's. Instead of enthusiasm and admiration, Donne was anguished and dismayed. In his eyes, the "new stars" visible with the spyglass, like the revival of an atomistic conception of nature, marked the end of a world already complete and orderly in and of itself.[79] Age-old and reassuring certitudes had been dashed, and chaos threatened to undermine the consolidated image of the universe, taking with it the role of man and even the order constituted by his social relations. If the hierarchical distribution of the heavens were overturned, sooner or later this process would sweep away "all coherence . . . [a]ll just supply, and all relation," to the point that "prince, subject, father, son, are things forgot."[80] In Donne's mind, the new astronomy called "all in doubt": conceptions regarding the universe but also those concerning political, moral, and social order.

This was the first time that a series of discoveries, sensational as they may have been, emerged from their specialized field—astronomical research—to spark such sweeping and dramatic reflections.[81] Yet this was not the first time Donne had tackled the issue, as he had already done so in a short but vitriolic anti-Jesuit satire titled *Conclave Ignati*. In this case, however, rather than indulging in catastrophe, his tone bordered on derision. In the *Conclave Ignati* the future dean of St. Paul's Cathedral in London was "merely pleasantly amused," although clearly his objective was one and the same: demonstrating the deleterious consequences of the new astronomy.[82]

• •

Published anonymously, undated, and with no indication of where it was printed, the *Conclave Ignati* was written between the assassination of Henry IV of France on May 14, 1610, and January 24, 1611, when it was entered in the Stationers' Register, the booksellers' record of new publications.[83] The English version, titled *Ignatius his Conclave* and likewise published anonymously, was listed on May 18, 1611, and was the most successful and widely circulated edition, as demonstrated by its numerous reprintings.[84] In the organization of the *Ignatius*, all of which revolves around a series of scenes taking place in Hell, the new astronomy seems to play a rather marginal role. There are a couple of fleeting references: at the beginning, where Galileo and Kepler are mentioned, and toward the end, where only Galileo is cited. Moreover, both scientists maintain the status of real, living men and are not transformed into figures of literary fiction, as is instead the case with nearly all the other historical figures involved in the plot, including Copernicus himself.

In the context of a fantastical narrative, brief mention of the latest astronomical discoveries, along with a reference to François Ravaillac, who murdered the French king, are the only windows onto reality.[85] These elements are extremely significant, not only because they help us date the composition of the work but above all because they tell us which ties the author was trying to establish between the Jesuits and the new astronomers. In his "impudent satire," Ravaillac's gesture, implicitly attributed to the sinister plotting of the Society of Jesus, and the astronomers' discoveries are the final, tangible outcomes of a conspiracy to destroy the tranquility of humankind and the world.[86]

The astronomical question is introduced in the very first pages, in which Donne explains his peculiar story: a cosmic journey during a trance, in the throes of an "ecstasy" in which his soul, separated from his body,

> had the liberty to wander through the all places, and to survey and reckon all the roomes, and all the volumes of the heavens, and to comprehend the situation, the dimensions, the nature, the people, and the policy, both of the swimming Ilands, the *Planets,* and of all those which are fixed in the firmament. Of which I thinke it an honester part as yet to be silent, than to do *Galileo* wrong by speaking of it, who of late hath summoned the other worlds, the Stars, to come neerer to him, and give him an account of themselves. Or to Kepler who (as himselfe testifies of himselfe) *ever since* Tycho Braches death *hath received it in his care, that no new thing should be done in heaven without his knowledge.*[87]

In the notes in the margins of this passage, we find references to Galileo's *Sidereus nuncius* and Kepler's *De stella nova in Cygno,* the latter published in 1606. Therefore, Donne's reflections clearly stemmed from his recent reading of these works. That said, it is equally clear that his attitude was neither indulgent nor open toward them.[88] Just the opposite, in fact, for while dwelling on the enthusiasm of both astronomers, he also ridiculed their exuberance and arrogance, almost as if he were saying, "Well, now, Galileo presumes to direct the motion of the stars, while Kepler professes that nothing new can happen in the heavens without his knowledge."

Donne's treatment of the new astronomers fully reflects the leitmotif of the entire work: using the instruments of satire to defuse the daring and pernicious ideas of all innovators. In fact, at the entrance to Hell, in addition to Lucifer we find only the souls of those who had "title"—in other words, those who proposed "any innovation," offended ancient beliefs, "induced doubts, and anxieties, and scruples," or felt "a libertie of beleeving" and thus went against "established opinions, directly contrary to all established before."[89] Despite the attempts of Ignatius of Loyola, who in Donne's mockery does everything in his power to convince Lucifer that the Jesuits were the only ones who had everything lined up properly, others also asserted their right to be considered "innovators." This was why they took part in the same conclave and demanded to be admitted to Hell. Ignatius and his followers are indubitably painted as the main figures disturbing the peace, champions of the policy of violence—above all in France and England—against the sovereigns who had embraced the Reformation and spurned the temporal power of the pope. Yet they were not the only ones to undermine the established truths of tradition. Paracelsus, Machiavelli, and Christopher Columbus also step forward as legitimate candidates on a par with the Jesuits.[90] And it is no accident that the first to ask Lucifer to welcome him into Hell is an astronomer: Copernicus, who had demolished Ptolemy's doctrine, a truth spanning the centuries that was the foundation of many Christian beliefs.

The way Donne introduces Copernicus in his *Ignatius* further confirms the poet's smug derision toward astronomers. Rather than immediately revealing his identity, Donne describes him as "a certain *mathematician*, which until then had been busy to find, to deride, to detrude Ptolemy." And in the following grotesque scene he puts words in his mouth that clearly make him seem arrogant. The author of *De revolutionibus*, who "with his hands and feet (scarce respecting Lucifer himself) [beats] the dores [of Hell]," is shouting: "Are these [doors] shut against me, to whom all the Heavaens were ever open; who was a soul to Earth and gave it motion?" With a knowing wink to the reader, Donne comments: "By this I knew it was Copernicus."[91]

We find this atmosphere intact in the final reference to Galileo, where his discoveries turn out to be the clever device through which Donne ends his journey. Exhausted by these pressing demands but by no means willing to relinquish his kingdom of Hell or share it, Lucifer suggests that the only way out for Ignatius and his Jesuits would be to retire to the Moon and set up a new empire there. The solution is promptly described:

> I will write to the bishop of *Rome:* he shall call *Galilæo* the *Florentine* to him, who by this time hath thoroughly instructed himselfe of all the hills, woods, and Cities in the new world, the *Moone*. And since be affected so

> much with his first *Glasses,* that he saw the *Moone* in so neere a distance that he gave himselfe satisfaction of all, and the least parts in her, when now being growne to more perfection in his Art, he shall have made new *Glasses,* and they received a hallowing from the *Pope,* he may draw the *Moone* floating like a boate upon the water, as neere the earth as he will. And thither (because they ever claime that those imployments belong to them) shall all the Jesuites bee transferred, and easily unite and reconcile the *Lunatique Church* to the *Romane Church;* without doubt, after the Jesuites have been there a little while, there will naturally grow a *Hell* in that world also: over which, you Ignatius shall have dominion, and establish your kingdome and dwelling there. And with the same ease as you passe from the earth to the *Moone,* you may pass from the *Moone* to the other *starrs,* which are also thought to be worlds, and so you may beget and propagate many *Hells,* and enlarge your *Empire.*[92]

In Donne's work, enthusiasm borders on frenzy. Galileo never said he had seen "woods, and Cities" on the Moon, nor did he make any of the other fanciful observations ascribed to him. The caricature aims at ridiculing the impact of those discoveries, presenting Galileo and his coeval astronomers as arrogant men who discredit the truth of the heavens and foment chaos on earth. According to Donne's way of thinking, these discoveries may well have been "real" facts, but there was no question that, along with all the other innovations, they drove humankind away from religion. Developments in astronomy, philosophy, political theory, and man's *Weltanschauung* revealed the decline of the world and of society rather than true progress toward wisdom. In short, this is the same hostility toward any kind of intellectual innovation that we also find in his *Anatomy of the World,* albeit in a more poetic vein.

Donne's satire was destined to make inroads, at least on a literary level. Just two years later, during the 1613–1614 season, the audience at Blackfriars Theatre in London would be amused by the *Duchess of Malfi,* a grim but true story that unfolded during the Italian Renaissance. Its author, the playwright John Webster, was referring to none other than *Ignatius his Conclave* when, to express female fickleness, he had one of his characters make this clearly anachronistic statement: "We had need go borrow that fantastic glass / Invented by Galileo the Florentine / To view another spacious world i' th' moon, / And look to find a constant woman there."[93]

* *

Donne was a close friend and correspondent of the English ambassador to Venice.[94] This means we cannot reject the idea that, knowing the poet's insatiable curiosity, Wotton may have sent him a copy of the *Sidereus*.[95] Furthermore, Donne established close ties with various courtiers who could have informed him. They didn't even need to put the *Sidereus* at his disposal: all they had to do was mention Wotton's accompanying letter with a short but accurate summary of the work.[96]

Nevertheless, we are convinced that Donne learned about the most recent views on natural philosophy and astronomy from a source that was not exactly close to James I: the Northumberland Circle. Donne had long been friends with Percy, and, in fact, when the embarrassing situation of the poet's secret marriage to Anne More emerged, he had no problem asking the earl to handle the delicate task of informing the father of the bride.[97] That said, when Donne was writing *Ignatius his Conclave* and *An Anatomy of the World*, Percy had already been imprisoned in the Tower for several years. Not only did this fail to convince Donne not to visit Percy in jail, but it also did not keep from freely accessing the earl's immense library at Syon House.[98] Percy had an impressive and very up-to-date book collection that, along with the works of Diogenes Laërtius, Hero of Alexandria, Diophantus, Boetius, Alhazen, and Witelo, included those of John Napier, William Gilbert, Tycho Brahe, Paracelsus, Giordano Bruno, Della Porta, and Kepler.[99]

It would be rather surprising if, during his frequent visits to Syon House, Donne did not have the chance to interact with Harriot or his closest collaborator, Lower, likewise a regular at the earl's home, given that he had married Percy's stepdaughter. Donne certainly knew Lower, as documented by at least a couple of letters specifically stating this.[100] As to Harriot, in addition to living in a comfortable annex at Syon House since 1597, he was also responsible for Percy's library, which he had established.[101]

Naturally, this concatenation of places and figures may seem like little more than an inventive idea. But if we add the few certain elements we have on hand to the picture we have just sketched out here, what emerges is more than mere hypothesis. Let us examine it from this perspective.

Both the *Ignatius* and *An Anatomy of the World* show that Donne's preoccupations stemmed from an awareness of the consequences—disastrous, in his mind—that the new astronomy would have for the destiny of humankind. The English poet had no doubt. Copernicus, Kepler, and above all Galileo were the ones chiefly responsible for the "decline of nature," the loss of harmony in the world and its order, going back thousands of years. To Donne, however, those culpable were not only the supporters of heliocentrism but also the philosophers who explained natural phenomena in atomistic terms. As he stated in *An Anatomy of the World*, the "old" world was threatened by the revival of the idea that it had once more "crumbled out again to his atomies."[102]

But when he wrote these heartbroken reflections, whom was Donne trying to warn? From which English environments did he need to distance himself in order to prevent such ideas from prevailing?

During the early seventeenth century, the scholars forming the Northumberland Circle in England were the only ones to declare themselves Copernicans and vaunt advanced astronomical expertise. Harriot was in touch with Kepler as of October 1606, and for the next three years they exchanged ideas and observations about issues that interested them both, from optics to atomism, keeping each other informed on the twists and turns of scientific debate in their respective countries. When the *Astronomia nova* was published in 1609, Harriot and his followers promptly embarked on in-depth discussion and unreservedly accepted the idea that the planets move along elliptical orbits rather than circular ones.[103] No other English astronomer did so until the mid-seventeenth century.[104] Moreover, these scholars always welcomed celestial discoveries with true enthusiasm. Lastly, it was only in the Northumberland Circle that the new astronomy became established with an updated atomistic worldview.[105]

In this light, there should be few doubts as to the milieus that kept Donne up to date on the latest developments in natural philosophy and astronomy. At the same time, the English poet rightly had his own reasons for not publicizing such heterodox and compromising frequentations. He had little sympathy for the supporters of those who, in his eyes, raised doubts about an established way of understanding the universe that he shared. Nevertheless, perhaps fearing that this difference of opinion—significant as it may have been—would not be enough to avoid the suspicion of political connivance with his intellectual acquaintances, he preferred to allow the reader to discover their identity. After all, the savviest would have no problem understanding those he was targeting.

Donne's approach thus concealed not only legitimate philosophical reservations but also less noble political opportunism. After all, his desire to make inroads at court would not have allowed him to reveal his ties with Harriot, accused by many—James I included—of atheism.[106] This is demonstrated by the fact that in 1615 Donne was appointed royal chaplain and in 1621 dean of St. Paul's Cathedral, one of the highest positions in the Church of England.[107]

EIGHT
Conquering France

News of the celestial discoveries spread swiftly across France, but it wasn't the likes of Henry Wotton who told people in real time that the *Sidereus* had been published. The various "ultramontanes"—particularly interested in such news—were instead instantly informed by both Padua and Venice.

By April 18 the book had already appeared in Paris, where Pierre de L'Estoile borrowed it from the Protestant Christophe Justel. A theologian and canonist, Justel had just finished writing an important collection of ecclesiastical decrees, dedicating them to Jacques Leschassier, "who, like a new Theseus," had guided him through that intricate labyrinth.[1] As it turned out, Leschassier happened to own a copy of the *Sidereus* that had just arrived in Paris. In fact, this was the copy Paolo Sarpi had sent on March 16, and it was now in L'Estoile's hands.[2] This homage does not seem to have interested its recipient a great deal; in his diary, he defined it as "a very curious book." He jotted down every detail of its title, also copying Sarpi's letter in its entirety, but his description is reticent as to the book's contents, noting only that he had read it hastily and confessing—not embarrassed in the least—that he hadn't understood a thing.[3]

A group of scholars that had converged around Nicolas Fabri de Peiresc, a councilman with the *parlement* of Aix-en-Provence, an amateur scientist, and a passionate collector, instead showed far greater interest and insight. Peiresc had traveled to Italy between 1599 and 1602, spending a great deal of time in Padua. At Gian Vincenzo Pinelli's home there, he had met Galileo and attended some of his public lessons.[4] It was a letter from Padua, sent by Lorenzo Pignoria on May 3, 1610, that alerted him to the publication of the *Sidereus* and the stir it had caused.[5]

In July, hoping to obtain the book, Peiresc wrote to the jurist Giulio Pace in Montpellier (with whom he had studied), asking if he could borrow the book for "seven or eight days." He specified that the favor was not so much for himself, as he was waiting to receive the copy he had ordered from Venice, but for "those who have all power over me"—in other words, the head of the Provençal *parlement*, Guillaume du Vair, with whom he shared a passion for astronomical studies.[6] Pace sent the book in a matter of days, stating that he had read it with great pleasure and had found confirmation for everything he had always suspected, "except that I think the Moon is illuminated by our sea, and not by the Earth as Galileo sustains."[7] Nonetheless, as a faithful follower of Aristotelian thought, he continued to manifest a certain level of "incredulity" fueled by the lack of solid proof.[8] Verifying the new celestial discoveries personally—and meticulously—thus became fundamental.

Meanwhile, working with his collaborators Guillaume du Vair, Jean Lombard, and Joseph Gaultier for over a year, Peiresc had spared no effort to obtain a good telescope, but in vain.[9] From Paris, the poet François Malherbe, whom he had commissioned to find the instrument, responded that it was not easy to obtain one "like those from the land where they were invented," adding that "not all Dutch spyglasses are equally good."[10] Following publication of the *Sidereus*, Peiresc resumed his search, sending new requests to Italy, Holland, and Paris. His brother Palamede sent him about forty lenses from the capital, but they were unsuitable for building a high-quality spyglass.[11] A disappointed Peiresc even thought about asking Galileo himself, but the futility of such an initiative was immediately clear. As he wrote to Pace: "I met a gentleman who spoke to [Galileo] after the book was published, but he was never able to see his instrument. So fancy him even selling one. We must resign ourselves to obtaining the invention without his help."[12]

Despite these difficulties, Peiresc continued to work on his telescope and observe the heavens. By early November, he had a far more powerful *lunette*, which he had fitted with a support and sights like those on harquebuses.[13] With that instrument he was able to confirm the accuracy of Galileo's description of the Moon's irregular surface. These initial results led him to "believe that the observations of the four Medicean planets and the other fixed stars that have recently been discovered are no less truthful than those of the Moon, although our telescope has not yet arrived that far."[14] A few days later, on November 20, he informed Pace of further indirect proof of the existence of the Jovian satellites: publication of Kepler's *Narratio*, which he had heard about but did not own.[15]

On November 24 he finally managed to see them. Peiresc wrote that the satellites were "exactly where our calculations placed them in completing their orbits," and this allowed him to recognize each one accurately. Consequently, he felt authorized to boast of his "right of distinction" *(droyt de distinction)*, which was independent of the "right of discovery" *(droyt d'invention)*. Based on the latter, Galileo had rightly coined the appellation "general and common" for the "Medicean" stars, but since Peiresc had managed to identify each of the four celestial bodies, the Frenchman claimed the privilege of naming each of them individually after four Tuscan rulers: Cosimo I, Francesco, Ferdinando, and Cosimo II.[16] Subsequently, these names would be corrected to include the two French queens from the Medici family, Caterina and Maria:

> Having learned that the queen [Maria de' Medici] wanted to have her name among these planets, I decided to call the one that follows the largest orbit around Jupiter (and is the third in order of magnitude) after Caterina and the one immediately after that (which is the largest and most beautiful of all) after Maria; the third (in other words, the second in terms of magnitude) will be given the name "Cosimo the Elder" or "the outermost" [*Cosmus Mayor ou Exterior*] and the last one "Cosimo the Younger" or "the innermost" [*Cosmus Minor ou Interior*]. Thus, there will only be illustrious figures from the Medici family and I believe the Grand Duke will have no cause for complaint, nor can Galileo [claim to be] the sole author, and I shall call these stars "Franco-Medicean" [*sidera Francomedicea*].[17]

Peiresc continued his observations in Provence until April 17, 1612, completing them in Paris on May 15.[18] During this period, every day he collected information about the Medicean planets, amassing what has been defined as "the largest surviving early modern archive devoted to Jupiter and its satellites."[19] To delineate their positions more accurately, he used a sort of graph paper, in which each square represented the value of the diameter of Jupiter (an arc-minute), a measurement of reference of the relative distances.

Through painstaking mathematical calculation of the data available to him, he also attempted to set up tables for each of the four satellites.[20]

At the same time, he proceeded to perfect the telescope. In September 1611 he announced he had an instrument that could magnify four hundred times.[21] Over the months that followed, he used four telescopes, recording the data he observed in order to obtain comparative information on their respective reliability.[22]

Confident of the accuracy of his work, Peiresc thought he had by far surpassed Galileo. In January 1611 he noted that the latter had admitted "he had been unable to determine the period of the motions of the satellites, despite two months of observations."[23] Moreover, he felt that Galileo's results contained a macroscopic error. "The same disturbance of one day that can be noted in the calculation of the period of one of the Medicean stars can also be found in that of each of the other three, although their motions are quite different. These considerations readily lead me to think he antedated them by one day," he wrote.[24]

In November 1611 Peiresc started reading the *Dioptrice*, and the preface told him that Galileo had identified the odd figure of the "three-bodied" Saturn, further stimulating him to engage in new observations that would confirm the discovery. To his eyes, the planet indeed seemed to have two appendices, "the eastern hump being smaller and longer than the western one." Starting with this observation, and considering it to be an innovative contribution, he was determined to emphasize its priority. "Thank God Galileo has not published anything yet: we must be a step ahead of him, if possible," he wrote to Pace in December 3, 1611.[25] He had already confirmed his intention to carve out a role for himself as a leading figure on the scene of astronomical debate a few months earlier when, with a sigh of relief, he learned that there was no news of further publications by the author of the *Sidereus*. "This has made me hope he has not had anything printed for the fair [in Frankfurt]."[26]

Peiresc clearly planned to develop an initiative that would compete with Galileo's. Convinced that he had more accurately observed the new celestial phenomena, he strived to present himself as the author of a significant leap in their knowledge. Above all, he felt he had devised an extremely reliable picture of the motion of the Medicean planets, whereas Galileo had "not deigned to write anything, nor to publish what he had promised."[27] Peiresc's opposition was fueled by his immense mistrust of a figure so sly and unconventional as to make him suspicious in the first place: "Knowing that he is Florentine, can we dare hope for something other than what one would expect from shrewd persons?"[28]

As early as January 1611 Peiresc began cultivating the idea of writing a pamphlet with the results of his ongoing observations of the satellites. At the same time, he did not think the work should be published in his name, not only because his profession did not seem to be reconciled with astronomy ("tropt mal accordée a celle de l'astronomie") but also because he was afraid of making mistakes.[29] As his observations continued, so did his faith in the results, along with his desire to publish the work and dedicate it to the French queen, Maria de' Medici.[30] In July his plans seemed close to fruition, and his friend Malherbe offered to revise the work and oversee its printing.[31] The painter Jean Chalette also prepared the title page (Figure 20). The image portrays the French queen seated on Jupiter and surrounded by the four satellites, represented by the grand dukes Cosimo I, Francesco I, Ferdinando I, and Cosimo II. Taking pride of place over their heads is the Medici coat of arms, but the six *palle*—balls—are replaced by six stars.[32]

The idea of publishing the work was never pursued, however, and nothing further came of it. According to Gassendi, the project was abandoned because Peiresc did not want to challenge Galileo's claim to the discovery.[33] Yet perhaps it was due to the difficulties in calculating the precise ephemerides of these satellites.[34] As Peiresc noted in his letter to Paolo Gualdo in January 1615, "We saw and observed the sickle of Venus before his [Galileo's] and Kepler's books were printed, and many other celestial curiosities, and even the entire motion of the Medicean planets; *but since we recognized some irregularities, which required more assiduous and constant observation than our profession would permit,* all must be abandoned. If we knew the observations he made after publishing his *Sidereo nuntio,* and if we had the chance to compare them with those performed here, perhaps this would be of great use."[35]

Figure 20. Jean Chalette, *bozzetto* for the title page of *Astra medicea* by Nicolas Fabri de Peiresc, pen and India ink, ca. 1611

He then confessed to Galileo, "Although our spyglass unquestionably did not achieve the perfection of Your Excellency's, I nevertheless wanted to renew my devotion and send you a good number of observations made here, along with the calculation of their motion . . . *but having encountered a few small problems* . . . when I then saw the other subsequent observations on your part and those of Messer Simon Mario [Mayr] and others, further reflection seemed pointless and, out of great respect, I refrained from mentioning them to you."[36]

The fact that Peiresc's planned work was not published can thus be attributed to the lack of accurate observational data. The need for such data was even more compelling if we consider that—like Galileo—Peiresc and his friends quickly grasped that the motion of satellites could be used to determine longitude at sea, which implied the development of extremely accurate tables.[37]

One thing was sure, however: in the years that followed, the Provençal group set aside the antagonism that had marked its earliest experiences. Through Gualdo, in 1614 Peiresc was once again in touch with Galileo, and after 1633 he would not hesitate to intervene with great courage and enthusiasm to try to convince the Roman authorities to reduce the effects of their condemnation of Galileo.

In any case, the enthusiasm of that exhilarating period remained vividly etched in his memory. In a letter to the Dupuy brothers dated November 8, 1626, it was without any polemical intent whatsoever that he recalled his astonishment upon using the "nouvelles lunettes de Galilée," with their amazing power to "bring close to the eyes of observers the greatest celestial wonders as if seeing them from the sky rather than the Earth."[38]

• •

In France, news of the telescopic discoveries reverberated far beyond Provence and the south. A little more than a year after the *Sidereus* had been published, at La Flèche, in the Loire region, the discovery of the Medicean planets was publicly cited at the local Jesuit college during a commemoration for Henry IV held on June 6, 1611.

To provide some background, at the end of 1603 Henry had given the Jesuits permission to establish a center in La Flèche, and the institution was thus named Collège Royal Henri IV in his honor. As a sign of gratitude, the king had ordered that his heart and the queen's should be buried there under a "large marble plaque on which, in gilt letters, the main events in the lives of Their Majesties shall be described."[39] Henry was stabbed to death on May 14, 1610, and on June 4 his heart was buried in the church of the college, following a solemn procession in a scenario that was at once mournful and magnificent.[40] The grandeur of the ritual was due not only to the status of the deceased. The Jesuits also wanted to dispel any suspicion that they had secretly instigated regicide and, being expert communicators, they staged the ceremony with great pomp to demonstrate their unfailing devotion to the unfortunate sovereign.

A year later, the clergy at La Flèche commemorated the ritual with a flurry of events. The three-day commemoration saw countless religious functions, readings of "philosophical theses and literary exercises," and theater presentations.[41] One of the poems recited by pupils for the occasion was the "Sonnet on the Death of the King Henry the Great, and on the Discovery of Some New Planets, or Stars Wandering Around Jupiter, Made This Year by Galileo Galilei, Famous Mathematician of the Grand Duke of Florence," which read as follows:[42]

France had already scattered so many tears
For the death of her King, that the Realm of the wave
Big with deluges ravaged her flowers from the Earth,
Threatening the whole world with a second Flood.

When the Day Star, which makes the circuit
Around the Universe, moved by impending sorrows,
Which were speeding their wandering course towards us
Spoke to her of Fate, above her distress,

FRANCE, whose tears, for the death of thy prince,
Are injuring with their excess every other Region,
Desist from grieving over his empty Tomb,

For, God having lifted him all above the Earth,
In the Heaven of Jupiter he now shines
To serve to mortals as a heavenly torch.[43]

When the Jesuit Camille de Rochemonteix rediscovered and published the sonnet in 1889, he dismissed it as "overblown" and "odd," as France's lament over the tomb of the great king seemed mawkish and unnatural.[44] However, he did not say one word about the young author's attempt to link the death of Henry IV with the discovery of the Jovian satellites through the poetic image—perhaps quite contrived—of the dead king who, almost as if he had become a new star, continued to shine on mortals from heaven. Yet this is the most intriguing aspect of the sonnet, for the *Sidereus* had already circulated for some time at the college in La Flèche and even students there were aware of its discoveries well before June 1611. Little does it matter whether we can establish that, as has been theorized, the author of the sonnet might have been René Descartes, a student at the college and barely fifteen at the time.[45] What counts here is that, alongside Henry's assassination, those discoveries were perceived as the event that had left the deepest impression on public opinion in Europe, so much so that they were cited in a novice's composition.[46]

It is hard to determine if Descartes was effectively the author of the poem comparing Henry IV to the Jovian satellites. The episode evokes the opening scene of Roberto Rossellini's 1974 film *Cartesius*, in which a Jesuit priest who had just arrived from Florence illustrates the marvels of the telescope, delighting the young René. In all likelihood, news about the Medicean planets reached La Flèche not from Florence but through the Collegio Romano.[47] In any event, it is significant that those surprising discoveries were also celebrated in northern France.

• •

Like La Flèche, Florence also paid tribute to the tragically murdered monarch. When Galileo arrived in the city on September 12, the court was busy with the final preparations for an event that would continue to be discussed long after the fact: the production of the magnificent funeral effigy of Henry IV, who was related to the Medici family through his marriage to Maria (later known in France as Marie), the cousin of Cosimo II.

The ceremony was held on September 15 in the Basilica of San Lorenzo in the very center of town, adorned for the occasion with majestic scenery created by the court architect Giulio Parigi, including twenty-six enormous canvases devoted to the life of the "most Christian king." The aim of the pictorial cycle was to convey with visual immediacy "the greatness and virtue of the king, with exquisiteness of order and invention" while also honoring Cosimo II by glorifying the power acquired by the regent, Maria de' Medici.[48] The series of rites officially giving the queen full powers had been celebrated a short time earlier, one after the other. On May 13, the day before Ravaillac stabbed the king three times, the ceremony granting her regency of the kingdom (in the sovereign's absence) had been held at Saint-Denis in Paris; on May 15 she was given regency on her son's behalf; on May 29, two days after the regicide was executed by being drawn and quartered, the monarch's solemn funeral was held in the same basilica. Lastly, June 1 marked the most spectacular *coup de théâtre*. The king's heart was brought to the college in La Flèche with the highest honors, fully confirming his unequivocal Catholicism and adding "another stone to the solidity of the edifice of the Regent, whom all had credited with allowing the Order [of the Society of Jesus] to return to the land of France."[49]

This was the climate that reigned in Florence during those days, a very unhealthy atmosphere for those who, like Sarpi, viewed the marriage of Cosimo II and Maria Magdalena of Austria as further confirmation that "we have turned to Spain yet again."[50] As we will see, however, it was also rather insalubrious for Galileo, although this did not keep him from returning to his city, as he was well aware that the religiose devotion of Christina of Lorraine, mother of the young Cosimo and responsible for much of the grand ducal policy, certainly rivaled that of the new queen of France.

The *Sidereus* had been met with great curiosity in Paris. With a legitimate sense of pride, Galileo himself noted the keen interest Henry IV had shown in his work. In a letter dated June 25, 1610, addressed to the grand ducal ambassador Vincenzo Giugni, he quoted a long passage from a missive sent to him from Paris on April 20, just two months earlier. Although no names are mentioned, it is easy to grasp that the writer was close to the king and had been asked to relay to Galileo the question of the worthiest way to celebrate one considered "the great star of France." Galileo recommended that as soon as the occasion of a new and astonishing discovery presented itself, the monarch's name be inscribed in the heavens alongside that of the Medici: a most welcome tribute and, indeed, rightful acknowledgment "of the true virtue and heroic merits of the greatest, most powerful, warlike, prudent, fortunate, magnanimous, and good prince to have appeared on this earth in many centuries."[51] When the letter reached Florence at the end of June, Giugni gave it the full attention it deserved, even underlining some of its passages. He highlighted the word "fortunate," the only jarring note on the list of virtues ascribed to the Gallic Hercules.

Henry never got the chance to admire the Galilean sky, and his murder was a tragedy for all those across Europe who had vested their hopes in his leadership to help the world emerge from the escalation of religious wars and finally restore peace and order. To get an idea of just how terrible this loss was, we can turn to Traiano Boccalini, one of the best-known European writers of his day. Boccalini used an image that fully conveyed people's dismay and grief over the death of their Hercules. As soon as the god Apollo hears the devastating news, his face is immediately obscured by a cloud: "In testimony of his inward grief, he presently vailed his face with a thick dark cloud, from which for three whole daies he showered down great store of tears: And all the Letterati, both Spaniards, English, Flemish, Germans, and Italians, did with abundance of tears bewaile the unfortunate mischance of so great a King."[52]

In the endless flurry of dispatches that princes, ambassadors, littérateurs, and court dignitaries exchanged after that fateful May 14, there were several in which anxiety and indignation over the assassination were entwined with the sensational news of another turn of events: one that came from another part of the world and was different, of course, but which in terms of astonishment and sheer unexpectedness certainly rivaled the news of the king's death. One example is the letter Belisario Vinta sent to the Tuscan ambassador Orso Pannocchieschi d'Elci in Madrid on May 23. For other current information, he referred his correspondent to other messages. In short, what his recipient was about to read did not come under the category of "common" dispatches. Above all, there were two "extraordinary" pieces of information he wanted to convey: the death of Henry IV and the equally unexpected discovery of a new sky. In his message he noted,

> Important and terrible news arrived when a courier came through from Paris on his way to Rome, that while the King of France was going through

> Paris in his carriage, he was mortally wounded, either with a knife or a dagger, by a Walloon who pretended to give him a petition, but so far, not only have we not received a courier, but not even the slightest message from Paris.[53]

Then he immediately went on to say:

> If a Messer Galileo Galilei, noble Florentine and chief Mathematician at the University of Padua, who has discovered and observed in the sky new stars and has named them Medicea Sidera, should send Your Excellency some of his demonstrations and compositions printed about these stars and planets, and also certain spectacles of his own invention to contemplate and observe them more easily, so that you may present them to His Majesty or to his literati and, in particular, to the Lord Constable, kindly accept them and help carry out his desire, because he is a mathematician and philosopher of great repute and great fame, and is also a very dear friend of mine, and all this also brings honor and glory to our Most Serene Lord.[54]

An earthly announcement of death and a celestial one of life were thus entwined. On earth what surprises us is the "terrible news" of the murder of Europe's most beloved king, while in the heavens we are astonished by the "birth" of new stars called the Medicea Sidera to further the "honor and glory" of the noble Florentine family. The tragic demise of the "great star of France" coincided with a new beginning etched in the heavens, with the observation of celestial bodies that had never been seen before. The contrast between the two pieces of news is jarring, but the wholly terrestrial significance ascribed to the spyglass is equally clear. In fact, it was so important that it was rightly mentioned in numerous diplomatic dispatches. As we have seen, however, in many cases this message of life was ultimately filled with uneasiness and disorientation, opening the door to surprises, threats, and upheavals that no one—not even its inventor—could ever have predicted.

• •

At court, the French queen promptly tasked her master glassworkers to reproduce "Galileo's spyglass," but the outcome was quite disappointing. Therefore, in early July the grand ducal ambassador Matteo Botti urged

Vinta to have at least one well-made spyglass sent from Florence. This was a request that not only would have met with the favor of the entire court but also would have enhanced the political and cultural prestige of the Medici family, not least because the Dutch spyglasses filling Parisian shops were highly regarded.[55]

"Galilei's large spyglass," as it was called in the correspondence between Florence and Paris, was being transformed into one of the most sought-after gifts. But the wait—in this case not by some lowly prince but no less than the queen of France—was quite long, confirming the enormous difficulties Galileo encountered in crafting high-quality lenses or obtaining enough to meet the numerous requests he received. It was not until August 23 that Vinta finally sent one to Paris, noting, "By courier . . . I am sending and addressing to Her Majesty, the Most Christian Queen, one of Galilei's large spyglasses and as soon as it arrives here, which we hope will be soon, Her Majesty the Queen can tell us if she prefers said spyglasses in one form rather than another, which will be made specially to her taste."[56]

Twenty days later, on September 13, Andrea Cioli informed Vinta that it had been delivered. "This morning, through channels of which I am not aware, Messer Galilei's large spyglass arrived for Her Majesty the Queen . . . and it will be up to the Marquis to present it to her."[57] The coveted instrument had finally reached its destination and was shown to the regent a few days later.

The event, eagerly awaited for months, had finally come to pass. "Galilei's large spyglass" was universally considered "a thing of princes."[58] Naturally, the device was a must at the palace of one of Europe's most powerful nations, the one closest to the only sovereign in the world to whom such an important piece of the sky had been dedicated.

Yet the outcome was discouraging, to say the least. Botti's words informing Vinta of the turn of events betray all the uneasiness of one who is professionally responsible for relaying the truth about what actually happened. He did so in a hushed tone, almost apologetic that he had to send such unpleasant and inopportune news: "When conversing with me, as she often does at length, Her Majesty confessed . . . that Galilei's spyglass arrived, but it shows little more than the others."[59] There was no denying that the famous "large spyglass" was just as much of a flop as the encounters Galileo had attended in April, first in Florence and Pisa, and then at Magini's house in Bologna. There was only the slightest difference between the ordinary spyglasses sold at the market stalls in Paris and Galileo's, sent specially from Tuscany. It would take nearly another year—until August 1611—before another spyglass could be delivered, this one of decidedly higher quality. The picture Ambassador Botti painted for Galileo of its presentation couldn't have been rosier: "Having presented to Her Majesty the Queen your instrument, I showed her that it is much better than another one sent earlier, which perhaps was not as well conditioned. Her Majesty liked it very much, and she even kneeled on the ground, in my presence, to see the Moon better. She enjoyed it infinitely and was very pleased by the compliment I offered her in your name, which was accompanied by much further praise, not only on my part but also on Her Majesty's, who demonstrates that she knows and admires you, as you deserve."[60]

Yet the letter didn't end there. Botti informed Galileo that his discoveries had been favorably received also at La Flèche, the most famous college of the Society of Jesus in France. There too, as had been the case a few months earlier among the Jesuits of the Collegio Romano, "great observations were made about what you wrote in this regard, and everything has been approved as being completely true."[61] Galileo was also conquering Paris and France.

NINE

Milan: At the Court of "King" Federico

Now we must return to Italy. This is essential if we want to proceed with our experiment in cartography and the cross-referencing of texts in an attempt to offer an overall vision of the circulation of the telescope in Europe. After Prague, England, and France, after the words and images we have collected, and after establishing a dialogue among them, pursuing Johannes Kepler, Rudolf II, Martin Hasdale, Michael Maestlin, William Lower, Thomas Harriot, John Donne, Francis Bacon, Nicolas Fabri de Peiresc, and the countless other figures involved in this complex story, we must head back to Italy, traveling through Milan. In this case, however, our guide is not an ambassador's dispatch but a painting by one of the best-known artists of the era.

Napoleon took the picture with him from Milan in 1796, along with other works by Brueghel. But unlike the paintings on the elements of water and fire, which were brought back to Italy in 1815, the *Allegory of Air* (Figure 21 and Plate 3) was never returned to the Pinacoteca Ambrosiana and is still at the Louvre along with its pendant, the *Allegory of Earth*.

Air has pride of place in the middle of the painting. She is on a cloud, and in her left hand she is holding a gleaming armillary sphere, while in her right is a parrot, a magnificent white-tufted cockatoo. In the sky, with the outlines of the chariots of Apollo and Diana off in the distance, flocks of birds are flying with several winged putti. Others, boasting the oddest plumage, are on the ground or perched on branches: herons, owls big and small, flamingos, birds of paradise, goldfinches, waxwings, toucans, gray parrots, a golden eagle. In the foreground on the right, sprites are playing with all sorts of astronomical and mathematical instruments, such as nautical compasses and dividing compasses, rulers, sundials, plumb rules, and

Figure 21. Jan Brueghel the Elder, *Allegory of Air*, oil on copper, 1621

astrolabes. And next to Air, there is a figure intently peering at the sky with a telescope.

Commissioned by Cardinal Federico Borromeo, it was the last painting of the four elements to be executed, and it is the one that, in this book, interests us the most. Federico waited a long time to receive it. In 1616 he wrote to Brueghel, "I would like that, in what remains to be done, you attempt to do your utmost, for your greater praise and my satisfaction."[1] With this series of allegories, the Flemish artist made the most of his extraordinary skill as a naturalistic painter, imitating nature itself "not only in the colors, but also fluidity, which is the greatest ornament in both nature and art."[2] Yet *Air* goes even further to acquire a distinctive meaning. The cardinal wrote, "I continue to await your *Spring*, the remaining of the four elements, in which I expect greater diligence from you than in the rest. I think this is what art itself demands, to continue to perfect the last things and their value. Though all are truly excellent and will be placed together with my other paintings in an Academy of Drawing, which I would like to establish."[3]

The date was December 3, 1616, and this passage is from another letter to Brueghel. As Stefania Bedoni has emphasized, Borromeo—unlike other great collectors such as the Barberini, Orsini, and Borghese families—was no ordinary art enthusiast. He intervened in the various phases involved in the commissioning and preparation of works and would "choose and hang them in the gallery according to specific arrangements; he commissioned them knowing what he wanted, participated in the artist's creation, and monitored the execution of the painting."[4] We also know that the spaces in which he arranged the artwork played a special role in his mind. In *De pictura sacra* he repeatedly noted that the disproportionate portrayal of objects was unbecoming and showed no decorum, meaning that this did not reflect their worth. For example, some artists painted St. John in the desert and wanted to depict him "in a dark and badly rendered part that can barely be seen, and then fill the most refined and noble parts of the field with animals, plants, rocks, and perspectives."[5] Furthermore, "other painters do worse when, as they think about painting a panel [to be displayed in church], they choose any mystery to represent, but in the middle, in the noblest and most conspicuous space, and with greater diligence than they have used elsewhere, they portray a lascivious woman in an inappropriate act without any need to do so, considering the story itself."[6]

Given that in painting—as in writing—it was essential to respect a hierarchy and order among the different parts, so that each object was well proportioned within the space in which it was represented, it is highly symbolic that Borromeo wanted the spyglass in the foreground. In effect, it seems more than likely that his youthful interest in skygazing was rekindled by the publication of the *Sidereus nuncius*. His interest in astronomy was well known, and not only among his closest collaborators. Under the date of May 2, 1614, the ledgers of the archdiocese of Milan list the cost of "a spyglass sent to His Excellency from Venice."[7] In his preparatory notes for a biography of the cardinal, his personal secretary Giovanni Maria Vercelloni recalled that "he was wholly enamored of and enchanted by the beauty of the starry sky, and at night he would often take the spyglass outside to admire and contemplate the stars in my presence, and particularly at Pobigo, a solitary place in the middle of the woods."[8] Francesco Rivola's *Vita di Federico Borromeo* notes that "one of the things that most inspired his love for Creation was viewing the stars, which he said were many divine eyes with which he observed divine majesty."[9]

• •

"Prencipe letteratissimo et ammirator delle cose nove celesti"—a most literate prince and admirer of new celestial things—was how Johannes Faber, physician and member of the Accademia dei Lincei, described Borromeo.[10] He had met him through the Jesuit Johannes Schreck, who in turn had had the chance to frequent the cardinal while in Milan during the winter of 1616.[11] In March Borromeo had turned to Faber to have him deliver one of own his spyglasses to Schreck, in Rome at the time and setting out for China.[12] And in August, when Galileo offered an even more detailed description of Saturn, it was again up to Faber to relay the surprising discovery to Borromeo, who was "most curious about such news."[13] He wrote, "It is my hope that, through this visual instrument, we will be able to enrich philosophy and mathematics."[14]

The observations of Saturn delighted the cardinal, and he discussed them a short time later with Kaspar Schoppe, who was passing through Milan. Even earlier, however, he had already demonstrated that he was perfectly aware of the celestial discoveries. As we can glean from a letter by Mark Welser (September 28, 1612) discussing sunspots, astronomy was a topic of conversation among Federico's entourage: "Your Excellency demonstrated that you enjoyed the discussions on sunspots sent to you several months ago. However, since the author [Christoph Scheiner] has had several papers printed again on this subject, expanding it to related topics, I felt it my duty to send them to you as well. And, if nothing else, they might serve for the honest entertainment of the esteemed Doctors at your Ambrosian College."[15]

This is also confirmed by the testimony of the Jesuato and Galilean disciple Bonaventura Cavalieri. The cardinal had shown him the various telescopes he owned, and particularly one measuring eight *braccia*, which he considered the finest in circulation. According to Cavalieri, "He says that, with it, he can see the body of the stars, by which methinks he is also inferring that he sees them as large as perhaps yours is able to magnify Jupiter, or more; if this is true, I think it is important."[16]

Then we have Christoph Scheiner, who in *De maculis solaribus disquisitio* cited the name of the "Reverendissimus et illustrissimus cardinalis Borromaeus" as one of the first in Italy to observe sunspots, mentioning him along with the Veronese physician and philosopher Andrea Chiocco, the Benedictine abbot Angelo Grillo, and Magini, who were followed on the list by Stevin, Praetorius, and Kepler.[17]

These documents confirm that Federico's interest in astronomy and telescopic observation was well known far outside Milanese circles. But we can also add his own words.

> Of all of the representing spectacles [*vetri rappresentanti*] I liked, because it should truly be liked by any reasonable man, not to mention one who is also literate, the *occhiale* that is commonly referred to as *canocchiale* is praiseworthy, as it has allowed us to see half the world much better, which we could not see before; in fact, it has discovered new worlds, meaning new stars, which are larger than the terrestrial world or slightly smaller. And I had one so excellent that for whichever star, I discovered the body of that star, concealing its splendor: which splendor, in this case, does not aid vision but prevents and obscures it. I do not know why this *occhiale* has this particular effect, except perhaps due to the singular material from which it is made. I have never found others like it, despite my great diligence; but perhaps we will discuss this elsewhere. . . . There are also those small *occhiali*, which can be used to see tiny things that are made immense, which is a marvelous thing. With these, I have been pleased to see so clearly how the smallest insects are made. In this I have discovered many times that this saying is quite true: *magis et minus non variant speciem* [bigger or smaller does not alter the kind]. Because I have learned that, unlikely as it may seem, little animals much smaller than the eye of a needle are the same species as the large ones that can be seen without the *occhiale*. This demonstrates the great mastery of nature; in other words, that it does not

> make such small things with any less industry and diligence, hard work, care, and artifice than the large ones.[18]

For Borromeo, the spyglass and the microscope represented experimental confirmation of his optimistic vision of nature, further proof of a harmonious world and an orderly nature going far beyond the natural senses of humankind. As Pamela Jones noted, "Borromeo's discovery of these newly visible worlds inevitably must have reinforced the cataloguing inclination so closely linked with his love of God's world and of Brueghel's paintings."[19]

Federico accepted the challenge posed by science not only in astronomy and cosmology but also on a philosophical and naturalistic level. Furthermore, he was well aware that this challenge was unlike the others, played out entirely in bookish knowledge and in which the cleverest were those who managed to deploy armies of *auctoritates*, ancient and modern, against enemies. This time, however, the battle had shifted to more insidious and less tested ground, where the key move entailed obtaining a powerful spyglass rather than burying oneself in a library. Under Borromeo—and to a far greater extent than at other cardinals' courts—Milan became an active research center for astronomy. Further proof lies in the plan to compose a treatise titled *Occhiale celeste*, which was to contain his own information about the new sky, and for which an extensive series of preparatory notes remains.

• •

We are not dealing here with ordinary curiosity. Borromeo's aim was far more ambitious: personally stepping into the debate by publishing a pamphlet aimed at confirming or refuting Galileo's discoveries. In effect, the *Occhiale celeste* is a composite text that is also quite surprising because of its numerous telescopic observations. We find Saturn, Mars, and Jupiter, but also descriptions of nebulae and comets, and in particular "the phenomenon now appearing [*de illa quae nunc apparet*]," referring to one of the comets of November/December 1618 that inspired Borromeo's idea—wholly Tychonic—of the nonexistence of solid and incorruptible orbs.[20] In his words, "It seems that the corruptible and generable sky can be seen and, as a result, these are not solid orbs, because the sky of Mars intersects the sky of the Sun."[21]

Naturally, these observations also offered philosophical and theological food for thought, and engaging with Galileo was what interested him most. In fact, one almost gets the impression that when recording his own data, Federico closely followed the lead of the *Sidereus nuncius*. Even when its author is not quoted directly, Federico looks toward him constantly, for example when rejecting the wholly Galilean idea of the analogy between sunspots and terrestrial clouds.

With "the instrument I built" the cardinal observed "three-bodied" Saturn, wondered about what appears on the Moon, its bright and sinuous areas, and also mused about the extent of the Milky Way. These discoveries further confirmed the order and harmony of the universe, representing a tangible sign of the unfathomable but benevolent power of God. His corroboration naturally contains a reference to the "newly found small eye" or, in other words, the microscope, which, thanks to the power of its glass *(virtus vitri)*, demonstrated the infinite perfection of nature through the knowledge of things invisible.[22]

The *Occhiale celeste* was effectively a collage put together at different times. Although no dates are listed, he worked on it continuously for a long time. The serried comparison with the *Sidereus* was indubitably one of the key motivations behind it. Federico carefully examined Galileo's statements with respect to his own observations.[23] For example, this is how he described the lunar surface: "The body of the Moon is mountainous, cavernous in thousands of places. The spots are not cavernous, but *pars rarior orbis* [constituting the more rarefied part of the sphere] and that which shines in the sky. . . . The Moon has different brightnesses, which are the parts most in relief crossed by the Sun's rays, as they are the least dense. Yet up there they are reliefs, but now they have not appeared and when they appear they are seen as brighter parts."[24]

He was also quite intrigued by the telescope, how it worked, and its history. The cardinal wondered if it had been invented by Heinrich Cornelius Agrippa, illustrated an empirical method to "learn how much it enlarges," and even asked "if the spyglass can continue to be improved."[25] Borromeo expressed great interest in celestial discoveries, a direct consequence of his conception of sidereal spaces as the full and openly revelatory manifestation of divine power and wisdom: "The sky is not a God but, rather, a book in which we see His greatest wonders painted on its pages and sides, and on the front of which we can almost find the name of the creator who made this."[26]

Therefore, he procured ever more powerful telescopes; he incessantly observed and described, annotated and compared; he sought corroboration in ancient and modern authors. The cardinal pursued a path that would create continuity with the past, even though at times what emerged were uncertainties and questions that did not always find reassuring answers, and which could not always be rejected or set within a proven philosophical structure. On one point, however, he was unswerving: Copernican cosmology. That the Earth moves was a proposition that had to be rejected on both philosophical and theological grounds.

Federico was well aware of Galileo's stance, as is evident from the passages in which he says that these theories must be rejected based on the authoritativeness of the Bible: "On the truth of the Holy Scriptures. Take inspiration from Galileo. Examine his reasons. Refute them."[27] But also: "Refute Galileo's question of the motion of the Earth through Psalm 135: *He established the earth above the waters, for his mercy is eternal.*"[28]

This was how Borromeo read and interpreted the *Sidereus*, inserting those discoveries, which he had promptly observed and verified, within his *philosophia christiana*. He never rejected any discoveries, celestial or terrestrial, a priori. His was an attempt to explain them within the framework of a traditional vision, thereby limiting any aspects that did not match this view.

• •

Borromeo's discussion with Curzio Casati, professor of mathematics at the Scuole Piattine, clearly shows that the link between the new astronomy and the Christian cosmos was the leitmotif of his scientific interests.

It was the summer of 1610, and Casati gave Federico the *Prima pars introductoriae constructionis astronomiae*, an introduction to the study of celestial motion.[29] In it, he noted that just a few days earlier he had talked to the cardinal about "which of the three main opinions on the structure of the world might best agree with the words of the divine Revelation." Casati was an ardent supporter of Copernicus, and of the three options—Ptolemaic, Tychonic, and Copernican—the last seemed "so consistent with celestial phenomena that nothing more probably can be suggested on the subject."[30] Furthermore, in addition to being scientifically superior to rivaling theories, the heliocentric concept also proved to be entirely "consistent with what is proffered by divinity."[31] Although it aroused astonishment and incredulity, the Copernican doctrine was confirmed by both facts and sacred texts. He wrote to Federico saying,

> Although I am sure you will be doubtful at first, surprised by this truth and even horrified by such an extraordinarily strange discovery that presents nature and the arrangement of the world upside-down, canceling the dogmas of all philosophers (albeit not of true philosophy) consolidated by an age-old adoption and long-established common belief; notwithstanding, if you were willing to examine the problem with the utmost care and assess it diligently on the grounds of the arguments copiously provided here, then I find it hard to believe that you would put so much stock in the human authority of philosophers that you would then choose to toss out reasons confirmed by divine words.[32]

To justify his viewpoint, he offered a detailed Copernican reinterpretation of the biblical passages commonly thought to clash with heliocentrism. Casati concluded that "if one were to attempt to corroborate the opinion of terrestrial motion in light of the testimony of the Holy Scripture, then he would succeed much better and could confirm his thoughts with conjectures far more probable than those of other doctrines."[33]

He finished the *Introductoria constructio* in 1609, but its dedication bears the date of August 18, 1610. Despite its timing, there is no mention of the telescopic discoveries, and yet it is hard to believe that none of this ever came up in his conversations with Borromeo. As documented by a subsequent letter from Cavalieri, Casati had attended Galileo's lectures in Padua and professed to be "completely devoted to his doctrine."[34] It seems unlikely that by August 1610 news of the *Sidereus nuncius* had failed to reach the ears of

Galileo's "devoted" pupil or those of the inquisitive and well-informed Federico. We should also bear in mind that at this point the spyglass must have been circulating in Milan for some time.[35] Curiosity about its astronomical applications was keener than ever, as we can glean from the words of the Spanish governor Juan Fernández de Velasco y Tovar, who in June 1611 insistently requested that the Florentine government provide him with "one of the large spyglasses of Messer Galilei," along with "instructions on how to use it."[36]

Around this time, the Jesuit Cristoforo Borri, professor of mathematics at the College of Brera, was discussing Galileo's discoveries in the *Tractatus astrologiae,* one of the texts he used for his lessons during the academic year of 1611–1612.[37] Like Casati, he pondered the theological correctness of the various cosmologies. In his opinion, the impossibility of reconciling the heliocentric view with the Bible made the Copernican system untenable.[38] In turn, there was nothing to prevent acceptance of the Tychonic system and Brahe's theory of fluid heavens, which did not go against the most widely embraced exegesis, as demonstrated by the duly quoted opinions of countless *auctoritates.*[39] Anticipating a choice that would later be shared by many Jesuit scientists, Borri unhesitatingly embraced the Tychonic cause. His decision was based on his awareness of the failings of the traditional Ptolemaic doctrine, an awareness that, in his words, he had developed at the very start of his career as an astronomer: "When I began to apply myself to mathematical sciences and to study the common description and distribution of the celestial bodies seven years ago, I noted the confusion of all the epicycles and eccentrics conceived by Ptolemy, and my mind loathed them so much that I could never be convinced to believe them."[40]

Aside from this explanation, however, one of the main reasons that had driven him to abandon the traditional image of the heavens was his own experience with telescopic observation, as it had clearly shown him a completely new reality. Regarding the Moon, he wrote:

> It is unquestionable that the Moon is not perfectly round, but its surface is made irregular by the presence of many mountains and valleys. And it is likely that the other celestial bodies are covered with mountains, like the

> Moon. As to the latter, the circumstance requires no proof, as it is evident to the senses thanks to the telescope said to have been invented recently by the Florentine Galileo Galilei, professor of mathematics at the University of Padua. I immediately obtained such a spyglass, and observed the mountains and valleys of the Moon; however, until now I did not dare discuss this publicly as I did not wish to be accused of impudence. But after noting that this and other phenomena (which we will discuss in due time) were observed by Galileo himself—who, in fact, also published them—I shall not hesitate to confirm and circulate this truth.[41]

Borri maintained that he had had a spyglass before the *Sidereus* was published, and that he had observed the rough lunar surface even before reading the book. Moreover, in illustrating the facts corroborating the presence of mountains and valleys on the surface of the Moon, he merely reproposed—to the letter—the arguments already advanced by Galileo.[42]

His discussion was not limited to the Moon alone, however, as he also investigated other discoveries. For example, regarding the strange figure of "three-bodied" Saturn, he alluded to Kepler's *Dioptrice*, and for Venus, which "waxes and wanes like the Moon," he again cited Kepler, along with Galileo.[43] As to the Jovian satellites, he instead stated that he had personally observed them: "Precisely as in the case of the lunar mountains, their existence needs no proof as, thanks to the telescope, it constitutes an appreciable fact. Galileo saw the satellites before anyone else; I myself observed them most scrupulously. It has not yet been possible to determine their real motions and periods."[44]

The *Tractatus astrologiae* eloquently documents the Milanese Jesuit's interest in Galileo's new sky. Based on what Borri himself recounts, however, there are various doubts about the power and effectiveness of his spyglass. He knew about the phases of Venus and Saturn's surprising form thanks solely to Kepler's testimony. Moreover, while he maintained that he carefully studied the Moon and the Medicean planets, his explanation relies largely on Galileo's observations. Yet this does not mean he was lying or boasting when he said he had used a spyglass before the *Sidereus* was published. More simply, the limitations of the instrument he used led to a significant gap between his results and Galileo's far more accurate ones.

All this aside, however, it is worth noting that the discoveries promptly became the topic of heated debate in the city's cultural circles, drawing in not only Borromeo, Casati, and Borri but many others as well. Mathematicians and astronomers such as Muzio Oddi from Urbino and the Milanese Ludovico Barbavara, who corresponded with Kepler and was a great expert in trigonometry, as well as the young but already "great mathematician" Giacomo Rho and the Barnabite Redento Baranzano, who later wrote a work supporting Copernican theory, would certainly not have been unaffected by the astonishing "sidereal message."[45]

It is no accident that the city's leading figure, our "king" Federico, would continue to show a keen interest in the *Sidereus nuncius,* also mentioning it in works on vastly divergent subjects. For example, in the Italian version of a text (later published in Latin) in which he hoped governments would commit to supporting science and culture, he praised the foresight of the Medici, the wise patrons of the discovery of the Jovian satellites: "And in the most modern times, examination of the stars with an attentive spirit and measurement of the height of the heavens was the reason the name of that other magnanimous prince is continuously celebrated along with those of the stars."[46]

Plate 1. Jan Brueghel the Elder, *Landscape with View of the Castle of Mariemont*, 1611

Plate 2. Jan Brueghel the Elder and Peter Paul Rubens, *Allegory of Sight*, oil on panel, 1617

Plate 3. Jan Brueghel the Elder, *Allegory of Air,* oil on copper, 1621

Plate 4. Ludovico Cigoli, *Assumption of the Virgin*, fresco, 1610–1612, detail

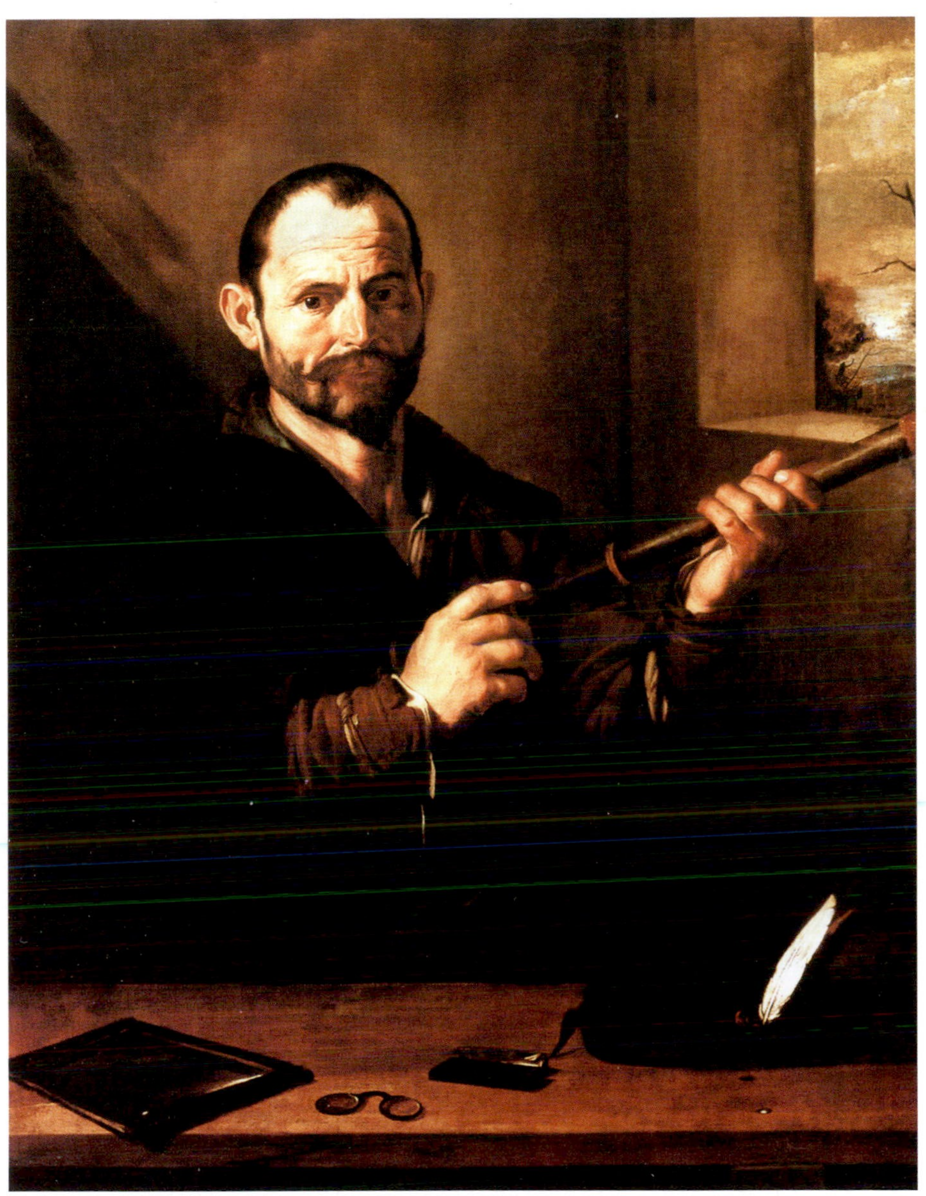

Plate 5. Jusepe de Ribera, *Allegory of Sight*, oil on canvas, ca. 1614

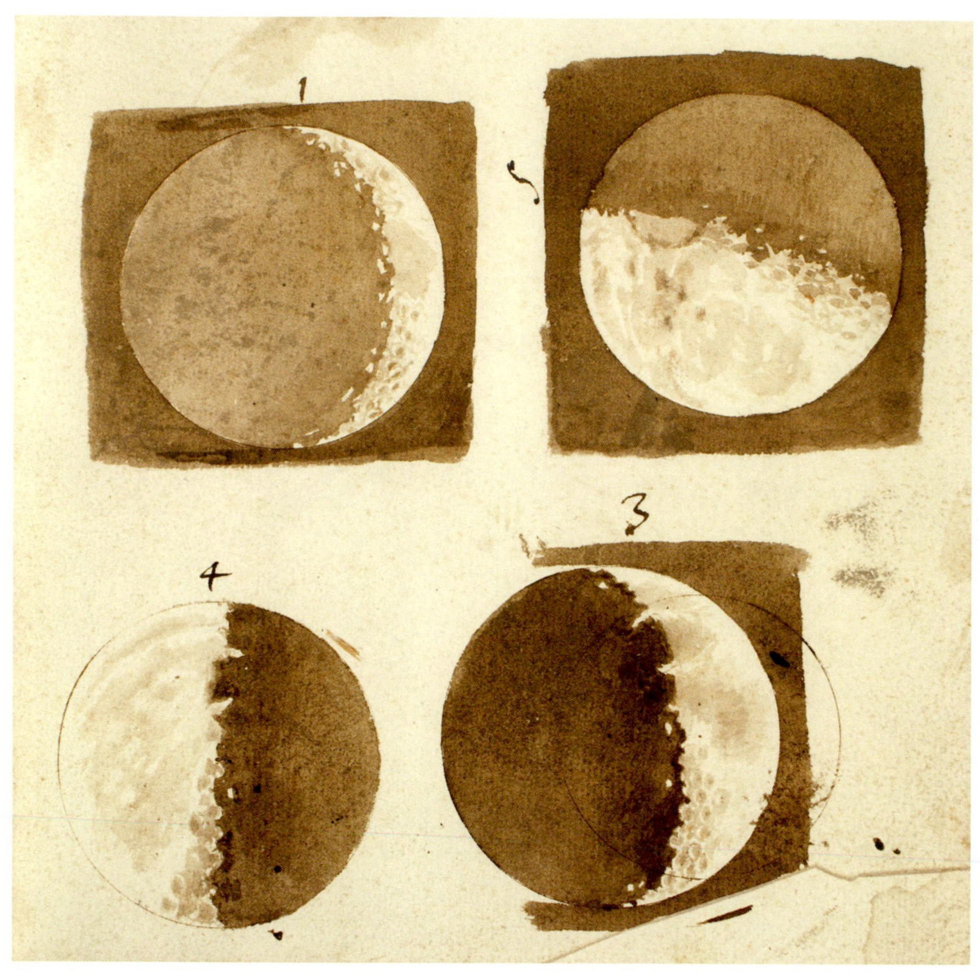

Plate 6. Galileo Galilei, drawings of the Moon, watercolor, 1609, detail

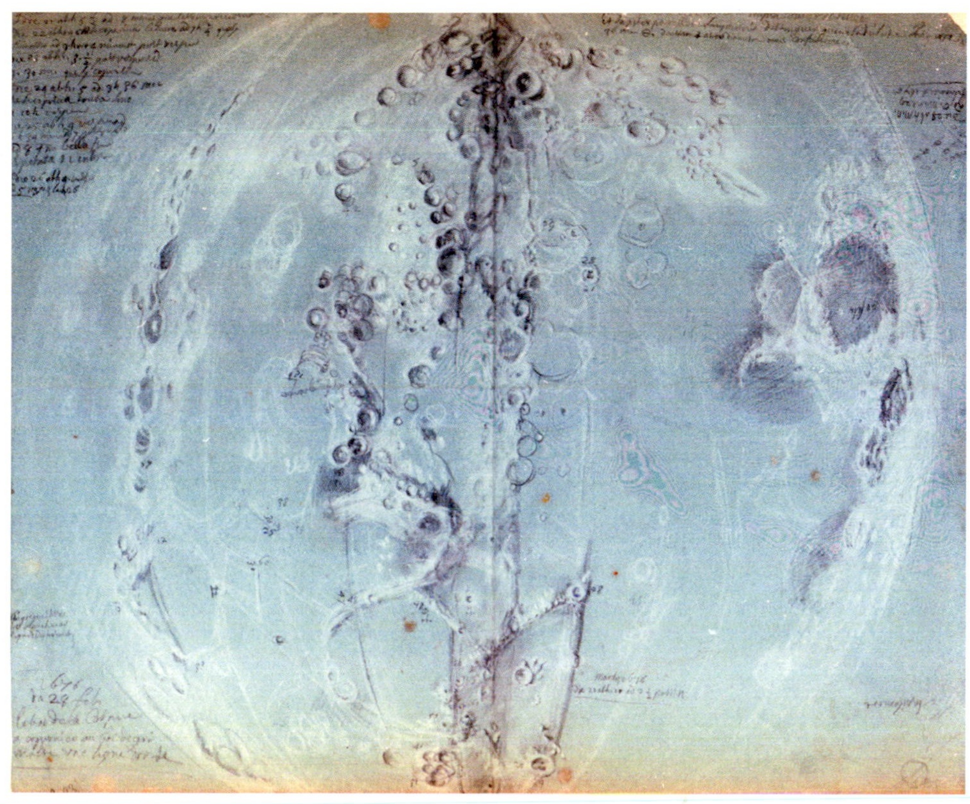

Plate 7. Gian Domenico Cassini, drawing of the Moon, pencil on paper, 1671–1679

Plate 8. English school, *Royal Observatory from Crooms Hill,* oil on canvas, ca. 1690

TEN
The Dark Skies of Florence

At this point, we must set out once more for Padua and Venice, and have Galileo step in personally. In this chapter, he will be our main narrator. He will be the one to guide us and talk to us, following his return to Padua at the end of April 1610, after his first trip aimed at spreading propaganda and offering demonstrations. As we have seen, things did not go well, at least not as well as he had hoped. In Florence, Pisa and, above all, Bologna, he encountered great hostility, just as he had in Padua among his colleagues at the university. In effect, at the end of March his trusted friend Enea Piccolomini had already warned him that in Pisa, some "obstinately refuse to believe what you say you have seen and wish to show to anyone."[1]

When Galileo reached Florence in mid-April, at Easter time, his arrival had been preceded by the first copies of the *Sidereus,* which instantly sold out.[2] The anticipation of seeing the celestial novelties through his telescope grew day by day, and the news of his visit sparked so much curiosity that it excited not only those close to the grand dukes, Cosimo II and Christina, but also ordinary court officials and dignitaries, men from Florence's various literary academies, physicians, jurists, and theologians. Reports of the meetings organized in Florence and Pisa rapidly spread by word of mouth, along with questions, doubts, and testimonials of admiration. For example, there is the terse message sent from Pisa on April 11, 1610 by Camillo Guidi, secretary to Grand Duchess Christina: "But what do you think of these four new planets found by Messer Galilei? I can say that last night he showed me three of them with his spyglass, which is wondrous. We will discuss this in person."[3]

There must have been many other letters like this, written by countless eyewitnesses and people who, in turn, had heard about the event indirectly and informed others. One of them was Cardinal Luigi Capponi, who wrote to Martin Hasdale from Rome in the middle of May to say, "Regarding Galileo's invention, I heard from Florence that he was called by Their Highnesses and also showed many distinguished gentlemen that the four stars around Jupiter do not remain in the same place; so that many scholarly persons fully agree with him."[4]

Galileo found himself on an unprecedented whirlwind tour, seeking the approval he needed for the return to Florence he so eagerly desired. The confrontation was quite harsh in some cases, and while there were high-ranking figures willing to help Galileo, others loathed him. The latter included leading religious figures as well as academics and members of the court. Nevertheless, among those who backed him we find the lawyer and poet Alessandro Sertini, Secretary of State Belisario Vinta, and, above all, his utterly devoted ex-student, the alchemist prince Antonio de' Medici, who not only became one of Galileo's main protectors but was also a valuable collaborator in building and improving his telescopes, as demonstrated by the supplies of lenses he frequently ordered from Venetian glassworks.[5]

During this period, discussion revolved chiefly around the objections that Giulio Libri, a philosopher and teacher from Pisa, had raised in the presence of the grand duke, seeking at all cost to "pluck and remove [Jupiter's satellites] from the sky," based on ideas that—as Galileo commented to Kepler on August 19, 1610—were closer to magic spells than solid reasoning.[6] As we can easily imagine, Galileo's discoveries were not warmly received by those who, because of their special position at the court, were obliged to attend the meetings the Medici had organized. One of them was Antonio Santucci, mathematician at the University of Pisa and grand ducal cosmographer (the enormous Ptolemaic armillary sphere that can be admired at the Galileo Museum in Florence was his); during this period he was finishing a little book published the following year with the title *Trattato nuovo delle comete* (New Treatise on Comets), although, belying its name, there was very little "new" about it, as it continued to embrace the existence of solid orbs.[7] A short time later, Santucci would make a name for himself by railing against Galileo in the dispute over hydrostatics. Following Santucci's death in 1613, the *provveditore* of the University of Pisa, Girolamo da Sommaia, painted this portrait: "To debate with Galileo, one must first have read Copernicus carefully, for otherwise one would seem ignorant, as was the case with Pomarance [Santucci, so called because he was born in Pomarance, a town near Pisa]."[8]

For Galileo, this meant convincing as many people as possible, but his counterparts' skepticism often gained the upper hand. In some cases, it was not merely a matter of disbelief, given that many of his opponents donned theological cloaks. Reasons defending the Aristotelian-Ptolemaic system and those of faith swiftly converged, so we do not even need to look ahead to Galileo's letter to Castelli—dated December 21, 1613—to grasp the objections raised against the new sky. We can merely recall that the unpublished *Contro il moto della Terra* (Against the Motion of the Earth) by the Aristotelian Ludovico Delle Colombe, listing numerous biblical quotations refuting the verity of the Copernican interpretation, was composed between late 1610 and early 1611. Moreover, this was by no means an isolated position, as the same considerations were soon raised in various cultural circles in Florence.

To get an idea of all this, it is quite instructive to read the correspondence of a figure who has been entirely forgotten today but was well known at the time: Bonifacio Vannozzi of Pistoia, protonotary apostolic, who had been in the service of Cardinal Enrico Caetani as papal legate in Poland and then secretary to Pope Paul V. Between August and September 1610, Vannozzi wrote from the Tuscan court to a fellow citizen, the magistrate and man of letters Gerolamo Baldinotti:

> Regarding Galileo I am of one mind with you, and any good theologian will laugh at those who maintain that the Earth really moves, *since it can never be moved* [*non inclinabitur in seculum*], and that the Sun stands still, since it moves by itself [*motu suo agit*]. Such things have been said on other occasions as hypothesis, not as truth. To say that the Moon is earthlike, with valleys and hills, is as if to say that there are flocks that graze there and cowherds who cultivate it. We must stand by the Church, which is the enemy of anything new, according to the teachings of St. Paul. These are thoughts from brilliant minds, but they are dangerous, and I prefer to be a theosophist rather than a philosopher, as you too seem to be, and I kiss your hands.[9]

The stakes could not have been higher, and they had to do with the idea of *truth*. This was the bone of contention that Vannozzi clearly understood, and Galileo would prove inflexible and uncontrollable about it. If these "new thoughts" were declared "as hypothesis"—as Monsignor Vannozzi desired and as Cardinal Bellarmine would reiterate several years later—they would not be an obstacle. But if the discoveries were demonstrated "as truth," then it was better to leave things as they were. The consequences would be explosive, starting with the absurd idea that the Moon might be inhabited like Earth. These were "thoughts from brilliant minds, but they [were] dangerous" and, as everyone knew, the Church was "the enemy of anything new." Accepting the truth of a new sky meant not only destroying the order and beauty of the firmament, that "majestical roof fretted with golden fire," as Shakespeare defined it, but also undermining the world's very essence, threatening all harmony and measure between man and nature as God had conceived and brought them into being.[10]

It is easy to grasp that this letter holds the crux of the conflict between science and faith that would lead to Galileo's battle with Bellarmine and the Church of Rome. The *Sidereus* had been printed only a few months earlier, and Vannozzi's invitation to his correspondent could not have been clearer: "We must stand by the Church."[11] The message he sent an experienced politician of the likes of Andrea Cioli, future first secretary to the grand duke, was of the same tenor: "I truly admire the talent of Messer Galileo, but in this matter I would like to stand by the majority and believe what has been believed and seen so far. Since as to the detail of the Moon and of its valleys and mountains I do not wish to reject what our elders believed, nor anything that can alter the Holy Scriptures in any way or the meaning they have been given for centuries, I wish to walk with the masses, without worrying about knowing too much."[12]

In other words, not knowing but instead following the opinions of tradition and the Church was better than knowing too much and running the risk of "altering" the meaning of the Scriptures. Vannozzi clearly agreed with Cioli, given that in the same letter the monsignor confessed that when he had tested "Galileo's telescope [*occhiale*] . . . to avoid displeasing anyone, the other day it was best for me to say that I had seen his new stars, though I never saw them."[13]

The letter by the unknown Vannozzi is by no means a minor detail, and not only because no one even knew it existed until a few years ago.[14] It is important for a number of reasons. As early as August 1610, it provides a cross section of the opposition and conflict Galileo faced upon leaving Venice. At the same time, it confirms what recent studies have increasingly highlighted: namely, that opposition to him emerged immediately and involved leading figures as high up as the ruling family, starting with Prince Giovanni de' Medici.

Giovanni was one of the key members of the Medici family. A valiant *condottiere* but also a patron of artists and literati, and an architect and engineer in his own right (he designed works such as the façade of the church of Santo Stefano dei Cavalieri in Pisa and the Chapel of the Princes at San Lorenzo in Florence), he had long been a passionate scholar of Jewish philosophy, astrology, and alchemy. During the endless winters spent in Prague, he had whiled away the hours talking to Rudolf II about "magic, and necromancy, and the marvels of chemistry."[15] Francesco Sizzi dedicated his anti-Galilean treatise (*Dianoia*, 1611) to him, and Ludovico Delle Colombe his response *Discorso intorno alle cose che stanno in su l'acqua* (Discourse on Floating Bodies, 1612). Moreover, the prince was an "expert in many sciences," knowledgeable "in many languages," and a participant in the heated discussion on buoyancy—attended by Galileo—at Filippo Salviati's house in the fall of 1611.[16] Along with Ludovico Delle Colombe and Francesco Sizzi, his "servants" and protégés included Orazio Morandi, the father general of Vallombrosa but also an authority on occult sciences such as magic and hermetic alchemy, and Raffaello Gualterotti, whom the prince updated on his progress in learning Kabbalistic "ciphers" and alchemical "recipes."[17]

Then there is the paradox within the paradox. In *his* Florence—and for the first time—Galileo found himself facing the various "spirits" of his worst enemies, who had formed alliances, one headed by Prince Giovanni and the other by Archbishop Alessandro Marzimedici. The two led an uncommon team, coalescing Dominican theologians and preachers including

Niccolò Lorini, Tommaso Caccini, and Raffaello Delle Colombe, scholastic philosophers such as Ludovico Delle Colombe, Francesco Sizzi, and Giulio Libri, and men who engaged in magical and astrological correspondence such as Orazio Morandi. They formed a solid front against their common enemy and were ready to join forces to prevent at all costs the new Columbus of the sky from succeeding.

The skies of Florence that would soon welcome and "celebrate" Galileo were an explosive mix that combined the secret correspondence of astrologers, the apocalyptic vision of the most rigid preachers and theologians, and the dogmatic frameworks of Aristotelian philosophers. They could not have been darker or more threatening.

• •

But let us take things one at a time. On his journey between April 23 and 26, 1610, Galileo stopped in Bologna. Here too—as we have seen—things didn't quite go according to plan. Based on what Horky and Magini had to say, those "telescopic demonstrations" were less than successful.[18]

Galileo arrived in Padua in late April. Waiting there for him was a stack of letters that had piled up in the meantime: one from his brother Michelangelo telling him that copies of the *Sidereus* had reached Munich and that the elector of Cologne wanted instructions to make a telescope; one from Hasdale in Prague, recounting the lively discussions the book had sparked at court; and another from the astronomer and astrologer Ilario Altobelli in Ancona, who had had a copy sent from Cardinal Carlo Conti, his "lord," and begging Galileo to send posthaste some "adequate lenses" *(vetri congrui)* so that he could observe—and he asserted he was convinced of this—other satellites around Saturn and Mars.[19] But there was also a surprise in the mail, a short treatise of just eight folios Kepler had sent him from Prague on April 19, and which the following month would be published with the title *Dissertatio cum Nuncio sidereo.*[20]

Galileo read it all in one go, and his ensuing words and actions went straight to the mark. We have no need to seek further proof that his ideas had accelerated in an unprecedented way. His strategy is clear: he couldn't waste another minute, and he had to take the initiative if he wanted to quash any opposition.

So in early March Galileo decided to deliver three public lectures at the university about the Medicean planets and his other celestial observations. This was not the first time he had lectured there, as he had done the same thing after observing the *stella nova* of 1604, inviting the entire university to discuss this extraordinary event. His account to Vinta of May 7 reads like the tale of a military leader who, in one fell swoop, had eliminated all his most fearsome enemies:

> The whole university attended my lecture, and I left each one capable and satisfied, so that finally those same heads who had been the bitterest opponents and contrary asserters of the things I wrote . . . unanimously said that not only were they persuaded, but were even prepared to defend and sustain *my doctrine* against any philosopher who would dare to challenge it. So that the threatened writings will be in vain, like the entire concept these men have so far tried to raise against me, perhaps hoping that, fearing their authority or astonished by the crowds of their gullible followers, I would retire to a corner and keep quiet.[21]

The enemy's last trenches had been overrun and destroyed, the war was over, and victory complete. So begins this important letter. Proof comes from a detail that can easily be overlooked at first: that at those packed lectures in Padua discussions revolved not only around the discoveries but also around "doctrine" (and we can easily imagine which of *his* doctrines Galileo was referring to, given that he had claimed to be a faithful disciple of Copernicus for over fifteen years).

But can we be sure about his report to Vinta? The allusion to cosmological discussions and the unanimous approval of his ideas makes it rather hard to believe. We know of no other testimony confirming that, in a stronghold of Aristotelianism such as Padua, his "bitterest opponents" were suddenly persuaded and even "prepared to defend and sustain [his] doctrine against any philosopher who would dare to challenge it." In fact, on several occasions in the years to come, Galileo would recall just the opposite: that the only ones who refused to accept his celestial discoveries were his Paduan colleagues. Yet none of this emerges in his letter to Vinta, in which Galileo is more euphoric and brilliant than ever. He elegantly and confidently portrayed himself as the victor. After all, he had just read Kepler's *Epistola,* and the opposition he had encountered in Florence and Bologna magically seemed to have vanished. The emperor's mathematician had sided with him, and that's what counted. The message he wanted to convey was precisely that of one who believes—or wants to convince others—that the worst is over and the truth will soon be recognized.

Galileo's letter of May 7, rightly quoted time and again, is a continuous crescendo of requests in which self-celebration reaches unheard-of levels. "I have plenty of particular secrets, which are as useful as they are curious and admirable," he writes, "and it is only their abundance that hurts me and always has; because if I had had just one, I would have greatly esteemed it, and advancing with that, with some great prince I might have encountered the fortune I have neither had nor sought so far." Then, taking up the opening of the *Sidereus,* he comments, "Great and very wonderful things I have [*magna longeque admirabilia apud me habeo*]: but they only serve or, rather, are put to work by princes, because it is the latter who commence and wage wars, build and defend fortresses, and, for their royal leisure, spend great sums, and not me or private gentlemen."[22]

The recognition he had won among his adversaries in Padua and Kepler's pronouncement helped bolster his reputation at court and anticipated the incredibly detailed list of requests (title, stipend, free time) he felt were essential if he were to return to his homeland, along with an equally extensive list of works he had already started and others he planned to undertake.

> The works I must complete are mainly two books, *De sistemate seu constitutione universi* [the future *Dialogo sopra i due massimi sistemi del mondo*], an immense conception and full of philosophy, astronomy, and geometry; three books, *De motu locali* [the future *Discorsi e dimostrazioni intorno a due nuove scienze*], entirely new science . . . ; three books on mechanics, two on demonstrations of its principles and foundations, and one on problems . . . I also have various pamphlets on natural subjects, such as *De sono et voce, De visu et coloribus, De maris estu, De compositione continui, De animalium motibus,* and more. I am also thinking about writing several books addressed to the soldier, training him not only in concept, but also teaching

> him, with very excellent rules, everything worth knowing and relying on mathematics, such as notions on castrametation, order, fortifications, assault and capture, drawing maps, measuring by sight, notions regarding artillery, the uses of various instruments, etc. Moreover, the use of my Geometric Compass, dedicated to His Highness, must be reprinted, as there are no copies left; said instrument has been widely embraced by the world, as truly no other instruments of this type are made, and I know that so far several thousand have been built.[23]

This is not the same wistful, humble tone of the subject who a few years earlier had hoped to wrest a benevolent but grudging yes from his prince. Rather, it was the lofty and solemn attitude of one who knew he could demand a role and had the power to negotiate—point by point—his new public position as philosopher and mathematician at the service of the Medici state.

This was the first time Galileo made such a request, an aspect that has not always been given the attention it deserves, but which emerges clearly when the various texts are cross-referenced, as we have attempted to do at the beginning of the chapter. In effect, until then, his Venetian correspondence with the grand duke and Vinta never mentioned any desire to return to Florence as a philosopher. It is entirely plausible that he did not express such a request until early May 1610 and that it was inspired by the fact that he had just read the *Dissertatio,* in which the prevalent image was certainly not that of Galileo the philosopher but Galileo the observer, a technical astronomer, albeit an excellent and very capable one. In fact, it is by no means irrelevant that, with all the air of throwing down the gauntlet, the first person Galileo would inform of his appointment as the grand duke's primary mathematician and philosopher was none other than Kepler.[24]

• •

His status as a victor was reinforced by the publication in Venice of several verses written in his honor during this same period by Girolamo Magagnati, a rather curious figure who was well known in Florence and was an old friend of Galileo's. A poet and future member of the Accademia della Crusca, but also a businessman, refined artist, and glassmaker, Magagnati had spent most of his life between Venice and Murano, where he lived and owned a furnace that produced highly sought-after blown and colored glass. In addition to Galileo, another figure who felt quite at home here was Traiano Boccalini, who praised Magagnati in his *Ragguagli di Parnaso.*[25]

The pamphlet, which Galileo promptly sent Vinta on May 21, was entitled *Meditazione poetica sopra i pianeti medicei,* and it would be the first of a long series of compositions dedicated to erecting the myth of the scientist as a "new Columbus" and "new Amerigo" who had conquered the heavens. Only a few verses suffice to convey the heroic merits of Galileo's exploits:

> But you, O Galileo of the Ether, crossed
> Boundless inaccessible fields,
> And sank the curious plow
> Of errant spirit into the eternal sapphires;
> Turning over the Sky's golden clods
> You discovered new Orbs and new Lights.[26]

The recipient of such gifts was naturally the "Etruscan Kingdom" of Cosimo II, who would "see, printed in the celestial annals // amidst the stars, to blaze the bright and pure // glory of your family and your name // the breadth of the Universe's immensity." In fact, on the title page—and perfectly in keeping with the subject—the author transformed Cosimo into Cosmo, and for the occasion he decided to redesign the Medici coat of arms, embellishing it with a splendid crown bearing five gleaming stars representing Jupiter and its four satellites (see Figure 22).

This publication could not have come at a better time: exactly two weeks after Galileo's request to return to his homeland, and thereby confirming all its exceptional merits. Perhaps this time, however, there was no need to rely on a poet's imagination in the hope that Galileo's petition would be accepted. On May 22, the day after Magagnati's little work was sent to Florence, Vinta informed Galileo of the decision ratifying his return to Tuscany. The grand duke had agreed to his every request, including the rather anomalous title of "philosopher." His projects—both present and future—would be duly funded, just as the entire diplomatic network of secretaries and ambassadors scattered throughout the most important European coun-

MEDITAZIONE
POETICA
DI
GIROLAMO MAGAGNATI
SOPRA I PIANETI MEDICEI:
AL
SER.^MO DON COSMO II.
GRAN DVCA DI TOSCANA.

IN VINEGIA, MDCX.

Figure 22. Title page with the Medici arms: the five stars over the crown represent Jupiter and its satellites, 1610

tries would be at his disposal whenever he needed. Vinta wrote, "The Lord Masters have informed me that they will send you 200 scudi to Venice to defray the costs of the spyglasses and printing: and [the ambassadors] at the court of Caesar [Rudolf], and in England, France, and Spain have been informed that when you send spyglasses or books, they are to receive and execute all the instructions in your letters, as if the Grand Duke himself had written to them."[27]

The mathematician and philosopher Galileo, like court painters and musicians, became one of the grand duke's men, and his creations were treated like affairs of state. His success would be the prince's success. Moreover, the Medici were prepared to underwrite the updated reprinting of the *Sidereus* as well as the technological research needed to make increasingly powerful telescopes.

After all, Galileo had never stopped working to improve his spyglass. It was during these months that he sent Kepler his observations of Jupiter, recorded between March 9 and May 21, while the planet and its satellites were still visible from Earth because they were not too close to the Sun.[28] After that, Jupiter would not reappear until the end of July, and Galileo had decided that, before reprinting the *Sidereus*, supplemented by new and more accurate details, he would wait for its return "outside the Sun's rays" in order to observe in the early morning hours. "The time to observe it again to the east in the morning will be in less than two months," he noted, "and it will easily be visible before sunrise; in the meantime, I will continue to make my astonishing observations and descriptions of the Moon, the vision of which surpasses all marvels, especially now, since I have improved the spyglass so much that I am able discover its most beautiful details."[29]

Therefore, as long as the Sun permitted, he continued to train his telescope on Jupiter. Then he would continue his observation of the Moon and, above all, keep working on lenses. He had two relatively calm months ahead of him before Jupiter reappeared, and they were months of extremely intense work and constant progress. We need merely consider that in his letter to Vinta of June 18, he asserted that he had "improved" his spyglass and that on several occasions he had observed Mars and Saturn but had not seen anything interesting ("I do not see other planets around them").[30]

We know very little about Galileo's work during these months. Nevertheless, there is no question that one of his aims was ever better lenses. We can glean this from a letter Cardinal Francesco Maria Dal Monte sent him on June 4, congratulating him on the method he had devised to craft lenses from rock crystal and, eager to satisfy any curiosity of his, informing the scientist that in Rome they were ground "artfully and with marvelous ease."[31]

How many lenses did Galileo manage to make during these months? What difficulties did he encounter in producing high-quality lenses? These questions have not been answered. But we do know that at the end of July he again turned his spyglass to Jupiter and, above all, Saturn, noting—for the first time—that it wasn't a single "star." He promptly informed the loyal Vinta and, through him, the grand duke and grand duchess, but begged him not to leak the news until the updated edition of the *Sidereus* had been published.[32]

His discoveries seemed endless. After the Moon, the Milky Way, and Jupiter, it was the turn of Saturn and Venus. The biggest and most resplendent celestial body continued to hold out against Galileo's telescope, but the havoc he was wreaking could not have been more devastating. Within the space of merely a year, the portrait of the sky had become even richer. The war on Mars that Kepler had so heroically waged for years seemed to pale compared to these unquestionable results, and his *Astronomia nova*, invisible to most, lay forgotten in some corner.

• •

To the very end, Galileo preferred not to circulate the news that he would be leaving Venice. Even in late June, Asdrubale Barbolani, the Tuscan resident there, told anyone who asked if Galileo was planning to leave the Republic that he knew nothing about it, though, as Vinta reported, he was well aware that it could "cause him trouble" in the city if anyone were to find out.[33] Once the decision was announced, some—such as Galileo's very close friend Sebastiano Venier—even decided to break all ties with Sagredo if the latter kept in touch with Galileo.[34] Sagredo nevertheless remained close to the Tuscan, despite his conviction that Galileo's departure was an irreparable loss for both himself and Venice.

It turned out to be the worst decision for Galileo as well, and time would unfortunately prove Sagredo right. Nowhere else but in Venice would Galileo feel like the master of his own fate. And this is even truer if we compare La Serenissima with a city such as Florence, where "the authority of the friends of Berlinzone"—that is, the Jesuits—had long been so influential.[35] Sagredo wrote those words to Galileo in one of the most moving letters of their entire exchange. The missive is too exceptional not to cite it in full, and it reads as if it had been penned by Machiavelli, the revelation of a lucid and disenchanted outlook that, in many ways, resembles that of the eminent Florentine secretary.

> Here the freedom and lifestyle everyone enjoys is widely admired and, perhaps, is unique the world over. . . . Where else can you find the same freedom and autonomy as you would in Venice? Mainly thanks to the support Your Excellency had here, which grew day by day, just as the age and authority of your friends also grew. You are now in your very noble country, but it is also true that you left the place where you were well off. You are now serving your own natural prince, who is great and virtuous, young and remarkably promising; but here you had control over those who order and govern others, and you had only to serve yourself, like a monarch of the universe. That prince's virtue and magnanimousness gives me great hope that your devotion and worth will be appreciated and rewarded. But in the stormy seas of the Court, who can be sure not to be travailed and troubled, if not submerged, by the raging winds of emulation? . . . Who knows what the infinite and incomprehensible accidents of the worlds can do, aided by the imposture of bad and envious men who, sowing and then cultivating false and calumnious ideas in the prince's mind, might then win over his justice and virtue to ruin a gentleman? For a while, princes get a taste for some curious things; but then, attracted by more important ones, they turn their minds to other matters. Moreover, I think the Grand Duke may delight that, with one of your spyglasses, he can observe the city of Florence and nearby places; but if for some important reason he should need to observe what is going on all over Italy, France, Spain, Germany, and the Levant, then he will set your spyglass aside: and despite its worth, if he finds some other instrument useful for this undertaking, who will be capable of inventing a spyglass to distinguish the mad from the sane, good advice from

> bad, the intelligent architect from an obstinate and ignorant eminence? Who knows who will be the judge of this court of infinite millions of idiots, whose opinions are estimated in number rather than weight?[36]

We do not know how Galileo responded. In any case, he probably paid little attention to this letter, as he had already been in Florence for nearly a year by this time and his interests were completely monopolized by his battle with the dreaded "friends of Berlinzone."

Sarpi wrote nothing at all, a profound silence demonstrating just how disappointed and bitter he felt over the decision of one of his dearest friends, with whom he had shared nearly twenty years of endless conversations about nature and humankind, and whom he admired for his versatility and unrivaled intellect. After the spyglass episode and the way Galileo had presented it in the *Sidereus*, Sarpi felt betrayed yet again and ceased all epistolary contact.

In Sarpi's correspondence with his Protestant friends, there is not one word about Galileo's departure from Venice. It hurt too much to even talk about it. Moreover, there is nothing even in Sarpi's letters to Micanzio or Sagredo. It would be more than five months after that fateful September 1610 before the two would be back in touch, unless we consider the annotation that Francesco Griselini, one of Sarpi's most important biographers, published in 1785 and had transcribed from folio CXXIV of the *Schede Sarpiane* kept at the library in the convent of the Servi di Maria before it was devastated by the fire of 1769. It is a controversial document whose authenticity has been questioned by a number of historians.[37] This is the entire text.

> Now that, through the note of His Excellency and Distinguished Senator Messer Domenico Molino, I have learned that Messer Galileo Galilei is about to go to Rome, invited there by various cardinals to show his celestial discoveries, I fear that, if in those circumstances he demonstrates the scholarly reasons that have led him to prefer the theory of the Canon Copernicus regarding our solar system, he will certainly not curry the favor of the Jesuits and other friars. Since they have changed the physical and astronomical question into a theological one, I most regretfully foresee that,

> in order to live in peace and without being called a heretic and be excommunicated, he will be forced to recant his ideas on the matter. The day will come, and I am quite sure of this, that men enlightened by better studies will deplore Galileo's disgrace and the injustice against such a great man; but in the meantime, he will have to suffer it and complain only in secret.[38]

If we choose to believe these words and attempt to date them, we would place this comment in late February 1611, when the news circulated about Galileo's imminent trip to Rome. After more than five months of silence, on February 12 the two started to correspond again, and for the first time, Galileo told Sarpi about his most recent and sensational telescopic discovery: "It is now certain that Venus revolves around the Sun and not under it (as Ptolemy believed), where it would show itself only as half a circle; nor above it (as Aristotle fancied), because if it were above the Sun, we would never see it as a crescent, but always much more than half, and almost always perfectly round. And I am sure we will see these same changes with Mercury."[39]

The astonishing news, with all its obvious cosmological consequences, immediately reverberated in Sarpi's circle. But the most fascinating thing is that on February 26, Fulgenzio Micanzio not only responded to Galileo, enthusiastically hailing his marvelous discovery, but also commented that he and Sarpi had confirmed his observation: "Father Maestro Paolo and I often remember you in our conversations; and particularly in the past few days, as we fully observed with the spyglass that Venus is effectively a moon, and the more it approaches the Sun, the thinner it becomes, and essentially does exactly the same thing as the Moon, except that its horns are not as sharp, perhaps because it is not close enough."[40]

Then, however, the letter suddenly changes tack. Instead of continuing to discuss astronomical matters, Micanzio switches subject and begs Galileo not to abandon his studies of motion, the "other science as rare as it is unknown . . . and for the understanding of which you were called by God and nature." This is far more than an invitation, in fact, and this mention of the science of the motion of heavy bodies *(de motu gravium)* that had so enthralled him and Maestro Paolo reads more like a joint appeal signed by his two Venetian friends, a plea to Galileo not to let himself get dragged into cosmological debates now that the "reasonable experience" of Venus had confirmed "it is completely true that all the planets revolve around the Sun as the center of their orbits."[41]

It is at this point that Sarpi's very controversial statement may come into play. There is no question that the tone—particularly at the end, so shamelessly prophetic—raises doubts about its authenticity, given the categorical "I most regretfully foresee" and the phrase "the day will come," seemingly too dramatic and sentimental to have been penned by Sarpi, not to mention the comment about the overly edificatory "men enlightened by better studies."[42] But there is a part at the beginning that should probably be examined more carefully. Why would Griselini have included such a detailed biographical reference, namely, that Senator Domenico Molino (although that "Messer" is indubitably a term he personally inserted) supposedly informed Sarpi of Galileo's upcoming trip to Rome? Why invent such a specific detail, running the risk of later being contradicted by Molino's biography? And then there is the next sentence: "I fear that, if in those circumstances he demonstrates the scholarly reasons that have led him to prefer the theory of the Canon Copernicus regarding our solar system, he will certainly not curry the favor of the Jesuits and other friars." It seems to be the right comment, stern and detached, for Sarpi to make in response to an opinion Galileo had expressed in his letter of February 12, 1611, telling his friend in no uncertain terms that with the exception of the Paduan Peripatetics, no one—Roman Jesuits included—now doubted his discoveries. In Galileo's words: "Because the most renowned mathematicians of different countries, *and Rome in particular*, after ridiculing at length and at every opportunity and everywhere, in writing and orally, the things I wrote, particularly about the Moon and the Medicean Planets, have finally been forced by the truth to write to me spontaneously, confessing and accepting everything; so that now I have no one against me but the Peripatetics, more partial to Aristotle than he himself would have been, and above all those of Padua, whom I truly have no hope of conquering."[43]

And whom else would Galileo have written to about his undeniable success besides the man who, more than anyone else, had manifested his skepticism and opposition to Galileo's decision to abandon the Republic? Yet not even these words, so triumphant in tone, seem to have reassured Sarpi. If we return to Sarpi's previously cited passage, what is most striking is his concern for his friend and his possible tragic downfall. There is nothing prophetic about these words. They are so grim that the only explanation is Sarpi's profound understanding of the powers at play within the Church of Rome, and particularly the noxious cultural and political war machine the Society of Jesus had become, dead set against any form of adherence to doctrines and disciplines that might cast the slightest doubt on the centrality of the Roman theological tradition. This is probably the source of Sarpi's distress: his anguish over a world he felt was irreparably lost and would penalize his friend, who had decided to play a game that, despite all appearances, he was bound to lose before it even started.

• •

Now let us go back a few months, to Venice, so we can take a look at Sarpi's bleak prediction and understand if it was overly pessimistic.

By the end of August, everything was ready for Galileo's farewell to the Republic. After packing his instruments, laboratory tools, books, and little else, Galileo decided to stop in Bologna to meet Magini and determine once and for all if the rumors he had heard about the latter's participation in the anti-*Sidereus* campaign had any foundation. He expected to arrive in Bologna on September 5 and then continue to Florence a few days later. In the meantime, he had just received the 200 scudi the grand duke had promised him to build new spyglasses and print the updated edition of the *Sidereus nuncius*, "to make its dignity commensurate with its subject and the dedication." He would never publish it.[44]

Yet as early as March 19, just a week after it had first come out, Galileo had thought about reprinting his announcement, already impossible to find. He was considering a much richer version with the addition of new celestial observations, beautiful copper engravings of the entire lunar cycle, and "many celestial images with all the stars that are truly there." And it would be written in the vernacular rather than Latin, complete with "many compositions by all the Tuscan poets."[45] If possible, in March he also wanted to publish the tables of the periods of revolution of the new Jovian satellites. In June his project was still very much alive and, if anything, had expanded even further. The book would also contain all his adversaries' objections and doubts, along with his answers and solutions, "so that everything will be utterly irrefutable."[46] Perhaps he was thinking about getting it printed in autumn, as this would give him enough time to observe the satellites in the morning starting at the end of July, when Jupiter was free of the Sun's rays, so he could "put in the work many observations made in this condition." In the meantime, given the latest improvements to the telescope, he also hoped to add other marvels. At the end of July, in fact, he had confided to Vinta about the surprising configuration of Saturn, requesting that he not tell anyone "until I have published it in the work I am reprinting."[47]

The project for the new edition took a few complicated twists. In addition to Horky's *Peregrinatio,* Galileo commented on another "downright load of balls" *(solennissima coglioneria)*—and written by a Florentine to boot. On August 7, Alessandro Sertini, who was supposed to handle the printing, asked Galileo for more specific instructions on how to proceed, advising him to wait "to see what all those who are writing or want to write have in mind, in order to respond to them all together."[48] But there is more. With this new work Galileo intended to reply to the doubts and questions Kepler had raised in his *Dissertatio* and which could potentially be used against him by his opponents.[49] As if that were not enough, he planned to include one of Kepler's letters and a text by Roffeni, both of which refuted Horky.[50] The more time passed, the more the original idea changed. What was supposed to be an enriched reprint of the *Sidereus* was being transformed into a new book with another structure and far different substance.

Slowly but surely, Galileo began to abandon the project. A pressing invitation by Federico Cesi, founder of the Accademia dei Lincei and Galileo's sponsor in Rome, fell on deaf ears. On August 20, 1611, Cesi wrote to Galileo, "I must urge you to print the supplement to your *Nuntio sydereo* as soon as possible. You have not written anything yet about horned Venus and three-bodied Saturn. Do it, please, as soon as possible, so that your children won't find some brazen father anxious to adopt them."[51]

There is no doubt that Kepler's unilateral decision to announce, in the preface to his *Dioptrice,* the discoveries made after the *Sidereus* was published undermined the aura of surprise and interest Galileo hoped would surround the reprint. However, perhaps there were other reasons that ultimately made Galileo set the project aside. He perceived that, to convince his adversaries, he needed to focus on practical observations and the discussions they would elicit, and not on circulating books. Above all, the new discovery of sunspots, which incontrovertibly demonstrated that the Sun rotates on its own axis, adding yet another decisive element in favor of the Copernican constitution of the universe, could not be relegated to the appendix of a reprint of the first celestial message. He needed a book devoted entirely to the Sun if he wanted to underscore the importance of this extraordinary telescopic observation.

Consequently, as soon as he arrived in Florence, one of the first people Galileo contacted was the Jesuit mathematician Christoph Clavius. He had likely pondered this for quite some time, but now it could finally happen. It is curious that in the opening of his September 17, 1610, letter to Clavius, he uses the words "It is time for me to break my long silence"—virtually the same words he would write to Sarpi a few months later, on February 12, 1611: "It is time for me to break my rather long silence."[52] But his silence toward the two men was vastly different. The former was due to enactment of the provision banning contact—epistolary or otherwise—with Jesuits after they had been expelled from Venetian territory due to the Interdict. The latter was the result of Galileo's grave insult toward the Republic with his decision to relinquish the privileges and recognition the Senate had offered him, and which he had initially accepted. In the first case, there was no need to explain the reasons behind this silence; he merely needed to reassure his illustrious correspondent that nothing had changed and that, even during his time in Padua, Galileo's "devotion to his great virtue had never waned."[53] In the second, however, Galileo's awkwardness is evident, stemming from the fact that he felt judged for a move considered politically incorrect and morally unacceptable—a move that, in Sarpi's eyes, not even the most extraordinary astronomical success could justify.

Galileo had been back in his city only a few days, but his mind already seems to have been elsewhere, pondering his trip to Rome. If we look at his words of September 17 in the right light, we might easily be convinced that Florence was simply a necessary stopover in order to conquer Rome. And if we think about it, was this not all along the real reason for his decision to leave a place as politically incorrect as the Venetian Republic? Wasn't this the first—and mandatory—step to take if he wanted to go to the Collegio Romano to convince Clavius and the other Jesuit mathematicians and philosophers of the validity of his telescopic discoveries, and demonstrate through them that the Aristotelian-Ptolemaic cosmology was wrong and the Pythagorean-Copernican view was not?

We certainly cannot raise any objections in terms of logic, but the two phases of the project would have seemed too simple and straightforward even to the most incurable optimist. So far, Clavius had not been able to see much with the telescope at his disposal. In fact, according to what Cigoli told Galileo in early October, "these *clavisi* [Clavius's disciples], every one of them, don't believe a thing; and among them, Clavius, who heads them all, mentioned the four stars to a friend of mine and laughed about them, saying that a spyglass would have to be built first to create and then show them, and that Galileo should keep his opinion while he would keep his own."[54] The only thing Cigoli could do at this point was to join Galileo in Rome as soon as possible; Antonio Santini, likewise in touch with the Jesuit mathematicians, felt exactly the same way, although he said he was convinced that "once they learn to use the spyglass, as long as its visual power is real, then they will have no choice but to confess."[55]

In turn, Galileo seemed unruffled as he waited to leave for Rome. In his letter of September 17 he had already reminded Clavius that, thanks to recent improvement of the telescope, he had managed to observe "the new planets as clearly and distinctly as stars of the second magnitude seen with the naked eye."[56] Moreover, he asked him not to doubt "the truth of this fact, which, if not earlier, you will be able to ascertain when I arrive, as I hope to do shortly and stay there a few days."[57] In the meantime, news of Kepler's observation of the Medicean planets was spreading across Europe, and one of the first to circulate it was Mark Welser, who had always been close to the Society of Jesus.[58] But Galileo had to wait quite some time for a reply from the Collegio Romano. Clavius did not respond to his letter—sent three months earlier—until December 17, and he did so by putting down on paper the position of the "new Medicean planets" he had "most clearly" observed in Rome a few days earlier. He had also seen Saturn's oblong outline, though it was hazy, and then the Moon, which did not cease to surprise (or trouble) him with "its unevenness and ruggedness when it is not full."[59]

The long-awaited letter had finally arrived and Rome was opening its doors at last. Galileo's reply, dated December 30, marked the sublime end to an astonishing year, and it was a good omen for the new one about to begin. Alongside known celestial phenomena (Saturn, Jupiter, the Moon), enriched by new telescopic observations, Galileo announced another discovery for the first time. Over the last three months he had managed to observe Venus clearly, and its configuration, which completely resembled the Moon's, had cosmological consequences he could no longer keep under wraps. "Here we are, sir," he wrote, "having clarified that Venus (and Mercury surely does the same) moves around the Sun, unquestionably the center of the maximum revolutions of all the planets; furthermore, we are sure that these planets are themselves dark and only shine when illuminated by the Sun, which I do not think happens with the fixed stars, based on some of my observations, and that *this system of the planets is surely different than what has commonly been thought.*"[60]

The season of doubts was over. Kepler's observations, on one hand, and the very recent discovery of the phases of Venus, on the other, had finally overcome the impasse of a situation poised between uncertainty and incredulity for far too long. All Galileo had to do was get over the ailment that had left him bedridden for weeks, so he could carefully prepare his mission and emerge from this battle victorious.

ELEVEN
The Roman Mission

The trip took an entire week, from March 23 to 29. And never once did Galileo fail to observe the position of the satellites: on the twenty-third from San Casciano, the twenty-fourth and twenty-fifth from Siena and San Quirico, the twenty-sixth from Acquapendente, the twenty-seventh from Viterbo, the twenty-eighth from Monterosi.[1] His tenacity in studying the Medicean planets proves his oft-claimed desire to determine the periods, meaning the time it takes them to complete their orbits around Jupiter. This was a task that Kepler and the astronomers of the Collegio Romano considered "most difficult and almost impossible," but Galileo didn't agree. In fact, he expected to crown his "truly Atlantic effort" with an even more ambitious achievement: "predicting the sites and dispositions that the new Planets will have in the future, but also those they have had in each time past."[2]

On March 29, Galileo finally entered Rome, on a litter provided by the grand duke and accompanied by two servants.[3] Given the public nature of the discoveries he was about to defend, the visit had a political spin, because it was a key element in his plan for cultural hegemony. Conquering Rome, the "workshop of all the world's practices," as defined by the cardinal (and later grand duke) Ferdinando de' Medici, would be a watershed in the battle for full acceptance of Galileo's telescopic discoveries.[4] As Belisario Vinta noted, "As soon as the truth of the Jovian satellites is established and confirmed in Rome, one can consider it as clarified around the world. And being shared by His Holiness, this new theory of the planets will be universally accepted by all mathematicians and astrologers."[5]

At the same time, such a victory would have critical repercussions on the more general war against the conceptual frameworks of tradition. As Tommaso Campanella wrote in January 1611, Galileo had "cleared men's eyes and showed the new sky and the new Earth in the Moon," and this had implications that transcended astronomical interests.[6] Those discoveries dramatically belied an age-old credence in the heterogeneity of celestial over terrestrial matter. They sketched out a universe far vaster than anyone had ever imagined (to the point that one might even suspect it was infinite) and corroborated the Copernican theory that shifted man from the center of the world. All of this confirmed the end of not only an image of the cosmos but also of a cultural model dominated by an anthropocentric and finalistic viewpoint.

To gain credit, however, the proposal needed the support of institutions that could embrace and substantiate the new theories, embodying them in the various shared ways of "practicing science" and a common vision of natural reality and knowledge. In this sense, Rome was a far more significant junction than Kepler's Prague, as the scientist's influence was limited almost entirely to specialists. Instead, if the papal city had accepted and circulated Galileo's ideas, this would have confirmed their universal value, in accordance with the confessional inspiration of the Roman Church and in keeping with the very etymology of its name (*catholicon*, meaning "universal").

Yet another aspect reveals his sojourn's markedly political aims. The mathematician and philosopher of the Grand Duke of Tuscany had gone to Rome not simply to defend his own discoveries and assert their scientific value but for the glory of the family of patrons to whom he had dedicated the four new "planets." His journey had been organized with Vinta's help and under the close supervision of Cosimo II, who had instructed the Tuscan administration to cover expenses. Moreover, Galileo would be allowed to stay at the residence of the ambassador, Giovanni Niccolini, and Cosimo and Antonio de' Medici wrote him letters of recommendation to give to the cardinals Francesco Maria Dal Monte and Maffeo Barberini, and to Father Virginio Orsini, all of whom were tied to the Medici rulers.[7] The Florentine government's interest in this visit is confirmed by Niccolini's eagerness to send the grand duke daily reports "explaining what has been done . . . and everything that will follow."[8]

In short, the initiative seemed to be a sort of cultural-political mission aimed at safeguarding and expanding the Tuscan rulers' prestige. As Cosimo II noted, the matter *(negotio)* was highly significant from at least three distinct—but connected—standpoints: "for the praise of [Galileo], born in Florence, and for public utility and for the glory of our age."[9]

• •

On March 30, the day after he arrived, Galileo visited the Collegio Romano and spoke "at length" with Clavius and "and two fathers, very well versed in mathematics, and his students" (likely Christoph Grienberger and the Belgian Odo Van Maelcote). When he arrived, they were reading Francesco Sizzi's *Dianoia*, which greatly amused them.[10] This was not the only time Galileo met with the Jesuit mathematicians. In a letter to his confrere Johann Lanz (in Munich) dated April 30, 1611, Paul Guldin confirmed that the author of the *Sidereus* often went to the Collegio and exchanged opinions with Clavius's collaborators.[11] It is no coincidence that Galileo kept a copy of the diary of the Jovian observations made by the Jesuit astronomers between November 28, 1610, and April 6, 1611.[12]

Among his papers, we also find another very important document: the query of April 19, 1611, in which Bellarmine asked Clavius and his collaborators for an opinion on the new discoveries.[13] The response arrived on April 24. Clavius, Grienberger, Maelcote, and Giovanni Paolo Lembo confirmed the observations, expressing reservations solely about the nature of the Milky Way, as they were not sure it was "made up of little stars." Clavius's position about the Moon stands out. He did not think it was rugged, but simply held that it was not uniformly "dense" or, in other words, that it had "denser and rarer parts, like the ordinary spots seen with natural sight."[14] Welser had already sent Galileo a similar message that he attributed to his "friend" (perhaps the Jesuit Christoph Scheiner).[15]

This thesis offered easy refuge to those who wanted to back the notion of the basic difference between celestial and sublunar bodies. So Clavius wasn't the only one defending this point of view. In a poem he wrote those weeks in Rome, the Sienese Jesuit Vincenzo Figliucci considered the concept closer to reality, illustrating it with these words:

But should be deemed closer to the truth
What the most scholarly have always believed,
That in this impure and insincere body,
There are truly the Sun's induced rays,
Through which the entire light is then not reflected,
But its splendors distorted and broken.
And just as something dense makes splendor brighter,
That which is more rare darkens.[16]

A short time later, Galileo took it as an important point in his favor that Clavius was the only one who refused to recognize the lunar mountains, whereas the other three astronomers agreed with him. In fact, he said he was sure that if he had the opportunity to meet the elderly mathematician more often, he would surely be able to convince him of his reasoning.[17]

Like Bellarmine's question, the reply bearing the signatures of the four Jesuit professors is also among Galileo's papers. This tells us that it must have been delivered to him personally by the cardinal or one of his writers, though he may have received the document at the assembly of the Collegio Romano on May 13, 1611, often referred to along the lines of a Galilean "triumph."[18] An *avviso* sent to Urbino on May 18 describes the event: "On Friday evening of last week [May 13] at the Collegio Romano, in the presence of cardinals, the Marquis of Monticelli [Federico Cesi], promoter of the meeting, an oration was given in Latin, with other compositions praising Messer Galileo Galilei, mathematician of the grand duke, glorifying and extolling up to the stars his new observation of new planets unknown to ancient philosophers, also aided by the magnification of the spyglasses procured by Porta [Giovan Battista Della Porta], Neapolitan, so that by means of this public demonstration Galileo may return to Florence most comforted and say he was crowned by the universal approval of this university."[19]

There were also four cardinals at the meeting, and the Latin speech praising Galileo was made by Maelcote, who in 1604 had already spoken about the *nova* before the Collegio Romano.[20] The text is known as the *Nuncius sidereus Collegii Romani* and it publicly attests to the truthfulness of Galileo's discoveries in light of the observations made by the Jesuit astronomers.[21]

Maelcote compared his own "sidereal message" to a dispatch delivered by a "lame messenger" who, precisely because he was slower, trailed behind the original courier (clearly Galileo) to confirm his revelations and vanquish the doubts of those who were still unsure. The oration corroborated most of the celestial discoveries, peremptorily in some cases, such as when Maelcote claimed that there was no longer any doubt about the large number of stars Galileo had recorded. Nevertheless, he was deliberately cautious about the interpretation of the unevenness of the lunar surface and, above all, the cosmological consequences of the phases of Venus.[22] According to Maelcote, it was not up to him to determine if the variation in the planet's appearance was due to its motion around the Sun or if instead there was some other cause, for he was not acting as master or arbiter in the matter but was merely a "celestial messenger."[23] Despite his great prudence, however, the philosophers at the meeting began to murmur, a sign that they had instantly perceived the perils for the traditional world system.[24]

Maelcote's oration had ratified the merits of the telescopic discoveries. As proof of the Collegio Romano's excellent rapport with Galileo, it inaugurated its courses in autumn 1611 with an opening lecture approving the discoveries that went against the laughable beliefs of the ancients.[25] The College in Parma celebrated the opening of its academic year with a lesson "in praise of Signor Galileo."[26] In a stellar catalogue Grienberger published in 1612, he praised the telescope that, "under the guidance of the courageous and fortunate Galileo," had made it possible to learn about new stars in the remotest reaches of the universe.[27] Figliucci followed suit:

> You, Galileo, above the terrestrial mire,
> Were the first to open the path closed to us.
> You, with crystals that I in poems express,
> Discovered a sky rich in new stars.
> While others were on the ground, you raised your heart
> To eternal matter and crossed the sea of the sky.[28]

The climate created between Galileo and the astronomers of the Collegio Romano certainly seemed to show mutual respect and sincere warmth. In reality, however, not everything portended a peaceful convergence of viewpoints and intents. Under these idyllic appearances, there were shadows and tensions that would explode into heated rivalry within a few years.

• •

Even before he went to Rome, Galileo was not pleased with Grienberger's statement that Jesuit astronomers had observed these celestial discoveries before they had heard about his telescope. "As soon as I got to Rome," Grienberger wrote, "I went to one of our confreres, Giovanni Paolo Lembo, who, *before anyone had news of your telescopes,* observed the irregularities of the Moon as well as the stars in the Pleiades, Orion, and various others, using instruments built *not by imitating others but based on his own conjectures.*"[29]

We can also be sure that Galileo did not appreciate Grienberger's claim of an independent method for comparing the distances of satellites with reference to Jupiter: "With this telescope, not only have we observed the Jovian satellites for nearly two months, but we have started to record their various positions, and *before learning from you* about this method to record the distances, we employed the observations in relation to the diameter of Jupiter, like your way."[30]

Nor did he like the Jesuits' assertion that they had independently observed the phases of Venus. "*Even before you pointed them out,*" Grienberger's letter went, "we quite clearly noticed that it was not a flaw in the instrument, but that Venus, like the Moon, slowly loses its brightness as it approaches the Sun."[31] Jealous as he was of his priority of discovery, Galileo certainly did not look favorably on this indirect debasement of his scientific merit. If anything, he spied a dangerous signal of the Jesuits' typical attitude, which a few years later, the Lyncean Fabio Colonna would qualify as the tendency to want to "seize the achievements others have made" in order to demonstrate that "they are the repository of science."[32]

In the spring of 1611, however, the support of the Collegio Romano was far too important for Galileo to indulge in controversy, especially since his mission had much more ambitious objectives: confirming the Copernican system as the only possible option, given his discoveries. In various passages of the *Sidereus,* Galileo designated the Sun as the center around which the planets move.[33] Naturally, the Copernican meaning of these expressions did not escape attentive readers such as Kepler: "While Jupiter moves along its

12-year orbit, four satellites encompass it before and behind. What was absurd, then, in Copernicus's statement (as you neatly remark, Galileo) that while the earth performs its annual revolution, a single moon clings to it in the same way?"[34]

Galileo was again launching the heliocentric cosmology but with new grounds, and this is what he referred to when, as he prepared for his journey, he expressed his will to underscore the "consequences" of the discoveries aimed at "renewing" astronomical science and "bringing it out of the darkness."[35] There was naturally resistance on this point. In this sense, the letter his friend Paolo Gualdo wrote to him on May 1611 was much more than just a warning: "So far, I have encountered no philosopher or astrologer willing to subscribe to your opinion that the Earth turns, and theologians are even less willing to do so. Consequently, think carefully before you unswervingly publish this opinion of yours as true, because many things can be said by way of debate that are not always best to assert as true, particularly when everyone's opinion is against you, absorbed, as it were, *ab orbe condito*."[36]

These words clearly reveal that Galileo intended to put the world system on his meeting agenda. At the same time, however, they also tell us about the insurmountable difficulties this position was bound to face in Rome. In particular, the Jesuit mathematicians could not follow him in this direction, as the rules of the Society placed very specific disciplinary and theoretical constraints on them. Cosmological matters were part of the field of natural philosophy and strictly the domain of Jesuit philosophers. Moreover, the *ratio studiorum* (the Society's plan of studies) specifically stated that in important issues they should maintain the Aristotelian doctrine: "*In rebus alicuius momenti ab Aristotele non recedat*."[37]

We can thus understand the caution of the *Nuncius sidereus Collegii Romani* regarding the most relevant of the celestial discoveries (from a cosmological standpoint): the discovery of the phases of Venus. While the Jesuits wholly backed what the telescope had found, they could not fully embrace Galileo's agenda for his "Roman mission."

The inevitable consequences of the celestial discoveries could only spark unease and apprehension among Jesuit ranks. As Lanz wrote to Guldin, in Rome at the time, "Please ask Father Clavius and Father Grienberger if, in order to save the motions of these new satellites of Jupiter, Saturn, and Mars, one need merely place epicycles with centers coincident with the centers of Jupiter, Saturn, and Mars; *or if a new theory must be devised*."[38] In essence, the most expert astronomers of the Collegio Romano were being asked if, given the changed observational framework, the old structure of the sky could still be sustained and taught.

Clavius would soon recognize the need for a more coherent theoretical framework of the new data, and in the newest edition of his commentary on Sacrobosco's *Sphaera*, he urged astronomers to reconsider the arrangement of the celestial orbs in order to "save" what the *Sidereus* had revealed.[39] On this basis, over the next few years Jesuit astronomers embraced the cause of Tycho Brahe's system, which allowed them to admit that the planets revolve around the Sun, while also preserving the immobility and central position of the Earth.

As we have already seen, in 1611 Cristoforo Borri theorized that the Tychonic system was the only plausible alternative to Copernican theory, considered unacceptable because it went against numerous scriptural passages.[40] Even Lembo, one of the leading figures in the observations conducted at the Collegio Romano, described the inevitable cosmological implications of the phases of Venus and then went on to praise the Tychonic view.

> We must confess that Venus and Mercury move around the Sun, and that they move sometimes above it and sometimes below, sometimes before it and sometimes after, as can be noted from various opinions of the ancients, some of whom placed these two planets above the Sun but others below it. . . . Tycho Brahe, a most scrupulous and modern observer of the path of the planets and stars, reconciles both of these opinions. In the eighth chapter of the second book of *De mundi aetherei recentioribus phenomenis*, he established that they orbit around the solar body, saying: "For Venus and Mercury, the smallest orbits go around the Sun and not around the Earth, because it seems they behave like epicycles"; and he said that all planets, except the Moon, revolve around the Sun as their head or king.[41]

Figliucci, likewise noting that "others [Copernicus] wrote falsely that the Sun is the center of the World," embraced the Tychonic solution:

Regardless, Venus and the Sun
Are best placed in the same Heaven.
And if what the great Dane Tycho wants,
We do not repudiate as truth,
That Mars is sometimes accustomed to being
Under the Sun and with its motion hastens down;
Three Planets have a common Orb,
Hence and thence they are separated by two [orbs].[42]

The Jesuits thus started to take an intermediate stance between the Ptolemaic and Copernican systems. This tendency would be further reinforced and fanned by the desire to preserve to the extent possible the established view of a stable Earth at the center of the universe. During Galileo's stay, the General of the Society of Jesus, Claudio Acquaviva, issued a circular asking all Jesuit professors to again respect the "uniformity and solidity of the doctrine," avoiding any dangerously innovative theories.[43] Jesuit astronomers could not follow the path Galileo asserted was the necessary "consequence" of his celestial discoveries.

This was the first but unmistakable sign that complete convergence would be impossible: a clue pointing to the difficulty of a rapport that would be marked by great tension and bitter disputes in the years to come.

• •

But there was more in Rome than just the Society of Jesus, and Galileo was well aware of this. The aim of his initiative was not merely to convince these experts, for a far vaster consensus had to be gained. His mission would not be complete until he had "received from and given to all the utmost and whole satisfaction."[44]

Therefore, at the end of April he noted that he had met many "cardinals, priests, and various princes who wanted to see my observations and were all satisfied, just as I was at the meeting when I saw their marvelous statues, paintings, furnishings, palaces, gardens, etc."[45] A flurry of appointments

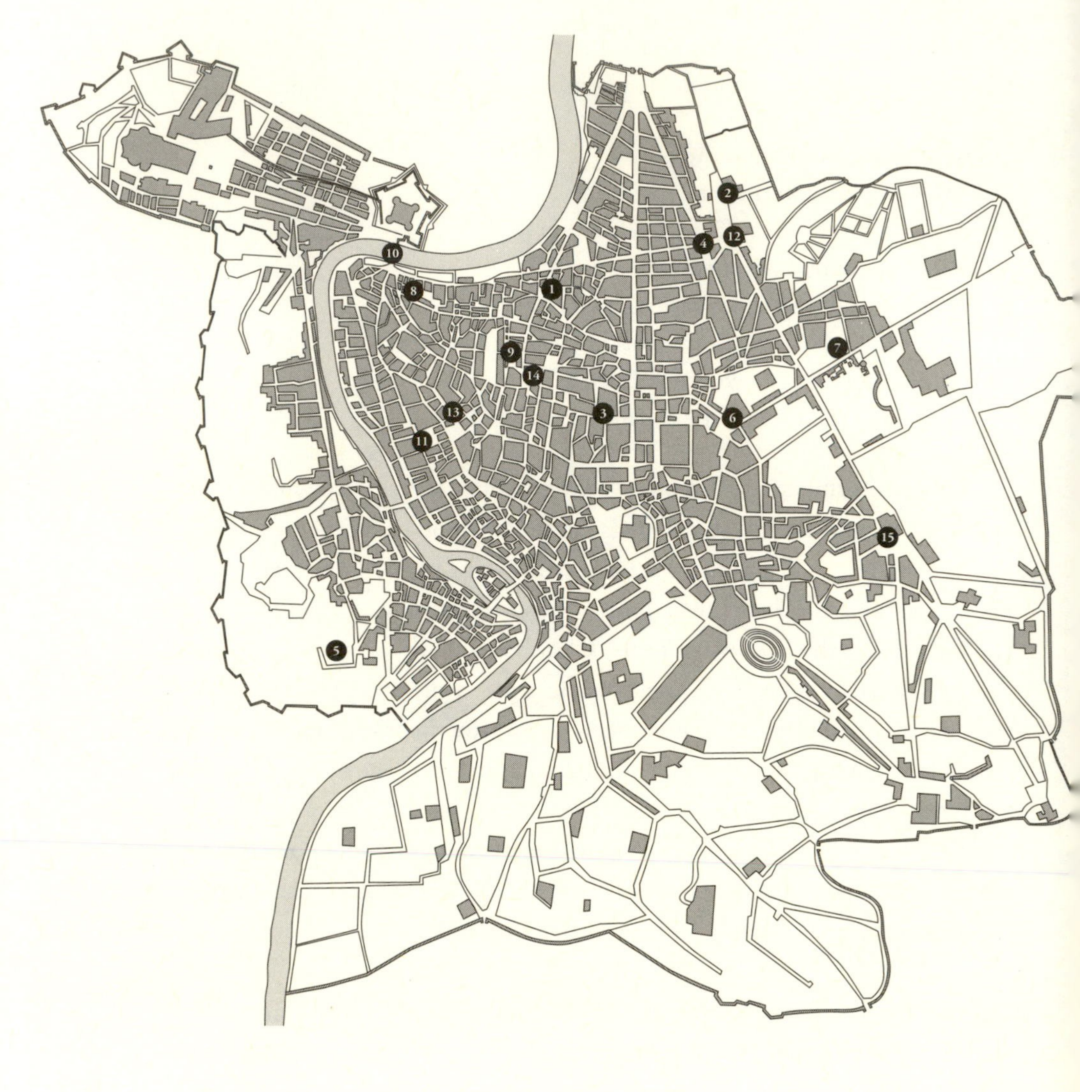

1
2
3
4
5
6
7
8
9
10
11
12
13
14
15

Map 4. Where Galileo stayed in Rome: March–June 1611

1. Palazzo Firenze. Residence of the Tuscan ambassador.
2. Villa Medici. Tuscan embassy in Rome.
3. Collegio Romano. G. went there a number of times and was publicly celebrated on May 13.
4. Palazzo Ferratini. G. attended a reading by G. B. Strozzi (April 6, 1611).
5. Vineyard of Msgr. Malvasia, Janiculum. G. conducted telescopic observations (April 14).
6. Quirinal Palace. G. is received by the pope (April 22).
7. Gardens of the Quirinal Palace. G. observed sunspots (April).
8. Palazzo Cesi, Via della Maschera d'Oro. Residence of Federico Cesi (G. becomes a member of the Accademia dei Lincei on April 25).
9. Palazzo Madama. Residence of Cardinal F. M. Dal Monte (G. goes there April 4–7).
10. Ponte Sant'Angelo. Residence of Cardinal Ottavio Bandini (G. is received there April 6).
11. Palazzo Farnese, Piazza Farnese. Residence of Cardinal Odoardo Farnese (G. is his guest in April).
12. Trinità dei Monti. Residence of Cardinal François de Joyeuse (G. meets him in April/May).
13. Campo dei Fiori. Residence of Count Virginio Orsini (G. brings a letter from the grand duke).
14. Via di Sant'Eustachio. Home of Cardinal Domenico Pinelli (on May 17, 1611, there is a meeting of the Holy Office asking for information about G.).
15. Church of Santa Maria Maggiore. Cigoli frescoes the dome, painting a Moon according to G.'s observations.

and conversations had allowed him to visit eminent figures at some of Rome's most beautiful residences in the city's most enchanting places: the garden of Trinità dei Monti, which Cardinal François de Joyeuse had purchased for the French college; the Palazzo Madama, where Cardinal Francesco Maria Dal Monte lived; Piazza del Monte di Pietà, with the *palazzo* of Maffeo Barberini; Ponte Sant'Angelo, near which Cardinal Bandini lived; the Palazzo Farnese, where he was the guest of Odoardo Farnese, who also hosted him at his estate in Caprarola.[46] It was at the Farnese residence that Galileo met the young Virginio Cesarini, future chamberlain of Urban VIII and recipient of *Il Saggiatore*, as well as Cardinal Carlo Conti, whom he contacted in the summer of 1612 for an opinion on the theological admissibility of terrestrial motion.[47] Galileo's encounter with Monsignor Giovanni Battista Agucchi, a man of letters but also an astronomy aficionado, likewise proved important; he was introduced to Agucchi by the mathematician Luca Valerio.[48]

His whirlwind of encounters those days included a meeting of the Accademia degli Ordinati at the residence of Monsignor Deti in the Palazzo Ferratini, near Piazza di Spagna. In the presence of cardinals Aldobrandini, Bandini, Tosco, and Conti, on April 6 Galileo listened to the speech on pride given by one of his friends, the Florentine littérateur Giovan Battista Strozzi, who, with "erudition and incomparable grace," made significant mention of the *occhiale:* "one of these new instruments, which, multiplying objects several times over, makes ants look almost like elephants."[49]

Galileo's relations with several members of the entourage of Pope Paul V became particularly important, such as the canon of St. Peter's, Tiberio Muti, and the cardinal-nephew Scipione Borghese, who was delighted to receive the scientist in mid-April.[50] All this was leading up to the mission's most significant event: Galileo's meeting with the pontiff on April 22. With immense satisfaction he wrote to Filippo Salviati, "This morning I went to kiss His Holiness's foot, and I was presented by our eminent and most illustrious ambassador, who said I had been treated with exceptional favor because His Beatitude would not let me say even one word on my knees."[51]

The audience had been promoted by the Tuscan ambassador, who, in the name of the grand duchy, would reap the intangible benefits—prestige and image—of the ruler's profitable investment in his "chief philosopher and mathematician." It was a way to crown a political operation and therefore represented confirmation that appreciation for the discovery of the new planets had indirectly won the Medici rulers points.

Three days after Galileo's visit to Paul V, there was another event on his calendar: initiation into the Accademia dei Lincei, established by Federico Cesi in 1603.[52] Galileo became the sixth member of the academy, preceded by the four founding members—Cesi, the Dutchman Johannes Van Heeck (Heckius), Francesco Stelluti of Fabriano, and Anastasio de Filiis of Terni—and also by Giovan Battista Della Porta. His entry into the Accademia permitted even more incisive development of his public relations tactics. From this standpoint, the role played by Cesi, member of an eminent noble Roman family, could not have been more valuable. Not only did he back the celebration of the *Sidereus* at the Collegio Romano, as already noted, but he also organized an important initiative to support the celestial discoveries. Thanks to his good offices, on April 14, in the vineyard of Cardinal Innocenzo Malvasia on the Janiculum hill, Galileo was able to provide further proof of the effectiveness of his telescope and show the Medicean planets to a number of exponents of Roman culture. According to an *avviso* for the court of Urbino dated April 16, 1611:

> On Thursday evening, with the patronage of the Marquis of Monticelli [Federico Cesi], nephew of Cardinal Cesis, who held a banquet and spoke along with his relative Messer Paolo Monaldesco [Monaldeschi] in the vineyard of Monsignor Malvasia past the Porta San Pancrazio, on a hill and outdoors, there was a meeting attended by Galileo, a certain Terrentio, a Fleming [Johannes Schreck], Messer Persio of Cardinal Cesis [the philosopher Antonio Persio], Galla, professor at this university [Giulio Cesare Lagalla], the Greek mathematician of Cardinal Gonzaga [Ioannis Demisianos], Messer [Francesco] Piffari, professor in Siena, and others up to a group of eight, some of whom came specifically to see this observation, and although they stayed there for seven hours that night, they did not agree in opinion.[53]

Using the telescopes, the guests saw "the *palazzo* of Duke Altemps, in the Tuscolano, so clearly that they could easily count all the windows, even the smallest ones, at a distance of sixteen Italian miles."[54] Nevertheless, opinions regarding the results of the observations diverged widely. The *avviso* sent to Urbino tells us that "although they spent seven hours there that night, they did not agree in opinion."

Galileo's mission involved a packed schedule of encounters aimed at winning over remaining skeptics.[55] Above all among the Lynceans, there reigned a feverish climate that Cesi described as follows:

> Every clear night we see new things in the sky, truly a suitable office for the Lincei: Jupiter with its four [satellites] and their periods, the mountainous, cavernous, sinuous, watery moon. There remain that horned Venus and three-bodied Saturn, which I should be able to see in the morning. As to the fixed stars, I shall say nothing else. Philosophers conclude that either the sky is fluid and is the same as air, or they maintain the spheres in this form of the planets, in keeping with the old declaration of Pythagoras and today's new observations. But there is a noteworthy difficulty if the Earth is at the center of the spheres.[56]

Galileo stayed in Rome longer than he had planned. Given his whirlwind schedule of meetings, he decided to take full advantage of the opportunity to convince as many people as possible. At the end of April, he asked the Tuscan secretary of state if he could stay longer "to fully satisfy each one, just as I have thus far done with so many others." Therefore, he asked if he could move from the Palazzo Firenze, residence of the ambassador Niccolini, to the other Tuscan venue in Rome, the Villa Medici, at Trinità dei Monti.[57] Plans for his return did not shape up until the end of May; when his departure was imminent, Galileo had Francesco Maria Dal Monte draft a letter confirming the success of his mission. The cardinal said it had been so successful that "if we were now in the ancient Roman Republic, I believe a statue would have been erected to him on the Capitoline Hill to honor his great merit."[58]

Convinced that he had "fully demonstrated his inventions, which were esteemed by all the capable and expert men of this city," Galileo headed home on June 4, 1611.[59]

• •

Certain of the validity of his discoveries, in Rome Galileo commenced a patient persuasion campaign devoted to establishing a new image of the heavens. His ultimate aim was a radical cultural change: replacing an old and worn-out model with a new ideal horizon that, starting with his discoveries and the Copernican viewpoint, could reform the entire approach to natural investigation and man's relationship with the world.

These were the very implications that preoccupied a number of people. Though Galileo had handily rebutted opposition to his theories, he could do nothing about what was going on behind the scenes. It was there, in the "secret rooms" far from public debate and noisy controversies, that his adversaries were taking action.

On May 17, 1611, the Holy Office held a session at the residence of Cardinal Domenico Pinelli in the Sant'Eustachio district of Rome. Present in addition to Pinelli were cardinals Arrigoni, Bellarmine, Taverna, Mellini, Rochefoucauld, and Veralli, as well as Andrea Giustiniani and Marcello Filonardi, respectively commissary and assessor of the Holy Office. A rather uncommon request was made during the meeting: to determine whether Galileo had been named in the trial against the Aristotelian professor Cesare Cremonini.[60]

Cremonini, who had taught philosophy in Padua since 1591, was being investigated on a number of counts. The first accusations against him went back to 1598: teaching Aristotle's *De anima* according to Alexander of Aphrodisias's doctrine of the death of the soul. The following year (and for the same reason) he was served a warning by the Inquisitor of Padua. His negation of the immortality of the soul then led to further judiciary action in 1608. It must be noted that Cremonini always managed to avoid sentencing and restrictive measures, deftly eluding the Inquisitors. The attitude of the Venetian authorities was decisive in this case, as they were quite determined to protect their philosopher's freedom. Clear proof came in 1604, when the names of Cremonini and Galileo were first associated in a condemnation submitted to the Venetian Inquisition, but thanks to opposition by the government of La Serenissima, nothing came of it.[61]

Now, seven years later, the names of Galileo and Cremonini came up again in an Inquisition document. This time, though, the top name was not Cremonini's. The need to investigate stemmed from Galileo's success during his springtime sojourn in Rome. The aim was to learn the true identity of that philosophy and mathematics professor whose innovative theories had been so sensational, and to discover if, during his years in Padua, he had somehow been involved in his philosopher colleague's miscreance. Possible involvement with Cremonini would make his theories even more suspect, but at the same time it would pave the way for examining his orthodoxy. The intent was to take a look, after which more detailed investigations might follow if necessary.

We have no way of knowing how things unfolded. But the Holy Office's request is crucial, as it portends the events that, starting in February 1615, would bring Galileo to the very center of the Roman Inquisition after charges were leveled against him by the Dominicans Niccolò Lorini and Tommaso Caccini.[62] In this case, one of the main accusations was no longer his rapport with Cremonini but his association with another "friend" who was equally dangerous: Paolo Sarpi.[63]

It was during this season, in December 1615, just before Galileo's next trip to Rome, that the Tuscan ambassador Piero Guicciardini would recall the episode of four years earlier: "The Consultors or the Cardinals of the Holy Office did not like his [Galileo's] doctrine, and some other thing; and, among them, Cardinal Bellarmine told me that great respect was owed to all the matters pertaining to these Serene Highnesses [the Medici], but that if Galileo had stayed here in Rome longer, he would have been forced to offer some justification for his ideas."[64]

Guicciardini's words confirm that Galileo's theories (along with an unspecified "other thing") had made the Holy Office suspicious. And if the mathematician had stayed in town much longer, this suspicion would have led to the opening of an inquiry. Evidently, the ongoing debates that had marked his stay in Rome and the opinions he had expressed while there must have seemed quite dangerous to the cautious custodians of Catholic orthodoxy.

When he left Rome, Galileo had no idea of the risk he had run, and the Holy Office certainly wasn't about to let on. After all, secrecy was an iron-clad rule of the Inquisitorial system.[65] Galileo's busy mission, animated by optimism and persuasive zeal, had ended up sparking an authoritarian and censorious response. Unfortunately, the years to come would lend credence to those early but veiled threats.

• •

Galileo's Roman mission did not end with his departure. It continued in other forms, leaving a tangible trace on one of the most important cultural undertakings of the papacy of Paul V: the pictorial cycle in the Borghese Chapel at Santa Maria Maggiore.

On the dome of the basilica Ludovico Cigoli depicted the Virgin standing on a Moon with the same features as the satellite observed through a telescope (Plate 4). The fresco (Figure 23) reproduced the same appearance Galileo had described: a surface that is "uneven, rough, and crowded with depressions and bulges. And it is like the face of the Earth" (Figure 24).[66]

The irregular shape of the terminator, and the presence of craters and luminescent points in the dark area (which Galileo interpreted as a sign that there were mountains), evoked the lunar morphology illustrated in the *Sidereus*. As Erwin Panofsky observed, Cigoli thus "paid tribute to the great scientist"; in the name of a long-standing friendship, he wanted to immortalize the "new" Galilean Moon.[67] The enormous fresco was an excellent opportunity to show the world one of the most surprising discoveries announced in the book and render it public outside the few copies of the *Sidereus* in circulation.

The work in the chapel had been commissioned following examination of a selection of drawings submitted by Gaspare Celio, Cherubino Alberti, and, naturally, Cigoli, whose work Paul V liked best.[68] It took approximately two years to complete, between September 1610 and October 1612.[69] When the fresco was finished, it met with some criticism, but it was praised by Cardinal Jacopo Serra, who had been appointed to oversee execution of the pictorial cycle.[70] The pope and Cardinal Scipione Borghese (an admirer and patron of Cigoli) also appreciated it, even sponsoring the artist's candidacy to become a Knight of Malta.[71]

How could a representation so divergent from tradition have obtained the consensus of Church authorities, and the pope in particular? Based on a well-established custom, the Moon was supposed to be portrayed as

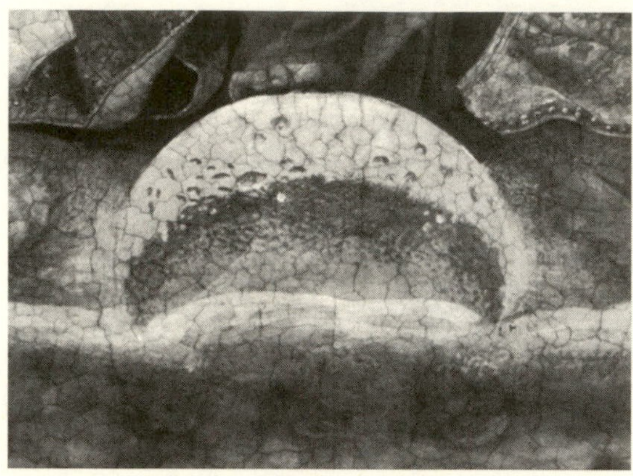

Figure 23. Ludovico Cigoli, *Assumption of the Virgin,* fresco, 1610–1612, detail

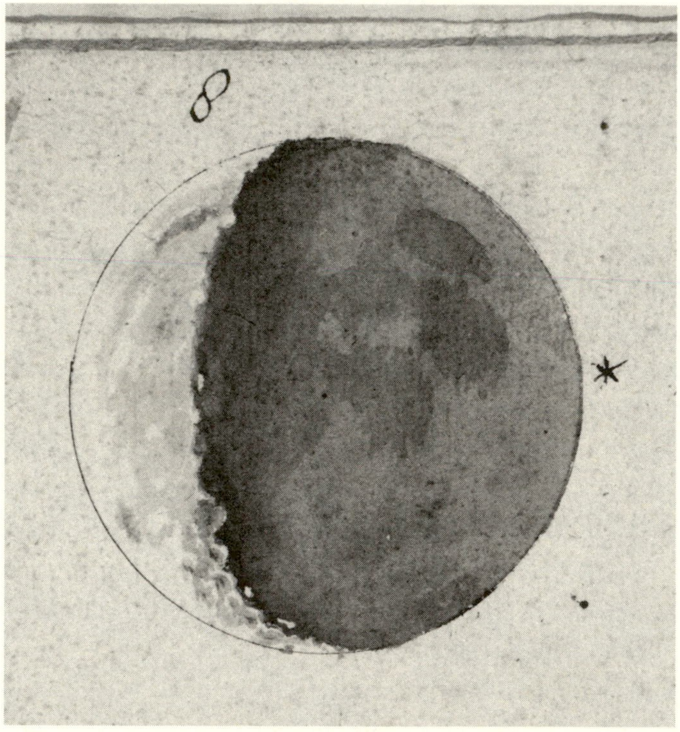

Figure 24. Galileo Galilei, drawing of the Moon, 1609

perfectly smooth and transparent, as it symbolized the Virgin and her purity as the "Immaculate Queen of the Heavens." As Stephen Ostrow noted, "When the moon as a symbol of Mary appears in painting, it is invariably as a pristine orb or crescent, for to depict the moon as less than perfect would be to tarnish the Virgin's image."[72] Regarding the Moon "debaxo de los pies" (beneath the feet) of the Virgin, the very devout painter Francisco Pacheco wrote that "although it is a solid body, I take the liberty of making it lucid and transparent above the landscape."[73] This meant that Cigoli's painting was a manifest violation of entrenched figurative and symbolic principles. So why was it so greatly admired by the pope's entourage and Paul V himself?

To understand the reasons, we must consider that the frescoes in the Pauline Chapel were dictated by a very specific plan that carefully laid out the subjects to be portrayed, detailing their spiritual meaning and symbolism.[74] The following instructions were provided for Cigoli's work.

> In the cupola will be painted the Vision from the Apocalypse, chapter 12: that is, a Woman clothed with the sun, and the moon under her feet, and upon her head a crown of twelve stars, facing Saint Michael the Archangel in the form of a Combatant, surrounded by the three hierarchies, each divided into three orders, [and] below [whom] emerges a serpent with its head crushed as in chapter 3 of Genesis. Around [are] the twelve Apostles. Such a Woman signifies the Church, . . . the Madonna literally signifies nothing less than the Church, which is the Madonna, who fights from the Beginning of the World, manifested to the Angels through the Incarnation, to the end of the World, Triumphing in heaven.[75]

Among the different portrayals of the Virgin in the Christian mind, the dome was to exalt the "Woman clothed with the Sun" evoked in chapter 12 of the Apocalypse.

A long exegetic tradition usually associated the Moon, trodden by the Virgin, with the changeability and imperfection of the terrestrial domain, over which the spirituality of the Church—identified with Mary—rose triumphantly. As the Jesuit Blas Viegas noted, "One must take the Moon as the symbol of fickleness because of its phases and because, quite often, it does not shine, due to the interposition of the Earth. . . . Therefore, the Moon symbolizes the things of the world, those that are subject to chance, variability, and instability."[76]

According to the Franciscan Jean de La Haye, "The Moon is the symbol and, indeed, the cause of fickleness. . . . It does not maintain the same figure, but looks different each time and never remains the same. Therefore, while the Sun has a positive meaning, the Moon has a negative one. . . . The meaning of the words *The Moon under her feet* is that this woman, in other words, the Church, gives no consideration to things subject to time, but crushes them under her feet, as they are fleeting and transitory."[77]

This perspective placed the accent on lunar imperfections. As the Portuguese João da Sylveira explained, "The moon, because it has certain blemishes in its body, and undergoes eclipse, and scatters darkness here and there, signifies the failings and faults of corrupt [human] nature. Therefore, the victorious Virgin Mary rightly crushes the moon beneath her feet."[78] Likewise, Johann Heinrich Bullinger, a Protestant from Zurich, noted, "The Moon is subject to change and undergoes variations, also in color. It waxes and wanes, and though it shines, it seems to be sprinkled with spots and receives its own light from the Sun. Therefore, the Church is crushing under her feet all temporal vicissitudes and all things from this century that are changeable, corruptible and flawed."[79]

In other words, an "imperfect," pockmarked Moon did not go against the symbolic meaning ascribed to the satellite in the exegesis of chapter 12 of the Apocalypse. Although Cigoli's intent was different—he wanted to celebrate the telescopic discoveries—the outcome fully respected the fresco project.[80]

The pontiff's endorsement corroborates this view. We can find further confirmation of this from an expert theologian, Andrea Vittorelli of Bassano del Grappa, who just three years after the chapel was opened wrote about 150 pages carefully analyzing the iconography of the frescoes. In the *Gloriose memorie della Beatissima Vergine* (Glorious Memories of the Blessed Virgin), dedicated to Paul V, Vittorelli interpreted the Moon as the symbol of terrestrial reality, variable and ephemeral: "The whole world (subject to various changes) [is] under Mary, because she did not value the world's grandness so appreciated by the lovers of mortal life." In keeping with this tradition, the Moon evoked both material and mental alteration: the folly of giving in to ephemeral mundane temptations. "The Moon not only signifies the flaw of corruption, but lunacy.... All ignorance [is] under the feet of Mary, who is filled with heavenly knowledge, and her glorious feet trample on the cause of all madness."[81]

Yet no mention is made of the Galilean significance of the fresco: not one word about vales and mounts, nor the moonscape the telescope had revealed. This reticence is striking, because Vittorelli was quite familiar with the *Sidereus*. A work on angelology, printed in Vicenza in 1611 and dedicated to Scipione Borghese, cited the book twice. Regarding the question of whether the zodiac was moved by a particular angel, Vittorelli wrote, "I've already said that in the starry sky, in addition to the numerous stars, there are many unknown to astrologers; I would now add that Mr. Galilei, Mathematician in Padua, used a spyglass to observe countless stars smaller than those of the sixth magnitude, which makes six more differences in magnitude."[82]

Galileo had argued that "stars of the fifth or sixth magnitude seen through the spyglass are shown as of the first magnitude."[83] Vittorelli seems to have read the *Sidereus* carefully, explicitly mentioning it in his notes. On the pages that followed, disputing the opinion of several "Jewish Rabbis" who ascribed an angel to each planet, he noted, "All this is uncertain, though not impossible; but if to the seven planets that have been known until now are added the four discovered by Signor Galilei, public mathematician at the University of Padua, with the benefit of a most perfect eye-piece of his invention (as he affirms in his little book titled *Sidereus Nuncius*), what will the Rabbis and others say about this important opinion?"[84]

It is strange that, well informed as he was, Vittorelli made no mention of the Galilean Moon portrayed by Cigoli. The metaphysical interpretation of the painting needed no astronomical references, of course, as its symbolism expresses an allegorical and emblematic world with prescriptive rather than cognitive aims and meanings. Yet this silence is striking and can perhaps be explained as opportunism. This was certainly not the right time to praise a new sky: the theologian published his book in 1616, when the accusations of heterodoxy leveled against Galileo had already reached the Holy Office.

Aside from Vittorelli's reticence and possible motivations, the lunar detail of the fresco was a public celebration of one of the aspects of the new world the *Sidereus* had started to reveal. Cigoli's painting can effectively be considered an appendix to the book, a back cover of sorts or even its dust jacket, with an astonishing image—that strange pitted Moon with spots and hollows—destined to evoke the overall meaning of this narration.

• •

Most of the faithful who visited the Pauline Chapel after it opened probably failed to grasp the fresco's singular lunar detail, but cognoscenti indubitably understood its astronomical reference. On December 23, 1612, Cesi wrote to Galileo, "Mr. Cigoli has beautifully finished the dome in His Holiness's chapel in Santa Maria Maggiore and, like a good and loyal friend, under the image of the Blessed Virgin he has painted the Moon as you discovered it, with its crenellated division and its islets."[85]

About one month later—it was Sunday, January 27, 1613—the "marvelous Capella [*sic*] Borghesia" (as Vittorelli called it) was inaugurated "with a solemn procession and an endless crowd." They carried the icon of the Virgin that, according to tradition, had been painted by St. Luke, placing it in front of the altar *cum luminaribus convenientibus* (with appropriate lights). Paul V had ordered that "in addition to the lamps burning perpetually," there should be "two large brass torchères with wax candles weighing four pounds each in front of the glorious image, and others weighing two pounds when said image is open."[86] That the fresco could not remain in the shadows clearly showed how much the pope appreciated the work.

In addition to exegetic reasons tied to symbolic aims, Pope Paul V may also have liked the image of the Moon as an emblem of the new discoveries. The way the pontiff received Galileo in April 1611, not permitting him to say even a word on his knees, was too recent to be forgotten.[87] The same can be said of the keen interest of Scipione Borghese, who was one of the first in Italy to receive an instrument from Flanders and who then promptly asked Galileo for one.[88]

Yet this had nothing to do with adhering to Galileo's scientific content, much less with embracing his cultural agenda. It would be an exaggeration to say that, through Cigoli's fresco, the Church "was quick to co-opt the new revelation" and "tacitly acknowledged Galileo's roughened Moon."[89] Instead, it would be more realistic to assert that the pontiff's approval stemmed from a desire to see the allegorical meaning of the figure (the "corrupt" Moon cited in chapter 12 of the Apocalypse) expressed in an original and fashionable image that tickled its patron's fancy because it was unprecedented.

Nevertheless, the fresco's significance was remarkable. That Moon was a tangible sign of Galileo's success. In fact, he personally participated in the preparatory work on the fresco. An autograph note of Sigismondo Coccapani, who collaborated with Cigoli on the Roman undertaking, eloquently confirms this: "The measurement of the dome of Santa Maria Maggiore based on the Roman palm is about 4000 and 700 palms without the lantern, according to Mr. Galileo Galilei. And the Roman surveyors said it was 3000, 200, 17 palms, after measuring it time and again."[90]

Though Cigoli was certainly knowledgeable in perspective, he had turned to Galileo for a more precise assessment of the area at his disposal.[91] Calculating the size of the dome was fundamental for splitting up the space he would allocate to his figures and sketching out the details of the entire composition. Galileo's technical information demonstrates that he was perfectly aware of what his artist friend was planning to paint.

• •

Publication of the "corrupt" Moon in the fresco at Santa Maria Maggiore gave the celestial discoveries an unexpected showcase. Despite this later partial success, however, Galileo's Roman mission did not achieve his intent to establish a different worldview. His design was too daring, the difficulties too great; the Holy Office's request for information irrefutably proves this.

The mild spring sky of 1611 had been a short-lived illusion, and a far different atmosphere was coming up on the horizon. A few years later, confirming the resolute stance of Paul V against the Copernican doctrine, the Tuscan ambassador Piero Guicciardini would say, "*This sky of Rome is very dangerous,* the most [dangerous] of this century, in which the Prince here abhors *belle-lettres* and clever people; he can't stand to hear about these discoveries nor these subtleties, and everyone tries to adapt his thoughts and character to the pope's."[92]

The Roman mission had seemingly ushered in a climate of openness to new ideas, a small, fleeting ray of sunshine—vague as it was—in the leaden sky of Counter-Reformation traditionalism. As of the end of 1613, theological debates on Copernicanism would quickly chill the tepid enthusiasm of that short season. As Guicciardini very realistically noted, Rome would no longer be the place "to come and discuss the Moon, nor to wish to . . . sustain or introduce new doctrines."[93]

TWELVE
In Motion: Portugal, India, China

India, "City of the Apostle St. Thomas," November 2, 1612:

> Somebody wrote to me from Italy that certain spectacles [*occhiali*] have been invented by means of which things fifteen and twenty miles away are seen clearly, and many novelties have been discovered in the sky, particularly in the planets. Your Reverence would do me a great favor by sending me these, along with a little treatise on such spectacles, if there is a demonstration of the things that can be seen. And if Your Reverence does not have the occasion or the money to send them, please send me *in writing and in figures* [*in scriptis et in figuris*], as clearly as possible, the manner of their construction, so that I may have them made in this land of many officials and an abundance of crystal.[1]

This is the first testimony of the circulation of telescopic discoveries in Asia. The writer, the Jesuit missionary Giovanni Antonio Rubino, was in the Portuguese colony of São Tomé, or Mylapore, on the Coromandel Coast, where the apostle Thomas is traditionally said to have died. Addressed to Father Grienberger in Rome, it expressed the missionary's great enthusiasm over the sensational discoveries he had just heard about. Since Rubino's astronomical interests were well known, the Collegio Romano notified him between May and June 1611 that the celestial discoveries had been confirmed there in Galileo's presence. Rubino then begged Grienberger to send him a telescope or at least a "little treatise" so he could figure out how to make one himself in India. While he waited for his request to be fulfilled, this is how he pictured his new instrument: "I imagine that the spectacles are made in the manner of pyramids, wide at the beginning and sharp at the end, and I don't know if they will be somewhat concave at the beginning; *and if this is so, long live the Perspectivists who maintain that vision takes place through*

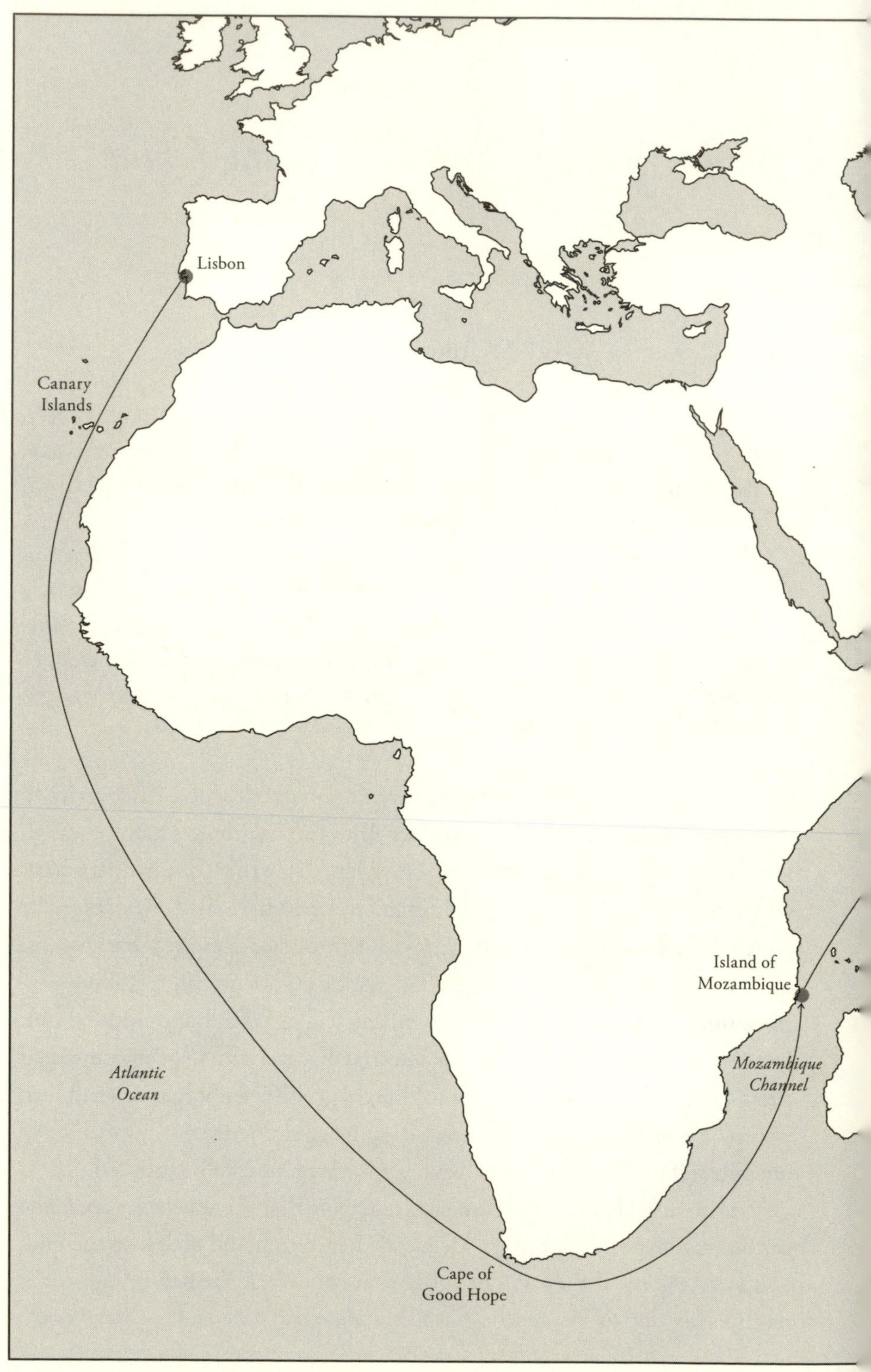

Map 5. Stops on the Jesuit missionaries' journey to China

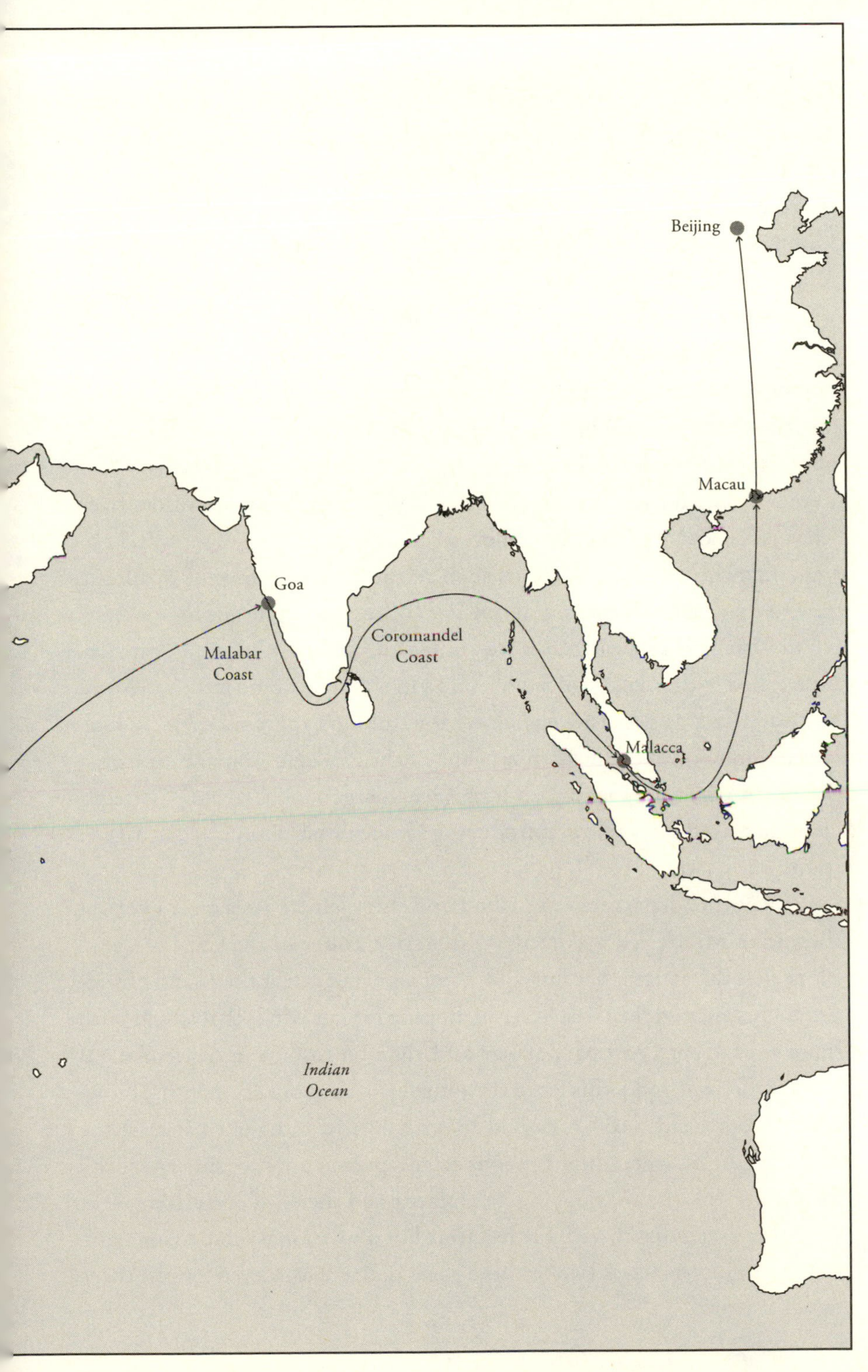

Beijing
Macau
Goa
Malabar
Coast
Coromandel
Coast
Malacca
Indian
Ocean

the extramission of rays [*et si ita est vivant in eternum Perspectivi, qui visionem fieri per extramissionem deffendunt*]. Your Reverence, please help me in everything, and send me everything written and well explained."[2]

Rubino had been in India for about ten years by this time and, like all Jesuits who went to Asia, before leaving he had spent the period from June 1601 until March of the following year in Lisbon, as the order's missions were under the jurisdiction of the Lusitanian Province.[3] He attended the College of Santo Antão, the center for the scientific training of Jesuits in Portugal, where, as of the end of the sixteenth century, an *aula da esfera* (course on the Sphere) had been set up to meet the request of Philip II to offer advanced mathematics courses for sea captains. These lessons, which were not only for Jesuit students, were taught mainly by professors from Clavius's school, and they dealt above all with the art of navigation and cartography: the use of instruments such as the astrolabe and the quadrant, how to draw nautical maps, and concepts of cosmography and astronomy.[4]

In Lisbon, Rubino attended the course taught by Grienberger, who had been called to the College of Santo Antão in 1599 and would return to Rome in 1602. This was a decisive encounter for Rubino's apprenticeship, as it allowed him to meet the mathematicians of the Collegio Romano and, once he got to Asia, correspond with both Grienberger and Clavius.[5] These exchanges played a key role in introducing Galileo's astronomical discoveries to the Far East.

The port of departure was Lisbon and ships left for Asia once a year, at the end of March. They skirted West Africa, rounded the Cape of Good Hope, left the equatorial calm, and, after a port of call at the Island of Mozambique, entered the Indian Ocean, hoping for fair winds that would push them toward the coast of Malabar and then on to Goa in September. If the weather was good, this leg of the journey took about six months. However, anyone headed to China had to stop over in India until at least the following April, when the monsoons would permit another ship to set sail for the port of Macau, stopping in Malacca and then—if everything went smoothly—arriving three months later. For a missionary departing from Rome, the journey took two or three years. Letters took even longer: three to seven years.[6]

It was through this network of communication that news about the celestial novelties circulated in India and, a short time later, China. It spread quite quickly, considering its long, complicated, and perilous journey. Just four years after Galileo's Roman mission, in Beijing the Portuguese Jesuit Manuel Dias Jr., known there as Yang Manuo, published a small book titled *Tian Wen Lüe (Epitome of Questions on the Heavens)*, which for the first time presented in Chinese a detailed description of the recent telescopic discoveries.[7]

• •

Dias was not an astronomer, in that he was not one of the Jesuit mathematicians, experts in science, that the Society trained at its European colleges and then sent to teach in the Far East.[8] Between 1596 and 1600 he had studied philosophy at the Jesuit College in Coimbra, where he learned the rudiments of cosmography and astronomy, based chiefly on the texts of Aristotle and Sacrobosco's *Sphaera*. He likely honed his knowledge of mathematics thanks to Grienberger's private course lasting just a few months and taught in Coimbra in 1599.[9] However, the contents of the *Tian Wen Lüe* reveal technical skills in astronomy rarely found in someone without advanced formal training in science. This tells us that Dias acquired those skills later, during his long stay in Asia, where he must have read specialized books and attended discussions with experts in mathematics and astronomy.

Dias left Lisbon on April 11, 1601, embarking on the galleon *Santiago* and landing in Goa, where he stayed until 1604.[10] This must have given him plenty of opportunity to meet Rubino, who reached the Indian city in September 1602 and immediately began to teach mathematics.[11] Nevertheless, it is likely that Dias continued his scientific studies during his six years at the Macau College between 1604 and 1610. The college, where he taught philosophy and theology, focused particularly on science, although Sacrobosco's *Sphaera* was always used as a reference to help students understand Aristotle's *De caelo*.[12] In any event, Dias took advantage of the presence of Sabatino De Ursis, a scholar from Puglia well versed in mathematics and astronomy, and later one of Matteo Ricci's closest collaborators in Beijing.[13] De Ursis had left Lisbon the same day as Rubino—March 25, 1602—but on a different ship, and reached Macau the following year, staying there until 1607.[14]

This means that Dias had the chance to study with a mathematician who had the appropriate skills to introduce him to the more technical sides of astronomy. The fact that he exploited the opportunity offered to him by the Macau College—where he would also cross paths with the Jesuit Giulio Aleni from Brescia, who arrived in 1610, had a solid background in astronomy, and was in touch with Magini—is also confirmed by the fact that in 1612 he and De Ursis were suggested for the Chinese translation of scientific books brought by European missionaries.[15] When he left Macau in 1610 for the Chinese interior, Dias already had an astronomical background that went beyond the basics of cosmography he had learned in rudimentary courses.

After stopping in Shaozhou, which he was forced to leave because of ongoing hostilities toward Jesuits, Dias arrived in Beijing in May 1613; De Ursis was already there. It was during this time that he started writing the *Tian Wen Lüe*, finishing it the following year in the nearby residence of Zhalan. We learn this from a Jesuit, probably De Ursis, who said he had spent twenty days in Zhalan with Dias, when the latter "finished his book of mathematics tien ven lio."[16] The text was published in Beijing in 1615 and required the collaboration of a number of scholars. Based on information in the *editio princeps* but missing from later editions, we know that two other Jesuits, João de Rocha and Pedro Ribeiro, as well as nine Chinese converts, helped the author proofread the work in Chinese.[17]

• •

The *Tian Wen Lüe* is an introduction to Western science, written in the form of a dialogue between a Chinese man asking questions and a European (that is, Dias) who answers them. But what stands out is its focus on the aspects that would most interest its readers, establishing an approach that, within just decades, became the foundation for a profitable exchange between the Jesuits and Chinese scholars.[18] Therefore, by recapping European cosmographic and astronomical knowledge, and taking up the tradition of Sacrobosco's *Sphaera*, Dias emphasized subjects both sides could discuss: solar and lunar eclipses, when the Sun rises and sets, the cycle of the seasons, and problems connected with the use of different calendars.

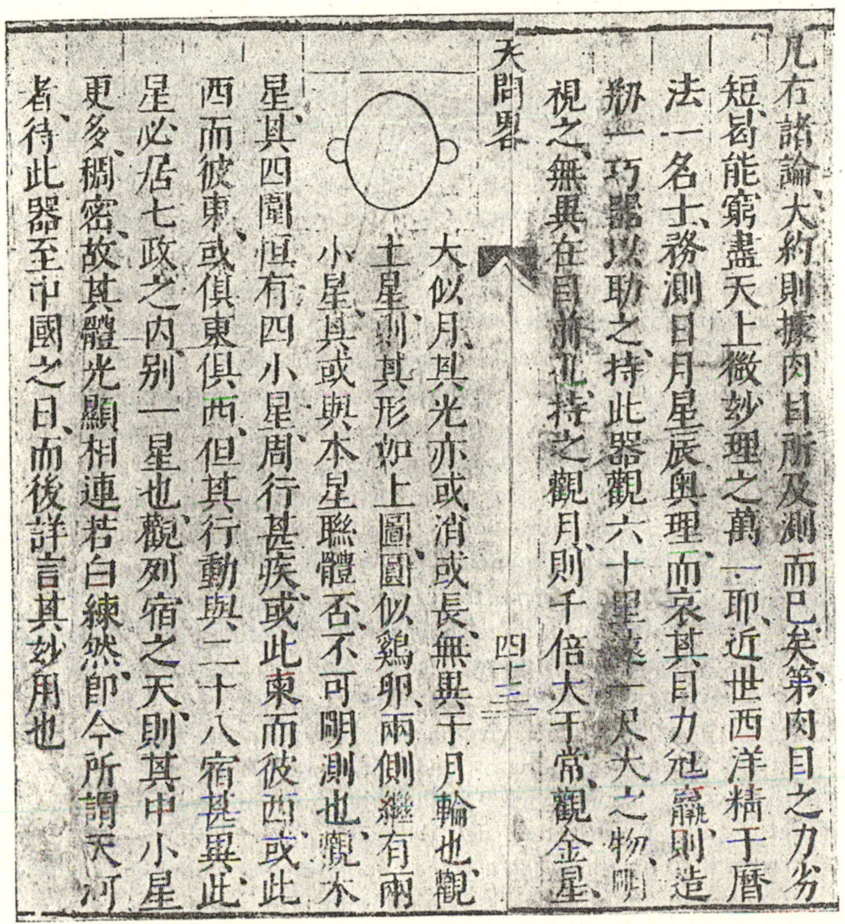

凡右諸論大約則據肉目所及測而已矣第肉目之力劣
短曷能窮盡天上微渺理之萬一耶近世西洋精于曆
法一名士務測日月星辰奧理而哀其目力尫羸則造
創一巧器以助之持此器觀六十里遠一尺大之物明
視之無異在目前也持之觀月則千倍大于常觀金星

天問畧

四十三

大似月其光亦或消或長無異于月輪也觀
土星則其形如上圖圓似鷄卵兩側繼有兩
小星其或與本星聯體否不可明測也觀木
星其四圍恒有四小星周行甚疾或此東而彼西或此
西而彼東或俱東俱西但其行動與二十八宿甚異此
星必居七政之內別一星也觀列宿之天則其中小星
更多稠密故其體光顯相連若白練然即今所謂天河
者待此器至中國之日而後詳言其妙用也

Figure 25. Manuel Dias, *Tian Wen Lüe*, 1615, the first Chinese text on Galileo's telescopic discoveries

In fact, in the part of the book devoted to the changing length of day and night, he included a table with variations calculated for all the provinces of China.[19]

It was here, in this presentation, that Dias decided to add a report on Galileo's telescopic discoveries. This was a *coup de théâtre,* as the subject had nothing to do with anything discussed previously. Yet it was the topic of his conclusions. Taking up just one page—the last page—it was embellished with a drawing of Saturn "with two little stars on its sides" (Figure 25).

> Most of the things explained so far are based on observations with the naked eye. But the vision of the naked eye is short. How can it measure the least part of the exceedingly small and admirable points of the firmament? Recently a famous Western scholar, well versed in astronomy, has undertaken to observe the mysteries of the Sun, the Moon and the stars. But grieved at the weakness of his eyes, he made a marvelous instrument to aid them. With this instrument, an object of the size of one *ch'ih*, placed at a distance of sixty *li*, seems to be right in front of one's eyes. Observed with this instrument, the Moon appears a thousand times larger. Venus looks as big as the Moon; its light waxes and wanes like that of the disk of the Moon. Saturn resembles the figure annexed here, like a hen's egg, with two small stars, one on each side, but whether they adhere to it or not is not exactly known. Jupiter appears to be attended constantly by four small stars moving around it very quickly; at times some of them are to its east, some to its west; and other times some are to its west, some to its east. Or all may be to its east, or all to its west. But their motion is quite different from that [of the stars] of the twenty-eight constellations; these stars are certainly part of the seven planets and are not [fixed] stars. With this instrument, the heavens of the constellations reveal a large number of small stars, close together, whose light is connected as if it formed a white chain: it is the one now referred to as the Celestial River [the Milky Way].[20]

What is striking about this passage is its wealth of information, not only about the observations reported in the *Sidereus*, but also later ones on the phases of Venus and the singular configuration of Saturn. Dias says nothing about how the telescope was built, however, simply because he had never seen one, as demonstrated by the final lines of his *Tian Wen Lüe*: "The day this instrument arrives in China, we will provide further details about its admirable use."[21] Moreover, he never mentioned the man who had made these discoveries, merely referring to him as a "famous Western scholar, well versed in astronomy," perhaps thinking that Chinese readers wouldn't care about his name in the first place. All these elements suggest that Dias reported this information secondhand, and this must certainly have been the case. Until the book was published in 1615, none of the Jesuits in Asia had had firsthand experience with the observations made by Galileo and then others in Europe.[22] One of the texts Dias could have used was the last edition of Clavius's commentary on Sacrobosco's *Sphaera*, published in 1611.[23] In it, Clavius not only confirmed Galileo's astronomical novelties but also pointed out that they seriously questioned the traditional cosmological order.

Clavius explained that, thanks to the telescope, many stars that had previously been inaccessible could now be seen clearly, "especially in the Pleiades, around the Cancer and Orion nebulae, and in the Milky Way." He then underscored his wonder that "with this instrument" the lunar surface looked "extraordinarily rough and uneven," adding:

> But on this matter, see the little book by Galileo Galilei, printed in Venice in 1610 and titled *Sidereus nuncius*, describing the various observations of the stars he had first made. One of the most important things seen with this instrument is that Venus receives its light from the Sun just as the Moon does, so that sometimes it appears to be more like a crescent, sometimes less, according to its distance from the Sun. In Rome I have observed these things in the presence of others more than once. Saturn has joined to it two smaller stars, one to the east and the other to the west. Lastly, Jupiter has four roving stars, which vary their places in a remarkable way among themselves and with respect to Jupiter, as Galileo diligently and carefully describes. Therefore, astronomers must consider how the celestial orbs may be arranged in order to save these phenomena.[24]

Clavius's commentary circulated widely within the Society of Jesus, but it is highly unlikely that this was Dias's source, and not only because the work still hadn't reached China when he was writing his *Tian Wen Lüe*. Dias didn't describe what had made the biggest impression on Clavius—the unevenness of the lunar surface—nor did he show any awareness of the dramatic cosmological consequences of Galileo's discoveries, and particularly the repercussions of the phases of Venus.[25] So how did he hear about all this?

As we have seen, Dias was in China when the first news of the telescopic discoveries reached India. We also know he had no direct contact with the mathematicians of the Collegio Romano. This means he was probably informed by a correspondent in Portugal or someone in Asia who had already heard about it.[26] We cannot reject the possibility that it was Rubino, whom he had met in Goa and who, in the aforementioned letter of November 2, 1612, wrote to Grienberger: "Perhaps I will go to China, as calls from there are insistent."[27] In any case, by the time Dias learned about the spectacular observations, the *Tian Wen Lüe* was almost finished. It was the very latest news, but he certainly wouldn't have wanted to omit it from a book presenting an overview of Western astronomical knowledge to Chinese readers.

• •

This is how Galileo made his debut in the Celestial Empire, although it would be a number of years before the Chinese learned the identity of that "famous Western scholar." It was in 1640 that the German Jesuit Johann Adam Schall von Bell, who arrived in Beijing in 1623, published his *Lifa xi zhuan* (History of Western Astronomers). Schall had met Galileo in person during the famous meeting organized in his honor by the Collegio Romano on May 13, 1611, and in 1626 he published a treatise on the telescope *(Yuanjing shuo)*, in which he not only provided a complete description of the instrument but also extensively discussed the celestial novelties, illustrated with a number of figures.[28] But Schall did not transliterate Galileo's name into Chinese, nor those of the most important ancient and modern astronomers, until his work of 1640. "The telescope was invented following the death of Di-Gŭ [Tycho Brahe] and the smallest stars in the sky were made visible through it. Thirty years ago Chia-li-lé-o drew a new map of the sky and announced what no astronomer had discovered in several thousand years. . . . So it was known that Jupiter is surrounded by four small stars revolving around it very rapidly; it was known that Saturn is surrounded by two other small stars, that Venus has phases, etc., all of which unheard of until then."[29]

Two years later, on January 8, 1642, Chia-li-lé-o died under house arrest at Arcetri. He was buried privately in Santa Croce, in a tiny room near the bell tower, and his tomb would remain hidden there for nearly a century.

Epilogue

The season that commenced on March 13, 1610, would prove to be far too short. As the celestial discoveries reached the ports of India and China and circulated in the Orient, in Europe the first steps were being taken to hinder interpretations of the phenomena glimpsed through the telescope, though not the instrument itself. Condemnation of Copernicus arrived swiftly—on March 5, 1616—and immediately weighed heavily on the cosmological implications Galileo had brought to the forefront. Exactly six years after publication of the *Sidereus* and just three years after Galileo's equally scandalous *Letters on Sunspots*, the Roman Congregation of the Index of Forbidden Books suspended Copernicus's *De revolutionibus orbium coelestium*, establishing an initial dividing line between the true and false constitution of the universe, between conclusions that could licitly be drawn thanks to telescopic observations and those that were banned, meaning that they could not be supported or taught as truthful. The mobility of the Earth was one of the latter.

Everything would have gone smoothly if heliocentrism had been accepted purely as a mathematical theory. After all, Bellarmine had been quite explicit on this point. On April 12, 1615, he spelled it out to the Carmelite friar Paolo Antonio Foscarini, who had published a troubling pamphlet on Copernican theology, and in the vernacular to boot. Foscarini's text was promptly banned. Wrote Bellarmine:

> Very Reverend Father, first I would like to say that you and Mr. Galileo are wise to speak *ex suppositione* and not in absolute terms, as I have always believed Copernicus did. Because so saying, supposing that the Earth moves and the Sun stands, still saves all appearances better than by positing eccentrics and epicycles, it is very well said and poses no danger; and this suffices for the mathematician. But wishing to assert that the Sun is truly at the center of the universe and turns on itself without moving from east to west, and that the Earth is in the third heaven and revolves around

> the Sun at great speed is a very dangerous thing that not only runs the risk of irritating all scholastic philosophers and theologians, but will also harm the Holy Faith by rendering the Holy Scripture false.[1]

The idea of the centrality of the Sun and mobility of the Earth had been considered so outrageous in the past that the ranks of the Church felt little cause for concern. But now, following the first—albeit partial—confirmations prompted by the discovery of the phases of Venus and sunspots, there was a serious risk that this would gain growing approval, the outcomes of which were hard to foresee. Embracing such concepts meant "[harming] the Holy Faith by rendering the Holy Scripture false." The cardinal's words were irrevocable, suggesting that Galileo act like a mathematician—meaning that he should speak *ex suppositione*—and not like a philosopher. In his letter to Foscarini, aimed at warning Galileo, Bellarmine continued, "Now consider, with your prudence, whether the Church *can tolerate* giving the Scripture a meaning that goes against the Holy Fathers, and all the Greek and Latin commentators."[2] The danger lay not only in teaching and transmitting theology as "the queen of sciences" but in the very idea of order and civilization. As John Donne had warned, the world and its system of ethical, social, and political relationships risked disintegration and destruction.

Bellarmine was well aware of the intentions of Galileo and the "sect" of *galileisti* that continued to grow around him in Florence day by day. Just two months before the cardinal's letter to Foscarini, the Dominican Niccolò Lorini had written to another authoritative member of the Holy Office, Cardinal Paolo Camillo Sfondrati, in a rather alarmed tone to report what was happening in the Tuscan city. All it would take to panic Bellarmine and the entire College of Cardinals was circulation of just one of those intentions, starting with what was probably the most shocking passage from "a text [Galileo's letter to Castelli of December 21, 1613] that everyone here has," asserting that the Bible is in "last place" when one is dealing with the knowledge of natural phenomena.[3] Then there were the closing words of the third and last of the *Letters on Sunspots*, in which the mathematician and philosopher of the ultra-Catholic Grand Duke of Tuscany unambiguously celebrated the truth of the "great Copernican system, to whose universal revelation we see such favorable breezes and bright escorts directing us, that we now have little to fear from darkness and cross-winds."[4]

The telescopic discoveries were being ferried back within rigid cosmological boundaries rendered insurmountable by specific theological precepts. The condemnation of March 5 was the first Inquisitorial censorship to be issued in a strictly astronomical ambit since the Council of Trent. It came more than seventy years after Copernicus's publication, a work that had been discussed and commented on throughout the sixteenth century, but nearly always among experts in the field and within the narrow confines that astronomers and mathematicians had established for themselves.[5] But it also came just six years after the astonishing discoveries the "celestial *occhiale*" had allowed everyone to see and which, as a direct result, could potentially wreak havoc. So they had to be brought back to normal and subjected to the strictest control. If, as the Jesuit Antonio Possevino had observed only a few years earlier, the world is "a great book" printed by God, in which the perfect concordance between religion and knowledge shines from every point, then by refuting the incorruptibility of the heavens and effectively moving toward a heliocentric universe, Galileo's discoveries made it possible to look at that book in a completely different and unprecedented way, overturning the entire traditional architecture of the celestial vault.[6]

According to the anti-Copernican decree of March 5, "This Holy Congregation has also learned about the spreading and acceptance by many of the false Pythagorean doctrine, altogether contrary to the Holy Scripture, that the Earth moves and the Sun is motionless, which is also taught by Nicholaus Copernicus's *On the Revolutions of the Heavenly Spheres* and by Diego de Zuniga's *On Job*."[7] These words give us enormous insight into just how oppressive the climate had become even on the home turf of astronomers and mathematicians. Reading between the lines, it becomes clear that the "looming presence" here was obviously Galileo, guilty of creating turmoil between religious belief and the knowledge of nature, and bringing total chaos to the heavens, which had always been considered untouchable, pure, and perfect, and where hosts of angels "live happily, contemplating God face-to-face."[8]

• •

After 1610 the celestial vault was no longer that exquisitely crafted palace in which men had always believed. It had become something that changes, deteriorates, and is transformed, just like everything on Earth. One might object that, in the past, phenomena such as new stars and comets had also called into question the image of an incorruptible and unchanging sky. But these were rare events explainable by terming them miracles that the human mind cannot grasp and that therefore maintained all their magical religious power. Instead, every day the telescope showed the whole world a sky that seemed more and more complicated and disorderly; the device risked being viewed as a sacrilegious instrument and desecrator of the ancient firmament. During this period, many were convinced of this and had no problem publishing their thoughts, such as the Benedictine monk Angelo Grillo, who wrote,

> We cannot deny that we can call brazen a glass daring to penetrate the bowels of the sky and the stars, and prying to discover if the Moon has stone disease [calculi], that is to say, if there are mountains and valleys inside her; and, in short, removing the veil of distance and discovering her imperfections, as if to say that her surface is not as clean, as smooth, and as flat as it looks, but is instead rugged, cavernous, and uneven, along with a thousand other blemishes; and shamelessly giving the lie to ancient astrology, manifesting new stars and new aspects not only to the intellect, but to sensibility itself. And finally making it conclude that this spyglass has become the school and teacher of the eye, and the keenest spy of heaven and earth.[9]

Grillo underscored all the audacity of this new instrument with a shudder: "In this sense, every man can be called a *cannonista*."[10] Spare as they are, these words portend troubling scenarios. The idea that "Galileo's large spyglass" would offer anyone and everyone the chance to see a new sky could have unexpected effects and, like a river in full spate, destroy every idea of order and hierarchy of fields of knowledge. After all, in his *Dissertatio cum Nuncio sidereo,* hadn't Rudolf's imperial astronomer been the one to point out Galileo's great debt to Giordano Bruno? Hadn't the Copernican Kepler been the one who recalled how much Bruno's "horrid" infinitistic and atomistic philosophy might benefit from the new telescopic discoveries? These reflections did not fall on deaf ears. Just the opposite. They were transformed into very serious accusations that were immediately taken up and used against Galileo, as we can see from this passage, which is long but too eloquent not to read in its entirety.

> Democritus thought that this immensity we see was empty and infinitely vast; moreover, it seems that perpetually infinite corpuscles, ingenerable and incorruptible as they constitute the first principles of all things, move about in it perpetually. They are minimal in size and vary in shape, indivisible and unalterable, moving solely through their own impetus; and by aggregating in multiple encounters with various orders, figures, and positions, and not through some intention nor for a given purpose, but randomly and fortuitously, they supposedly produce not only one but many worlds. . . . And since these worlds are generated and decay, infinite in their succession, as Democritus states, there exist many terrestrial globes corresponding to these many worlds. So why wouldn't the Moon be one of these globes, or the terrestrial globe of its own world, with mountains, valleys, seas, lakes, and everything its inhabitants need, like our world, as Galileo's telescope shows us? A few years ago a modern resurrected this claim from the underworld, asserting that each planet is a sort of terrestrial globe similar to ours, and that other skies and other spheres turn around each planet; and now Galileo's new observation about the four satellites of Jupiter and two of Saturn has been interpreted favorably by some, suspiciously by others, as confirmation of that opinion, as Kepler notes, so that one might think that the orbiting of those little planets is at the service of the life of the inhabitants of Jupiter and Saturn.[11]

These words could not have been more trenchant or venomous. They were written by Giulio Cesare Lagalla, physician and professor of philosophy at La Sapienza, in a text published in Venice in early 1612 (and, in a strange twist of fate, by the same printer that had published the *Sidereus nuncius*). Lagalla closely linked atomism, the infiniteness of the universe, inhabited worlds, and celestial discoveries: in other words, Democritus, Giordano Bruno, and the man he considered their latest disciple, who with his discoveries was resuscitating their noxious philosophies from the depths of the underworld. As if that weren't bad enough, other details—more theological in nature—were added a few years later to flesh out this already bleak portrait. Wrote Tommaso Campanella:

> Based on Galileo's opinion, it ensues that there is a multiplicity of worlds, lands, and seas, as Mohammed retains, and also men who inhabit them, if it is true that among the stars there are four elements as in our world; indeed, if each star is composed of all four elements, then each one is unquestionably a world. But since the Scripture talks about only one world, and only one human race, he seems to sustain opinions that go against them. Not to mention the heresy that would stem from this, that Christ also died on the other stars for the salvation of men.[12]

That is a page from the *Apologia pro Galileo,* written in 1616, before Copernicus's work was placed on the Index, and the passage is part of a list of eleven arguments "contra Galileum" that Campanella intended to confute. He would naturally go about it in his own way, starting with the conviction that the sacred philosophy of Moses transmitted to Pythagoras, after long centuries of oppression under the pagan Aristotle, would finally be liberated thanks to the telescopic discoveries of the "Christian Galileo."[13] But here we are not that interested in the frankly bizarre ideas of the friar and prophet from Calabria (whom Galileo completely ignored). What is striking, however, is that the perception of Galileo's public image, so well documented in Lagalla's work and Campanella's testimony, does not coincide with anything he had written, nor is it corroborated by his actual behavior. It was built on a combination of often vague and imprecise opinions that were based chiefly on his adversaries' fears, unease, and prejudices, as well as the philosophical and theological flights of fancy of admirers who could not have been more dangerous or unreliable. This blend of images, repeated *ad libitum,* was magnified, and ultimately came to be considered true and real. And it worked against him far more deeply than we tend to think.

• •

As the Italian author Primo Levi has taught us, books sometimes behave like nomadic animals. This fate did not elude the *Sidereus,* and the texts we have just discussed prove that fate couldn't care less about what its author actually wrote. Likewise, recognition of the visual power of the new instrument immediately took on meanings that went far beyond the science of celestial bodies. The image of the "long-sighted spectacle" *(occhiale dalla vista lunga)* was too enticing to remain relegated to an astronomical announcement.

All the discussions, debates, and ideas that the "celestial spyglass" spawned in those years offer undeniable proof. The almost instantaneous reversal of perspective, shifting from heaven to earth, from astronomy to the quintessentially human sphere of politics, is also fully and rightly a part of the history of the telescope. It is a consequence we might also define as necessary, initially merely whispered but then emerging clearly and powerfully, as we can see from this singular document by the Bohemian nobleman Wentzel von Meroschwa, published just ten years after the *Sidereus:* "Now, just as the new Mathematicians have discovered new stars in the firmament and new spots on the Sun through their spyglasses, so too the new Politicians have their lenses and their optics [their points of observation and their perspectives], whence they can see other parts of the division to add to those of antiquity."[14] The success of the spyglass was transformed into an intriguing political metaphor, an image that, far better than a thousand treatises, could convey the function of a new political viewpoint as a way to unmask tyranny and *raison d'état.* Galileo certainly did not foresee or want any of this, above all in its most radical and heterodox version, such as Meroschwa's.

Printed simultaneously in Latin, French, and German in 1620, the *Epistola* by Meroschwa (possibly a pseudonym) is a formidable Protestant work that strongly criticized the serious political errors made by the litigious Reformation coalition, confirmed shortly thereafter when the Bohemian army and its allies were routed in the Battle of White Mountain. Like the modern mathematicians who discovered things in the heavens that had never been seen before, modern politicians had to establish a new theory of politics: this is the slant of the metaphor. Like astronomy, politics, too, had to invent its own "lenses" and specific viewpoints, unknown to the ancient world, through which to examine reality. Through telescopic observation the astronomer discovered the true face of the celestial world, and through the art of dissimulation the politician deciphered the most recondite human behavior. In order to do so, to reveal the different practices of power, the latter needed his own *specularia*, his own "optical tubes."

Discovery of the secrets of nature through the "celestial spyglass" had paved the way for other impenetrable secrets. The *arcana imperii* ran the risk of being laid bare by a science of politics that had found its archetype in telescopic astronomy. So it comes as no surprise that in the tragic conflict between Catholics and Protestants, the metaphor of the spyglass was used as a cultural and political weapon by both warring sides. The year after Copernicus was condemned, a German Jesuit, Adam Tanner, published a book of over a thousand pages entitled *Dioptra Fidei* (The Spyglass of the Faith), in which he stoutly defended Catholic principles and dogma.[15] Moreover, just a few years earlier the famous Catholic theologian Leonardus Lessius had paid his own tribute to the *Sidereus* and the spyglass. Like Bellarmine, he naturally took a politically correct stance that neutralized the scope of the new science in the realm of knowledge, dwelling on unfathomable divine knowledge and power, on the one hand, and the absolute and inherent weakness of human knowledge, on the other.

Lessius wrote: "Through a spyglass [*fistulam dioptricam*] recently invented by a Dutchman, astonishing phenomena were discovered in the heavens a short time ago, things entirely unknown to our predecessors. . . . I have often observed all these phenomena with the spyglass, to my greatest admiration of divine knowledge and power, which has regulated the motion of the celestial bodies with an artifice so marvelous that the human mind cannot understand it in any way."[16]

This work was published in Antwerp in 1613. As we know, Bellarmine's letter to Foscarini was dated April 1615. Meanwhile, in Milan, Federico Borromeo was in the throes of his telescopic passion, determined to obtain and build ever better spyglasses. At the same time, to satisfy the requests of his confreres in Perugia, Grillo started to look for two good lenses, though he could not refrain from warning them about the possible inauspicious consequences of such great curiosity: "Antiquity tells us that the giants of Phlegra placed mountains on mountains to raid the stars; our adding lenses on lenses to find mountains in the sky, and pry into the secrets of the Moon and the stars, is just as reckless an attack as what was attempted with solid and excellent machines, and [we are doing so] with tiny and most fragile lenses."[17]

Those "tiny and most fragile lenses" were becoming the guardians of the secrets of nature, like the instruments used by royal secretaries, the loyal custodians of the intrigues of politics and states. Yet unlike the latter, hidden from subjects and adversaries, the former were available to anyone who wanted them, putting them in a position to inspire new ideas or restore ancient and forgotten opinions. In Grillo's words:

> With Galileo's spyglasses, *which have now become the secretaries of the Moon and the stars*, we have discovered new aspects and new stars; and because here it is not so much the intellect at play but the senses, a new school of slippery curiosity has been established; and the opinion of Copernicus has been raised, that the Earth moves like other spheres and that the Sun stands still in the middle of the universe to illuminate them; and that the Earth is to the Moon as the Moon is to the Earth, and that they illuminate each other. This is not far from the opinion of Pythagoras, who (if I recall) thought the Earth was a star. And so the ages renew the ages and opinions, and the years and human intelligence are more and more consumed by going around and around these circles.[18]

As we can see, Bellarmine's caution and preoccupation are also evident in the writings of other theologians, philosophers, and literati. At the same time, procuring good spyglasses was one of the main concerns of the Jesuit mathematicians and astronomers of the day. Everyone had the same goal: building better and better instruments, and becoming excellent observers of the sky in order to limit and oppose the change of mind-set those astonishing discoveries seemed to portend.

The telescopic passion that spread to every corner of the world did not turn into an overall embracement of the Galilean and Copernican cosmology (and not only in the Church). It was akin to what had happened in the first few decades after Columbus discovered America, which did not lead to the "jettisoning of many traditional preconceptions and inherited ideas."[19] In a word, the telescope could be accepted, but not a new sky. People were not willing to embrace a new sky if this meant questioning their beliefs in a traditional cosmos and literal interpretation of the Holy Scripture, nor did they want to contemplate the possibility of another world that this would necessarily entail. Just as Cortés described Aztec temples like mosques, many turned the telescope skyward, but remained prisoners of their image of the world.[20]

• •

The time has come to bring this book to a close, and in doing so, we would like to expand on the thoughts of the French author Marguerite Yourcenar.[21] *Hamlet* was the first tragedy of doubt, *Faust* that of pride and intellectual arrogance, and the *Sidereus nuncius* the first modern work to overturn the order of the heavens (and more).

Bringing the spyglass to the world stage meant shifting that world and turning it upside down. If the Moon viewed through the telescope looked as mountainous as Earth, if Jupiter had its own satellites just like Earth, if the Sun, with its spots, was as corruptible as Earth, then this made it all the more likely that Earth was a planet just like all the others and that, like Venus, it circled the Sun. So the distinction between the sublunar and celestial worlds was merely a figment of the imagination, because matter had to be a single uniform substance spread across the whole universe. But if that was the case, didn't this make the sky a corruptible place with bodies composed of the selfsame terrestrial matter? If the sky was subject to generation and corruption, could it continue to be the home of angels and saints? And could Hell, inhabited by demons and the damned, be at the center of the Earth if our planet, in turn, was in the sky and no longer the region furthest from heaven?

Galileo's astronomy and new philosophy raised huge theological problems, in which the telescope proved to be a decisive ally. This is the main reason that the season of freedom proved to be so short. And the journey to regain that freedom would be a long and difficult one.

Notes

References

Credits

Acknowledgments

Index

Notes

Abbreviations

Archives and Libraries

ABIB Archivio Borromeo, Isola Bella
ANTT Arquivo Nacional da Torre do Tombo, Lisbon
APUG Archivio della Pontificia Università Gregoriana, Rome
ASB Archivio di Stato, Bologna
ASDM Archivio Storico Diocesano, Milan
ASEPD Archivio Società di Esecutori di Pie Disposizioni, Siena
ASF Archivio di Stato, Florence
ASV Archivio Segreto Vaticano, Vatican City
ASVe Archivio di Stato, Venice
BAM Biblioteca Ambrosiana, Milan
BIC Bibliothèque d'Inguimbert, Carpentras
BL British Library, London
BNCF Biblioteca Nazionale Centrale, Florence
BNCR Biblioteca Nazionale Centrale, Rome
BNF Bibliothèque Nationale de France, Paris
BRF Biblioteca Riccardiana, Florence
BUB Biblioteca Universitaria, Bologna
BVR Biblioteca Vallicelliana, Rome
PH Petworth House, Sussex
SBD Studienbibliothek, Dillingen
UBG Universitätbibliothek, Graz

Printed Works

DBI *Dizionario biografico degli italiani.* Rome: Istituto della Enciclopedia, 1960–.
DNS (=*Dissertatio cum Nuncio sidereo*) *Kepler's Conversation with Galileo's Sidereal Messenger.* First complete translation with an

introduction and notes by Edward Rosen. New York: Johnson Reprints, 1965.

KGW Johannes Kepler, *Gesammelte Werke*, ed. Walther Von Dyck, Max Caspar, and Franz Hammer. Munich: Beck, 1937–.

OG Galileo Galilei, *Opere*, national edition, ed. Antonio Favaro. Florence: Barbèra, 1890–1909, reprinted 1968.

SN Galileo Galilei, *Sidereus nuncius or the Sidereal Messenger*. Translated with introduction, conclusion, and notes by Albert Van Helden. Chicago: University of Chicago Press, 1989.

1. From the Low Countries

1. We were stimulated to examine this painting and its relationship with the advent of the first Dutch spyglasses by the brief reference in Inge Keil, *Augustanus Opticus. Johann Wiesel (1583–1662) und 200 Jahre optisches Handwerk in Augsburg* (Berlin: Akademie Verlag, 2000), 267–268. See also Paolo Molaro and Pierluigi Selvelli, "The Mystery of the Telescopes in Jan Brueghel the Elder's Paintings," *Memorie della Società Astronomica Italiana* 75 (2008): 282–283.
2. On the reign of Albert and Isabella, and their political and cultural plans, see Jonathan I. Israel, *Conflicts of Empires: Spain, the Low Countries and the Struggle for World Supremacy, 1585–1713* (London: Hambledon Press, 1997), 1–21; W. Thomas, "Andromeda Unbound: The Reign of Albert and Isabella in the Southern Netherlands, 1598–1621," in *Albert & Isabella, 1598–1621: Essays*, ed. Werner Thomas and Luc Duerloo (Turnhout: Brepols, 1998), 1–14.
3. See Hugh Trevor-Roper, *Princes and Artists: Patronage and Ideology at Four Habsburg Courts, 1517–1633* (London: Thames and Hudson, 1976), 116–152.
4. See *Albert & Isabella, 1598–1621: Catalogue*, ed. Werner Thomas and Luc Duerloo (Turnhout: Brepols, 1998), 178–185.
5. This image of power is eloquently illustrated by a series of paintings portraying the archdukes near the buildings symbolizing their sovereignty. It is quite evident in the two portraits by Rubens and Brueghel, now at the Prado in Madrid, showing Albert in front of the castle of Tervuren and Isabella in front of Mariemont (ibid., 149–157).
6. See Klaus Ertz, *Jan Brueghel der Ältere (1568–1625). Die Gemälde mit kritischem Oeuvrekatalog* (Cologne: DuMont, 1979), 163.
7. See Joëlle Demeester, "Le domaine de Mariemont sous Albert et Isabelle (1598–1621)," *Annales du Cercle archéologique de Mons* 71 (1978–1981): 210, 277.

8. Trevor-Roper, *Princes and Artists*, 140; Ertz, *Jan Brueghel der Ältere*, 157–163.
9. Demeester, "Le domaine de Mariemont," 258–262.
10. See Matías Díaz Padrón, *Museo del Prado: Catalogo de Pinturas I, Escuela Flamenca Siglo XVII* (Madrid: Museo del Prado y Patronato Nacional de Museos, 1975), 52–53.
11. On the role of Maurice of Nassau in the development of the telescope, see Rienk Vermij's recent update, "The Telescope at the Court of the Stadtholder Maurits," in *The Origins of the Telescope*, ed. Albert Van Helden, Sven Dupré, Rob Van Gent, and Huib Zuidervaart (Amsterdam: Knaw Press, 2010), 73–92.
12. The original text of the letter, like most of the sources regarding the invention of the telescope, has been published with an English translation in Albert Van Helden, *The Invention of the Telescope* (Philadelphia: American Philosophical Society, 1977), 36.
13. See Dirk Van der Cruysse, *Louis XIV et le Siam* (Paris: Fayard, 1991), 53–69.
14. See Jan Julius Lodewijk Duyvendak, "The First Siamese Embassy to Holland," *T'oung Pao* 32 (1936): 285–292.
15. For a detailed reconstruction of the different phases of the conflict, see Jonathan I. Israel, *The Dutch Republic: Its Rise, Greatness, and Fall, 1477–1806* (Oxford: Oxford University Press, 1995), 399–420.
16. *Ambassades du Roy de Siam envoyé à l'Excellence du Prince Maurice*, in Van Helden, *The Invention of the Telescope*, 41.
17. See Vincent Ilardi, *Renaissance Vision from Spectacles to Telescopes* (Philadelphia: American Philosophical Society, 2007), 22–23, 51–55, 64–73, 136, 152.
18. See Sven Dupré, "Galileo, the Telescope, and the Science of Optics in the Sixteenth Century: A Case Study of Instrumental Practice in Art and Science," Ph.D. diss., University of Ghent, 2002, 234–284; Jurgis Baltrušaitis, *Le miroir: Révélations, science-fiction et fallacies* (Paris: Éditions du Seuil 1978).
19. See Giovan Battista Della Porta, *Magia naturalis libri XX* (Naples: Horatium Salvianum, 1589), 269. On the vast influence of *Magia naturalis* in Europe, see Laura Balbiani, "La ricezione della Magia naturalis di Giovan Battista Della Porta. Cultura e scienza dall'Italia all'Europa," *Bruniana & Campanelliana* 5 (1999): 277–303.
20. Ilardi, *Renaissance Vision*, 224–235.
21. See Rolf Willach, *The Long Route of the Telescope* (Philadelphia: American Philosophical Society, 2008), 94–95.
22. See Chapter 3.
23. *Ambassades du Roy de Siam*, 41.

24. Pierre Taisan de L'Estoile, *Journal du régne de Henri IV Roi de France et de Navarre* (The Hague, 1761), 3:513–514, now in Van Helden, *The Invention of the Telescope*, 44.
25. OG, 10:252 (our italics): Della Porta to Cesi, August 28, 1609.
26. See Johannes Georgius Walchius, *Decas fabularum humani generis* (Strasbourg: Lazar Zetzener, 1609), 249–250. See also Eileen Reeves, *Galileo's Glassworks: The Telescope and the Mirror* (Cambridge, MA: Harvard University Press, 2008), 9–10.
27. SN, 37; OG, 3:61. On the fact that telescopes of this size could magnify objects up to a maximum of three times, see Van Helden, *The Invention of the Telescope*, 11–12.
28. See "Minutes of the States-General, 2 October 1608," in Van Helden, *The Invention of the Telescope*, 36.
29. Ibid., 38.
30. See Jan M. Baart, "Una vetreria di tradizione italiana ad Amsterdam," in *Archeologia e storia della produzione del vetro preindustriale*, ed. Marja Mendera (Florence: All'Insegna del Giglio, 1991), 423–437.
31. See "Letter from the Committee of Councillors of Zeeland in Middelburg to the Zeeland delegation at the States-General in The Hague, 14 October 1608," in Van Helden, *The Invention of the Telescope*, 38–39. Regarding the identity of the "young man" with the Middelburg spectacle maker Zacharias Janssen, see Cornelis De Waard, *De uitvinding der verrekijkers: eene bijdrage tot de beschavingsgeschiedenis* (Rotterdam: W. L. & J. Brusse, 1906), 172–173.
32. See "Letter from Jacob Metius to the States-General, ca. 15 October 1608," in Van Helden, *The Invention of the Telescope*, 39–40.
33. See Simon Mayr, *Mundus Iovialis* (1614), in Van Helden, *The Invention of the Telescope*, 47. According to De Waard (see *De uitvinding der verrekijkers*, 168–170), the Dutchman at the Frankfurt fair was Zacharias Janssen, who had returned to Middelburg around October 13. Discovering that Lipperhey had filed for an exclusive patent, he promptly went to his provincial authorities to inform them that he too had made such an instrument. See also Van Helden, *The Invention of the Telescope*, 21–22n.
34. See Isabelle Pantin, "La lunette astronomique: une invention en quête d'auteurs," in *Inventions et découvertes au temps de la Renaissance*, ed. Marie Thérèse Jones-Davies (Paris: Klincksieck, 1994), 161.
35. See "Minutes of the States-General, 17 October 1608," in Van Helden, *The Invention of the Telescope*, 40.
36. See "Minutes of the State-General, 15 December 1608," in Van Helden, *The Invention of the Telescope*, 42.

37. See Engel Sluiter, "The Telescope before Galileo," *Journal for the History of Astronomy* 28 (1997): 224.
38. See Thomas and Duerloo, *Albert & Isabella: Catalogue*, 178–185.
39. See Alberto Clerici, "Ragion di Stato e politica internazionale. Guido Bentivoglio e altri interpreti italiani della Tregua dei Dodici Anni (1609)," *Dimensioni e problemi della ricerca storica* 2 (2009): 198.
40. The letter, at the ASV (Fondo Borghese, II, 114, fol. 127*r*), was first published by Antonius Hubertus Leonardus Hensen, "De Verrekijkers van Prins Maurits en van Aartshertog Albertus," *Mededelingen van het Nederlandsh Historisch Instituut te Rome* 3 (1923): 203–204. More recently, it has been reproduced at the end of Sluiter, "The Telescope before Galileo," 231–232.
41. See De Waard, *De uitvinding der verrekijkers*, 230. See also Jean-Charles Houzeau, "Le téléscope à Bruxelles, au printemps de 1609," *Ciel et terre* 3 (1882): 25–28.
42. ASV, Fondo Borghese, II, 114, fol. 127*r*.
43. See Sluiter, "The Telescope before Galileo," 226.
44. ASV, Fondo Borghese, IV, 52, fols. 227*v*–228*r*.
45. On Mancini, see Silvia De Renzi and Donatella L. Sparti in DBI, 68, under the specific entry.
46. ASEPD, XIX, Eredità Mancini, 167, fol. 247*r*. This information was kindly provided by Michele Maccherini, to whom we are grateful.
47. See Chapter 6.
48. See Sluiter, "The Telescope before Galileo," 226.
49. See Pierre Jeannin to Henry IV, December 28, 1608, in *Les Négociations de Monsieur le President Jeannin* (Paris: Pierre Le Petit, 1656), 518–519.
50. See L'Estoile, *Mémoires-Journaux*, ed. G. Brunet et al. (Paris: Librairie des Bibliophiles, 1881), 9:164.
51. Ibid., 168.
52. See John J. Roche, "Harriot, Galileo, and Jupiter's Satellites," *Archives Internationales d'Histoire des Sciences*, 32 (1982): 10.
53. See Chapter 2.
54. Jeannin to Henry IV, December 28, 1608, in *Les Négociations de Monsieur le President Jeannin*, 519.
55. See L'Estoile, *Journal du régne de Henri IV*, and *Le Mercure Francois, ou la suite de l'histoire de la paix* (Paris, 1611), in Van Helden, *The Invention of the Telescope*, 44, 46.
56. See Girolamo Sirtori, *Telescopium, sive ars perficiendi* (1618), in Van Helden, *The Invention of the Telescope*, 48.
57. See OG, 10:250.

58. Ibid., 10:253.
59. See John W. Shirley, "Thomas Harriot's Lunar Observations," in *Science and History: Studies in Honor of Edward Rosen*, ed. Erna Hilfstein, Pawel Czartoryski, and Frank D. Grande (Wrocław: Polish Academy of Sciences Press, 1978), 283. On Harriot, see Chapter 7.

2. The Venetian Archipelago

1. Paolo Sarpi, *Lettere ai protestanti*, ed. Manlio Duilio Busnelli (Bari: Laterza, 1931), 2:15.
2. Sarpi to Groslot de L'Isle, January 6, 1609, in *Lettere ai protestanti*, 1:58.
3. See Libero Sosio, "Paolo Sarpi, un frate nella rivoluzione scientifica," in *Ripensando Paolo Sarpi*, ed. Corrado Pin (Venice: Ateneo Veneto, 2006), 186–187.
4. Sarpi to Groslot de L'Isle, in Sarpi, *Lettere ai protestanti*, 1:22.
5. Paolo Sarpi, *Lettere ai gallicani*, critical ed., introductory essay and notes by Boris Ulianich (Wiesbaden: Steiner, 1961), 179.
6. Sarpi to Badoer, March 30, 1609, in *Lettere ai gallicani*. Regarding Badoer, see Boris Ulianich in DBI, 5, under the specific entry.
7. Sarpi, *Lettere ai protestanti*, 1:16.
8. Ibid., 19.
9. Sarpi to Lollino, January 20, 1603, in Paolo Sarpi, *Lettere*, ed. Filippo Luigi Polidori (Florence: Barbèra, 1863), 1:10–11.
10. William Gilbert, *De Magnete* (London: Petrus Short, 1600), 6. See Libero Sosio, "Galileo Galilei e Paolo Sarpi," in *Galileo Galilei e la cultura veneziana* (Venice: Istituto veneto di scienze, lettere ed arti, 1995), 282–283.
11. Sarpi, *Lettere ai protestanti*, 2:45.
12. See Galileo, *Le Messager céleste*, ed. Isabelle Pantin (Paris: Les Belles Lettres, 1992), 59 n. 14.
13. See Mario Biagioli, "Did Galileo Copy the Telescope? A 'New' Letter by Paolo Sarpi," in *The Origins of the Telescope*, ed. Albert Van Helden, Sven Dupré, Rob Van Gent, and Huib Zuidervaart (Amsterdam: Knaw Press, 2010), 221–222.
14. See also Galileo, *Le Messager céleste*, x.
15. See Chapter 3.
16. Pignoria to Gualdo, August 1, 1609, in OG, 10:250.
17. Bartoli to Vinta, August 22, 1609, in OG, 10:250.
18. OG, 10:250.
19. OG, 10:255.
20. Galileo to Landucci, August 29, 1609, in OG, 10:253.

21. See Barbolani to Cosimo II, May 23, 1609, in ASF, Mediceo del Principato 3001, fol. 97*v*.
22. OG, 10:253.
23. Ibid.
24. Ibid.
25. OG, 10:254, 257.
26. Bartoli to Vinta, September 5, 1609, in OG, 10:257.
27. Gerolamo I Priuli, *Cronaca*, August 21–25, 1609, in OG, 19:587. Regarding identification of the author as Gerolamo I Priuli, see Gaetano Cozzi, "Galileo Galilei, Paolo Sarpi e la società veneziana," in *Paolo Sarpi tra Venezia e l'Europa* (Turin: Einaudi, 1979), 181 n. 99.
28. OG, 19:588.
29. Fulgenzio Micanzio, *Vita del Padre Paolo dell'Ordine de' Servi e theologo della Serenissima* (Leiden 1646), in Paolo Sarpi, *Istoria del Concilio Tridentino, seguita dalla Vita del padre Paolo di Fulgenzio Micanzio*, ed. Corrado Vivanti (Turin: Einaudi, 1974), 2:1372–1373.
30. This is the case of Raffaello Gualterotti, who on March 6, 1610, wrote to Galileo from Florence, but the letter bears Sarpi's address (OG, 10:286–287). See Cozzi, "Galileo Galilei, Paolo Sarpi e la societa veneziana," 183.
31. Pignoria to Gualdo, October 15, 1609, OG, 10:260.
32. Bartoli to Vinta, September 26, 1609, OG, 10:259–260.
33. OG, 10:260.
34. Bartoli to Vinta, October 24, 1609, OG 10:261.
35. In the two letters to Belisario Vinta, dated October 30 and November 20, and one to the *Riformatori* of the University of Padua, dated November 4, there is no mention of this. Regarding the letter of December 5, see Chapter 3.
36. On the uncertainty of who the recipient was, see Galileo, *Le Messager céleste*, xiii.
37. See Chapter 3.
38. Sarpi, *Lettere ai gallicani*, 73–74.
39. Ibid., 73.
40. Cozzi, "Galileo Galilei, Paolo Sarpi e la societa veneziana," 185.
41. See Paolo Sarpi, *Opere*, ed. Gaetano Cozzi and Luisa Cozzi (Milan: Ricciardi, 1969), 225.
42. Sarpi, *Lettere ai gallicani*, 240.
43. Ibid., 238.
44. Ibid., 79. See Eileen Reeves, "Kingdoms of Heavens: Galileo and Sarpi on the Celestial," *Representations* 105 (2009): 68.

45. Maurizio Torrini, "Il Rinascimento nell'orizzonte della nuova scienza," in *Nuovi maestri e antichi testi. Umanesimo e Rinascimento alle origini del pensiero moderno*, ed. Stefano Caroti and Vittoria Perrone Compagni (Florence: Olschki, 2012), 315–330.
46. Corrado Pin, "'Qui si vive con esempi, non con ragione': Paolo Sarpi e la committenza di stato nel dopo-Interdetto," in *Ripensando Paolo Sarpi*, ed. Corrado Pin (Venice: Ateneo Veneto, 2006), 357.
47. Barbolani to Grand Duke Cosimo II, May 2, 1609, in ASF, Mediceo del Principato 3001, fol. 77*r–v*.
48. Bartoli to Vinta, November 7, 1609, in ASF, Mediceo del Principato 3001, fol. 186*r*.
49. Manlio Duilio Busnelli, *Études sur Fra Paolo Sarpi: Et autres essais italiens et français* (Geneva: Slatkine, 1986), 174. See also Vittorio Frajese, *Sarpi scettico. Stato e Chiesa a Venezia tra Cinque e Seicento* (Bologna: Il Mulino, 1994), 263.
50. Paolo Savio, "Per l'epistolario di Paolo Sarpi," *Aevum* 10 (1936): 12.
51. See Ulianich in DBI, 5, under the specific entry.
52. Sarpi, *Lettere ai protestanti*, 2:98–99.
53. "We should rightly pray to God that he rid the world of such a plague": Sarpi to Leschassier, August 3, 1610, in Sarpi, *Lettere ai gallicani*, 89.
54. Sarpi, *Lettere ai gallicani*, 81.
55. Sarpi to Groslot de l'Isle, May 10, 1610, in Sarpi, *Lettere ai protestanti*, 1:122.
56. Letter of June 8, 1610, in Sarpi, *Lettere ai gallicani*, 84.
57. Leschassier to Sarpi, June 29, 1610, in Sarpi, *Lettere ai gallicani*, 241.
58. About Da Mula, see Luisa Cozzi and Gaetano Cozzi in DBI, 32, under the specific entry.
59. Gloriosi to Schreck, May 29, 1610, in OG, 10:363.
60. OG, 11:350.

3. Breaking News: Glass and Envelopes

1. See Chapter 2.
2. See Chapter 2.
3. See William Eamon, *Science and the Secrets of Nature: Books of Secrets in Medieval and Early Modern Culture* (Princeton, NJ: Princeton University Press, 1994), 238.
4. SN, 37; OG, 3:60.
5. Therefore, Biagioli's sensationalism surrounding this letter seems rather excessive (see his "Did Galileo Copy the Telescope? A 'New' Letter by Paolo Sarpi," in *The Origins of the Telescope*, ed. Albert Van Helden, Sven Dupré,

Rob Van Gent, and Huib Zuidervaart [Amsterdam: Knaw Press, 2010], 203–208), as is his interpretation, which sounds quite unilateral.

6. See OG, 6:258.
7. See Chapter 2.
8. For a useful, recent historiographical survey, see Sven Dupré, "Galileo, the Telescope, and the Science of Optics in the Sixteenth Century: A Case Study of Instrumental Practice in Art and Science," Ph.D. diss., University of Ghent, 2002, 6–10.
9. SN, 37–38; OG, 3:60–61.
10. In fact, as opposed to what Galileo asserts in the *Sidereus nuncius*, in his *Cronaca* Priuli mentions an instrument capable of magnifying objects up to nine times. See OG, 19:588.
11. See Vincent Ilardi, *Renaissance Vision from Spectacles to Telescopes* (Philadelphia: American Philosophical Society, 2007), 82–95.
12. See Albert Van Helden, "Galileo and the Telescope," in *The Origins of the Telescope*, ed. Albert Van Helden, Sven Dupré, Rob Van Gent, and Huib Zuidervaart (Amsterdam: Knaw Press, 2010), 187–188.
13. See Simon Mayr, *Mundus Iovialis* (1614), in Albert Van Helden, *The Invention of the Telescope* (Philadelphia: American Philosophical Society, 1977), 47–48.
14. See Chapter 1.
15. Sirtori, *Telescopium, sive ars perficiendi*, in Van Helden, *The Invention of the Telescope*, 50.
16. See Ilardi, *Renaissance Vision from Spectacles to Telescopes*, 145.
17. Sometimes Galileo also acted as an intermediary for other customers. In September 1602, for instance, Paolo Pozzobonelli, a former student in Padua, received glasses from him. Pozzobonelli thanked him for "taking the trouble to send me good ones and also the gift your lordship gave me concerning their price" (Pozzobonelli to Galileo, September 12, 1602, in OG, 10:93).
18. See Ilardi, *Renaissance Vision from Spectacles to Telescopes*, 183–184.
19. OG, 10:259.
20. OG, 10:280. For evidence showing that Galileo was in Florence between the end of September and the end of October 1609, see Antonio Favaro, *Galileo Galilei e lo Studio di Padova*, 2 vols. (Florence: Le Monnier 1883; Padua: Antenore, 1966) 1:287 n. 5.
21. See Ammannati to Piersanti, November 21, 1609, OG, 10:268.
22. OG, 10:279.
23. See Favaro, *Galileo Galilei e lo Studio di Padova*, 2:49–51.
24. OG, 10:280.
25. SN, 37; OG, 3:60.

26. See OG, 10:270 n. 1. For a careful survey of the contents, see Giorgio Strano, "La lista della spesa di Galileo: un documento poco noto sul telescopio," *Galilaeana* 6 (2009): 197–211; Matteo Valleriani, *Galileo Engineer* (Dordrecht: Springer, 2010), 42–44. Galileo's shopping list had already been mentioned by Luigi Zecchin in a 1957 study, "I cannocchiali di Galilei e gli 'occhialeri' veneziani," reprinted in *Vetro e vetrai di Murano: studi sulla storia del vetro* (Venice: Arsenale, 1987–1990), 2:256.
27. See Michele Camerota, *Galileo Galilei e la cultura scientifica nell'età della Controriforma* (Rome: Salerno Editrice, 2004), 113.
28. For a map of the Venetian guilds of glassworkers, see Francesca Trivellato, *Fondamenta dei vetrai. Lavoro, tecnologia e mercato a Venezia tra Sei e Settecento* (Rome: Donzelli, 2000), 136.
29. Carlo Antonio Manzini, *L'occhiale all'occhio* (Bologna: Herede del Benacci, 1660), 8.
30. See OG, 11:314, 351, 521–522, 536, 539, 545, 549–550. On the correspondence exchanged between Galileo and Sagredo concerning lenses, see Zecchin, "I cannocchiali di Galilei e gli 'occhialeri' veneziani," and particularly Olaf Pedersen, "Sagredo's Optical Researches," *Centaurus* 13 (1968): 142–150.
31. On the various stages in manufacturing lenses, see Strano, "La lista della spesa di Galileo"; Piero Solaini, "Storia del cannocchiale," *Atti della Fondazione Giorgio Ronchi* 51 (1996): 838–857; Rolf Willach, "The Development of Lens Grinding and Polishing Techniques in the First Half of the 17th Century," *Bulletin of the Scientific Instrument Society* 68 (2001): 10–15.
32. OG, 10:301.
33. See respectively Galileo to Landucci, August 29, 1609, in OG, 10:253, and SN, 37 (OG, 3:60).
34. OG, 6:259. See also Isabelle Pantin, "La lunette astronomique: une invention en quête d'auteurs," in *Inventions et découvertes au temps de la Renaissance*, ed. Marie Thérèse Jones-Davies (Paris: Klincksieck, 1994), 165–166.
35. This is, for example, the interpretation provided by Biagioli, "Did Galileo Copy the Telescope?," 210.
36. See Thomas B. Settle, "Ostilio Ricci, a Bridge between Alberti and Galileo," in *Actes du XIIe Congrès International d'Histoire des Sciences* (Paris: Blanchard, 1971): 3B, 121–126. Settle's attribution of the textbook to Alberti was mistaken. More recently, the authorship of the work has been credited to Fontana. See Eugenio Battisti and Giuseppa Saccaro Battisti, *Le macchine cifrate di Giovanni Fontana* (Milan: Arcadia Edizioni, 1984), 24.
37. Tommaso Garzoni, *Dello specchio di scientia universale* (Venice: Andrea Ravenoldo, 1567), fol. 55*v*.

38. See Sven Dupré, "Ausonio's Mirrors and Galileo's Lenses: The Telescope and Sixteenth-Century Optical Knowledge," *Galilaeana* 2 (2005): 148–149, 152–170.
39. Ibid., 154.
40. See *Considerazioni d'Alimberto Mauri sopra alcuni luoghi del Discorso di Lodovico delle Colombe intorno alla stella apparita nel 1604* (Florence: Gio. Antonio Caneo, 1606), fol. 8*r*–*v*. About the circumstance that Galileo was actually behind the pen name "Alimberto Mauri," see Stillman Drake, *Galileo against Philosophers in His Dialogue of Cecco di Ronchitti (1605) and Considerations of Alimberto Mauri* (Los Angeles: Zeitlin & Ver Brugge, 1976), 61–71.
41. See Antonio Favaro, "La libreria di Galileo Galilei," *Bullettino di bibliografia e di storia delle scienze matematiche e fisiche*, 19 (1886): 262.
42. Ibid., 263. In Galileo's library there was also a copy of Della Porta's *Magia naturalis*, a work containing a description of the effects produced by a combination of concave and convex lenses. Galileo owned the 1611 Italian edition of the *Magia naturalis*, but this does not exclude the possibility that he had already read the Latin edition of 1589. See also Dupré, "Ausonio's Mirrors and Galileo's Lenses," 167–172.
43. See Bruce S. Eastwood, "Alhazen, Leonardo, and Late-Medieval Speculation on the Inversion of Images in the Eye," *Annals of Science* 43 (1986): 413–446.
44. On Kepler's theory of vision, see David C. Lindberg, *Theories of Vision from Al-Kindi to Kepler* (Chicago: University of Chicago Press, 1976), 178–208; Stephen M. Straker, "Kepler, Tycho and the 'The Optical Part of Astronomy': The Genesis of Kepler's Theory of Pinhole Images," *Archive for History of Exact Sciences* 24 (1981): 267–293; Gérard Simon, *Archéologie de la vision. L'optique, le corps, la peinture* (Paris: Éditions du Seuil, 2003), 203–222.
45. See Galileo, *Le Messager céleste*, ed. Isabelle Pantin (Paris: Les Belles Lettres, 1992), lxxxi–xxxii.
46. OG, 10:441.
47. For a careful study of the contents of the *Dioptrice*, see Antoni Malet, "Kepler and the Telescope," *Annals of Science* 60 (2003): 107–136.
48. This letter, to an unknown addressee, was discovered and published by Franco Palladino, "Un trattato sulla costruzione del cannocchiale ai tempi di Galilei. Principi matematici e problemi tecnologici," *Nouvelles de la République des Lettres* 1 (1987): 95–102.
49. For biographical information on Venturi, see Isidoro Ugurgieri Azzolini, *Le pompe sanesi o' vero relazione delli huomini, e donne illustri di Siena e suo Stato* (Pistoia: Stamperia di Pier'Antonio Fortunati, 1649), 679; Ettore Romagnoli, *Biografia cronologica de' Bellartisti Senesi, 1200–1800* (Florence: Edizioni S.P.E.S., 1976), 9:711–719.

50. Giulio Mancini to his brother Deifebo, November 4, 1609, in ASEPD, XIX, Eredità Mancini, 167, fol. 262*v*.
51. See Chapter 10.
52. Bonifacio Vannozzi, *Della suppellettile degli avvertimenti politici, morali, et christiani* (Bologna: Heredi di Giovanni Rossi, 1609–1613), 3:685.
53. See Palladino, "Un trattato sulla costruzione del cannocchiale ai tempi di Galilei," 101.
54. Sarpi, *Lettere ai gallicani,* critical ed., introductory essay and notes by Boris Ulianich (Wiesbaden: Steiner, 1961), 240.
55. See Chapter 2.
56. See OG, 10:280.
57. OG, 10:271.
58. OG, 10:273.
59. On Galileo's youthful "inclination to drawing," which he continued as an adult, see Vincenzo Viviani, *Racconto istorico,* in OG, 19:602. About Galileo's painting techniques, see Horst Bredekamp, "Gazing Hands and Blind Spots: Galileo as Draftsman," *Science in Context* 13 (2000): 423–462; Horst Bredekamp, *Galilei der Künstler. Der Mond. Die Sonne. Die Hand* (Berlin: Akademie Verlag, 2007), 101–121, 346–362. See also Reeves, *Painting the Heavens: Art and Science in the Age of Galileo* (Princeton, NJ: Princeton University Press, 1997), 138–183; Samuel Y. Edgerton Jr., *The Mirror, the Window and the Telescope: How Renaissance Linear Perspective Changed Our Vision of the Universe* (Ithaca, NY: Cornell University Press, 2009), 151–167.
60. See Ewan A. Whitaker, "Galileo's Lunar Observations and the Dating of the Composition of 'Sidereus Nuncius,'" *Journal for the History of Astronomy* 9 (1978): 155–169.
61. See Elizabeth Cavicchi, "Painting the Moon," *Sky and Telescope* 83 (1991): 313–315.
62. OG, 10:277–278.
63. OG, 10:278.
64. See Van Helden, "Galileo and the Telescope," 192–193.
65. See John D. North, "Thomas Harriot and the First Telescopic Observations of Sunspots," in *Thomas Harriot: Renaissance Scientist,* ed. John W. Shirley (Oxford: Clarendon Press, 1974), 158–160.
66. SN, 36; OG, 3:60.
67. OG, 10:262 n. 2.
68. OG, 10:261.
69. See Fernand Braudel, *The Mediterranean and the Mediterranean World in the Age of Philip II* (Berkeley: University of California Press, 1995), 1:360–365;

Luciano De Zanche, "I vettori dei dispacci diplomatici veneziani da e per Costantinopoli," *Archivio per la storia postale* 2 (1999): 25–35; Luciano De Zanche, *Tra Costantinopoli e Venezia: Dispacci di Stato e lettere di mercanti dal Basso Medioevo alla caduta della Serenissima* (Prato: Istituto di studi storici postali, 2000), 25–26, 95–96; Eric R. Dursteller, "Power and Information: The Venetian Postal System in the Early Modern Eastern Mediterranean," in *From Florence to the Mediterranean and Beyond: Studies in Honor of Anthony Molho*, ed. Diogo Ramada Curto, Eric R. Dursteller, Julius Kirshner, and Francesca Trivellato (Florence: Olschki, 2009), 1:605–608.

70. See Guglielmo Berchet, *Relazioni dei Consoli Veneti nella Siria* (Turin: Paravia, 1886), 66.

71. See *I diarii di Marino Sanudo (MCCCCXCVI–MDXXXIII) dall'autografo Marciano ital. cl. VII codd. CDXIX–CDLXXVII* (Venice: F. Visentini, 1882), 7:299. Sanudo's journal faithfully recorded the arrival of the various letters read at the meetings of the Venetian Senate in the years 1497–1532.

72. OG, 10:277.

73. The document is now preserved at the University of Michigan Library, Ann Arbor. Stillman Drake pointed it out for the first time in 1962. See Drake, "Galileo Gleanings XIII: An Unpublished Fragment Relating to the Telescope and Medicean Stars," *Physis* 4 (1962): 342–344. Drake provided a more accurate survey of its contents a few years later in his "Galileo's First Telescopic Observations," *Journal for the History of Astronomy* 7 (1976): 153–168. Drake's conclusions have been revised by Owen Gingerich and Albert Van Helden in "How Galileo Constructed the Moons of Jupiter," *Journal for the History of Astronomy* 42 (2011): 259–264. See also Paul Needham, *Galileo Makes a Book: The First Edition of "Sidereus Nuncius" Venice 1610* (Berlin: Akademie Verlag, 2011), 13–18.

74. See Drake, "Galileo's First Telescopic Observations," 165.

75. SN, 66; OG, 3:81. In this log Galileo recorded all the observations of Jupiter and its satellites performed from January 7 to March 2, 1610 (i.e., in the space of time covered by the *Sidereus nuncius*). Entries resume on March 9 and continue with interruptions until 1613. For a reprint of the autograph, see OG, 3(2):427–453.

76. For the conjecture that Galileo used loose sheets, see Antonio Favaro, "Le osservazioni di Galileo circa i Pianeti Medicei dal 7 gennaio al 23 febbraio 1613," *Atti del Reale Istituto veneto di scienze, lettere ed arti* 59 (1900): 524.

77. OG, 3(2):427. See Gingerich and Van Helden, "How Galileo Constructed the Moons of Jupiter," 263.

78. OG, 3(2):428.

79. On the vicissitudes of the editorial history of the *Sidereus*, see Galileo, *Le Messager céleste*, xvii–xxx; Gingerich and Van Helden, "From 'Occhiale' to Printed Page: The Making of Galileo's 'Sidereus Nuncius,'" *Journal for the History of Astronomy* 34 (2003): 251–267; David Wootton, "New Light on the Composition and Publication of the 'Sidereus Nuncius,'" *Galilaeana* 6 (2009): 123–140; Needham, *Galileo Makes a Book*, 63–75.
80. OG, 10:280–281.
81. OG, 10:281.
82. OG, 10:283–284. On the secrecy Galileo maintained in an attempt to defend his priority of discovery, see Mario Biagioli, *Galileo's Instruments of Credit: Telescopes, Images, Secrecy* (Chicago: University of Chicago Press, 2006), 77–134. For a critical discussion of Biagioli's thesis, see Franco Giudice, "Only a Matter of Credit? Galileo, the Telescopic Discoveries, and the Copernican System," *Galilaeana* 4 (2007): 391–413.
83. OG, 3(2):432–433.
84. Baglioni was a small publisher who had been a partner of Roberto Meietti, the principal publisher of the works on Venetian Interdict (see Mario Infelise, "Ricerche sulla fortuna editoriale di Paolo Sarpi," in *Ripensando Paolo Sarpi*, ed. Corrado Pin (Venice: Ateneo Veneto, 2006), 530–531. Baglioni had already published Galileo's *Difesa* against Baldassarre Capra in 1607. In 1611, he printed the *De radiis visus et lucis in vitris perspectivis et iride tractatus* by Marcantonio De Dominis, archbishop of Split. According to Giovanni Bartoli (not to be confused with the homonymous secretary of the Medicean ambassador to Venice), who wrote the foreword of *De radiis visus et lucis*, De Dominis had explained how the telescope worked twenty years before Galileo; see Enrico De Mas, "Il 'De radiis visus et lucis.' Un trattato scientifico pubblicato a Venezia nel 1611 dallo stesso editore del 'Sidereus Nuncius,'" in *Novità celesti e crisi del sapere*, ed. Paolo Galluzzi (Florence: Giunti, 1984), 160–162. Among the books published by Baglioni, we should also mention *De phoenomenis in orbe Lunae* (1612) by Giulio Cesare Lagalla. See Chapter 11 and the Epilogue.

4. In a Flash

1. See Galileo to Vinta, March 13, 1610, in OG, 10:288. On the *Sidereus nuncius* see Galileo, *Le Messager céleste*, ed. Isabelle Pantin (Paris: Les Belles Lettres, 1992), ix–civ; Maurizio Torrini, "'Et vidi coelum novum et terram novam'. A proposito di rivoluzione scientifica e libertinismo," *Nuncius* 1 (1986): 49–77; Philippe Hamou, *La mutation du visible. Essai sur la portée épistemologique des*

instruments d'optique au XVIIe siècle (Villeneuve d'Ascq [Nord]: Presses Universitaires du Septentrion, 1999), 1:29–111; John L. Heilbron, *Galileo* (Oxford: Oxford University Press, 2010), 147–160; David Wootton, *Galileo: Watcher of the Skies* (New Haven, CT: Yale University Press, 2010), 96–105; Galileo Galilei, *Sidereus nuncius: o mensageiro das estrelas*, ed. Henrique Leitão (Lisbon: Fundaçao Calouste Gulbenkian, 2010), 17–136; Eileen Reeves, "Variable Stars: A Decade of Historiography on the *Sidereus nuncius*," *Galilaeana* 8 (2011): 37–52.

2. OG, 10:289.
3. OG, 10:288.
4. Logan Pearsall Smith, *The Life and Letters of Sir Henry Wotton* (Oxford: Clarendon Press, 1907), 1:486–487.
5. Ibid.
6. OG, 10:291.
7. Manso to Galileo, March 18, 1610, in OG, 10:296.
8. Manso to Beni, March 1610, in OG, 10:293.
9. OG, 10:295.
10. See Mario Biagioli, *Galileo Courtier: The Practice of Science in the Culture of Absolutism* (Chicago: University of Chicago Press, 1993), 103.
11. See Galileo to Vinta, March 19, 1610, in OG, 10:299.
12. See Chapter 10.
13. OG, 10:299.
14. OG, 10:301.
15. OG, 10:302 (our italics).
16. OG, 10:301.
17. ASF, Mediceo 3004b, fols. 222*v*–223*r*. The dispatch was sent on May 22, 1610. See Massimo Bucciantini, *Galileo e Keplero. Filosofia, cosmologia e teologia nell'età della Controriforma* (Turin: Einaudi, 2003), 175.
18. For a different opinion, see Mario Biagioli, "Replication or Monopoly? The Economies of Invention and Discovery in Galileo's Observations of 1610," *Science in Context* 13 (2000): 547–551; Mario Biagioli, *Galileo's Instruments of Credit: Telescopes, Images, Secrecy* (Chicago: University of Chicago Press, 2006), 83–84.
19. Bartoli to Vinta, March 27, 1610, in OG, 10:306–307.
20. Piccolomini Aragona to Galileo, March 27, 1610, in OG, 10:305.
21. Sertini to Galileo, March 27, 1610, in OG, 10:305.
22. Lorini to Cardinal Sfondrati, February 7, 1615, in Galileo Galilei, *Scienza e religione. Scritti copernicani*, eds. Massimo Bucciantini and Michele Camerota (Rome: Donzelli, 2009), 250.

5. Peregrinations

1. See Vincenzo Sampieri, *Origine e fondatione di tutte le Chiese che di presente si trovano nella Città di Bologna* (Bologna: Clemente Ferroni, 1633), 92; Giovanni Niccolò Pasquali Alidosi, *Diario. Overo raccolta delle cose che nella Città di Bologna giornalmente occorrono per l'Anno MDCXIV* (Bologna: Bartolomeo Cochi, 1614), 25; Antonio Masini, *Bologna perlustrata*, 3rd imp. (Bologna: Erede di Vittorio Benacci, 1666), 297.
2. Massimo Caprara was the son of Girolamo and Margherita Barbazza; he married Caterina Bentivogli in 1621 and died in 1630. See Giuseppe Guidicini, *Alberi genealogici*, in ASB, II, fol. 37*r*; Ludovico Montefani Caprara, *Famiglie bolognesi*, in BUB, MS 4207.23, fols. 91*r* and 131*r*.
3. See Magini to Kepler, May 26, 1610, in OG, 10:359. According to Martin Hasdale's report, twenty-four people attended. See OG, 10:390.
4. See OG, 3:142, 196; Cittadini to Galileo, July 3, 1610, in OG, 10:389; Hasdale to Galileo, April 28, 1610, in OG 10:345. On Roffeni, see Denise Aricò, "Giovanni Antonio Roffeni: un astrologo bolognese amico di Galileo," *Il Carrobbio* 24 (1998): 67–96. On Bottrigari, see Marina Calore and Gian Luigi Betti, "'Il molto illustre Cavaliere Hercole Bottrigari.' Contributi per la biografia di un eclettico intellettuale bolognese del Cinquecento," *Il Carrobbio* 35 (2009): 93–120. On the Dominican missionary Paolo Maria Cittadini, see Giovanni Michele Cavalieri, *Galleria de' sommi Pontefici, Patriarchi, Arcivescovi e Vescovi dell'Ordine de' Predicatori* (Benevento: Stamperia Arcivescovile, 1696), 568. On Papazzoni, see Michele Camerota, "Flaminio Papazzoni: un aristotelico bolognese maestro di Federico Borromeo e corrispondente di Galileo," in *Method and Order in Renaissance Philosophy of Nature. The Aristotle Commentary Tradition*, ed. Daniel Di Liscia, Eckhard Kessler, and Charlotte Methuen (Aldershot: Ashgate, 1997), 271–300.
5. Kepler to Magini, March 22, 1610, in KGW, 16:295.
6. Hasdale to Galileo, May 31, 1610, in OG, 10:365. On Hasdale, see Chapter 6.
7. "D. Maginus honoratum convivium, et lautum et delicatum, Galileo paravit"; Horky to Kepler, April 27, 1610, in OG, 10:343.
8. Letter from Magini to Zuckmesser, reported by Hasdale to Galileo on April 28, 1610, in OG, 10:345.
9. Horky to Kepler, March 31, 1610, in OG, 10:308.
10. Horky to Kepler, April 6 and 16, 1610, in OG 10:311, 316.
11. Giovanni Antonio Roffeni, *Epistola apologetica* (Bologna: Heredes Joannis Rossij, 1611); see OG, 3:195–196.
12. Horky to Kepler, April 27, 1610, in OG, 10:343.

13. Martin Horky, *Brevissima peregrinatio contra Nuncium sidereum* (Modena: Iulianum Cassianum, 1610); see OG 3:142. See also OG, 10:358, 387. According to Vasco Ronchi, that night Horky first observed Mizar, which is actually a binary star, in Ursa Major. See Vasco Ronchi, *Galileo e il cannocchiale* (Udine: Idea, 1942), 266. In reality, the Bohemian stated he saw the "rider" of Ursa Major, or Alcor (not Mizar). He did not say he had seen them double: he simply stated that he had seen that nearby there were four small stars similar to the Medicean planets. See OG, 3:142. Mizar's binary nature was first observed by Galileo and Castelli in 1617. See Umberto Fedele, "Le prime osservazioni di stelle doppie," *Coelum* 17 (1949): 65–69.
14. See John Wedderburn, *Quatuor problematum quae Martinus Horky contra Nuntium sidereum de quatuor planetis novis disputanda proposuit. Confutation* (Padua: Marinelli, 1610), in OG, 3:172.
15. OG, 3:140–141.
16. OG, 10:343.
17. OG, 10:342–343. The description can be found again—almost verbatim—in another letter from Horky dated May 1610. See OG, 20:599.
18. BNCF, MS Gal. 48, fol. 34*v*. See also OG, 3:436. Kepler couldn't help but notice that some of Galileo's observations and Horky's statements coincided. See Kepler to Galileo, August 9, 1610, in OG, 10:416.
19. Hasdale to Galileo, July 5, 1610, in OG, 10:390.
20. Magini to Kepler, May 26, 1610, in OG, 10:358. When Magini republished this letter in 1614, he decided to omit the aforementioned passage. See Giovanni Antonio Magini, *Supplementum ephemeridum ac tabularum secundorum mobilium* (Venice: Haeredem Damiani Zenarii, 1614), 267.
21. Horky to Kepler, April 27, 1610, in OG, 10:343.
22. "No es Peregrinación aquel vagante, / Inquieto y solicito camino / Del que por ser curioso es caminante. / Ni el que por melancólico destino / O por necesidad o vanagloria, / O por intento vano es peregrino. / La peregrinación que de memoria / Y de albanza es digna en cielo y suelo, / Y la que se encarece en esta historia, / Es la de aquel que con piadoso celo, / Por voluntad u obligación, visita / Los lugares que acá señala el cielo." Bartolomé Cairasco de Figueroa, "Peregrinación," in *Templo militante. Flos sanctorum, y triumphos de sus virtudes* (Lisbon: Pedro Crasbeeck, 1613), 157; see also Juergen Hahn, *The Origins of the Baroque Concept of Peregrinatio* (Chapel Hill: University of North Carolina Press, 1973), 15–16.
23. Martin Horky, *Brevissima peregrinatio contra Nuncium sidereum* (1610), in OG, 3:131.

24. Ibid.: "Germaniam incolui, Gallorum urbes vidi; Italia, Philosophiae ac Medicinae amore, exuladii."
25. Horky to Kepler, January 12, 1610, in KGW, 16:268. The references to Lipsius concern the latter's letter to Philippe Lanoye dated April 1578. See Justus Lipsius, *Epistolarum selectarum centuria prima* (Antwerp: C. Plantinum, 1586), 107. Previously, in 1605 Horky stayed in Brzeg, where he discussed a thesis published in Liegnitz that year. See Martin Horky, *Disputatio ethica de beatitudine politica* (Legnia: N. Sartorii, 1605).
26. See the note by Carl Frisch in Johannes Kepler, *Opera Omnia*, ed. Carl Frisch (Frankfurt: Heyder & Zimmer, 1858–1871), 2:62. On Horky, see also Josef Smolka, "Böhmen und die Annahme der Galileischen astronomischen Entdeckungen," *Acta historiae rerum naturalium necnon technicarum* 1 (1997): 41–69; Christoph Strebel, "Martinus Horky und das Fernrohr Galileis," *Sudhoffs Archiv* 90 (2006): 11–28.
27. Martin Horky, *Ein richtiger und sehr nützlicher Wegweiser, wie man sich für der Pestilentz bewahren sole* (Rostock: Sachs, 1624).
28. Martin Horky, *Talentum astromanticum, oder Natürliche Weissagung und Verkundigung, auß deß Himmels Lauff, vom Zustand und Beschaffenheit deß Schald-Jahrsnach Christi Geburt 1632* (Leipzig: Ritzsch, 1632); Horky, *Das grosse Prognosticon, Oder Astrologische Wunderschrifft . . . auffs 1633. Jahr Christi, durch die Handt Gottes am Gestirnten Firmament deß Himmels auffgezeichnet* (Hamburg, 1633); Horky, *Chrysmologium Physico-Astronomicum, Oder Natürliche Weissagung und Erkundigung auß dem Gestirn und Himmelslauff von dem Zustand und Beschaffenheit deß 1653. Jahrs Jesu Christi* (Nuremburg, 1639); Horky, *Alter und Newer Schreib-Calender sambt der Planeten Aspecten Lauff und derselben Influentzen auff das Jahr nach der Geburt Jesu Christi MDCLIII Auß den rechten wahrhafftigen alten und newen Canonibus mit Fleiß gestellet* (Nuremberg: Endters, ca. 1649). A few years earlier he had published a pamphlet entitled *Eine newe Diania Astromantica, oder gewisser Beweiß, was zu halten sey von den schrecklichen Göttlichen Wunderwerck, so diß jetzige 1629 Jahr Christi an der Sonnen ist gesehen worden* (1629). In this regard, see Josef Smolka, "Martin Horký a jeho kalendáře [Martin Horky and his calendars]," in *Miscellanea. Oddělení rukopisů a starých tisků* 18 (Prague: Narodni Knihovna, 2005), 145–160.
29. In a letter dated December 3, 1609, Magini stated that he had taken on "a young German copyist who writes quite well": see Antonio Favaro, *Carteggio inedito di Ticone Brahe, Giovanni Keplero e di altri celebri astronomi e matematici dei secoli XVI e XVII con Giovanni Antonio Magini* (Bologna: Zanichelli, 1886), 118 n. 3. At the end of April 1610, Horky said he had been in Bologna

for six months (KGW, 16:307), while at the end of May he wrote: "Bononiae apud D. Maginum . . . septem annos lunares delitesco" (OG, 20:599). Therefore, he must have arrived in Bologna in November 1609.

30. Horky to Kepler, April 27, 1610, in KGW, 16:306.
31. OG, 3:139.
32. Horky to Kepler, May 24, 1610, in OG, 10:359.
33. OG, 10:358: "Es beisst ein Fuchss den andern nicht, undt ein Hundt beldt den andern nicht ahn."
34. See Horky to Kepler, June 30, 1610, in OG, 10:386.
35. Roffeni to Galileo, June 22, 1610, in OG, 10:376.
36. Magini to Santini, June 22, 1610, in OG, 10:379.
37. Ibid.
38. Cittadini to Galileo, July 3, 1610, in OG, 10:389. See also the letter from Roffeni to Galileo dated July 6, in OG, 10:391–92.
39. OG, 10:379, 384, 391–392, 418. In 1607, Baldassarre Capra was reported for plagiarizing Galileo's work on the geometric and military compass. See Michele Camerota, *Galileo Galilei e la cultura scientifica nell'età della Controriforma* (Rome: Salerno Editrice, 2004), 124–130.
40. Sertini to Galileo, August 7, 1610, OG, 10:412.
41. See Kepler, *Discussion avec le Messagere céleste,* ed. Isabelle Pantin (Paris: Les Belles Lettres, 1993), xli.
42. See OG, 20:599–600. The addressee of the letter is cited with the initials "D. H. M. Vall. Mon." We think he must undoubtedly be Orazio Morandi ("Dominus Horatius Morandi Vallombrosanus Monachus"). On Morandi, see Germana Ernst, "Scienza, astrologia e politica nella Roma barocca. La biblioteca di Orazio Morandi," in *Bibliothecae selectae. Da Cusano a Leopardi,* ed. Eugenio Canone (Florence: Olschki, 1993), 217–252; Brendan Dooley, *Morandi's Last Prophecy and the End of Renaissance Politics* (Princeton, NJ: Princeton University Press, 2002).
43. In the *Brevissima peregrinatio,* Horky mentions a "Segretario di Madonna Luna, Capitano della Via Lattea, Cerimoniere di Orione, testimone oculare dei quattro nuovi pianeti (Secretary of Lady Moon, Captain of the Milky Way, Master of Ceremonies of Orion, eyewitness to the four new planets"), OG, 3:142. As is the case with Morandi, the letter cited Sizzi, not in full but in the abbreviated form of "D. fr. sit.," i.e., "Dominus Franciscus Sitius"; see OG, 20:600.
44. See Francesco Sizzi, *Dianoia astronomica, optica, physica, qua Syderei nuncii rumor de quatuor planetis a Galilaeo . . . vanus redditur* (Venice: P. M. Bertanum, 1611); see OG, 3:203–250. The work was probably drawn up in collaboration

with Morandi; see OG, 10:411. On Sizzi, see Michele Camerota, "Francesco Sizzi. Un oppositore di Galileo tra Firenze e Parigi," in *Toscana e Europa. Nuova scienza e filosofia tra '600 e '700*, ed. Ferdinando Abbri and Massimo Bucciantini (Milan: F. Angeli, 2006), 83–107.

45. See OG, 3:208.
46. Roffeni to Galileo, June 29, 1610, in OG, 10:384: "Magini sent a copy of the letter he had received in Florence, where he fully realized how arrogant [Horky] had been and had wanted to write to his friends as if he had given his consent, which is totally false."
47. "Te [Morandi] numquam vidi *sed literas tuas legi*": OG, 20:600 (our italics). On the relationship between Magini and Morandi, see Favaro, *Carteggio inedito*, 54. Note that Roffeni also mentioned Horky's Florentine contacts as Magini's "friends" (see previous note). Magini would later complain that Horky had rifled through his papers and "took the liberty to examine even the letters I received from my friends": OG, 10:446.
48. Perhaps this is the very "secret" that Magini begged Sizzi to keep under wraps. See the latter's letter dated March 11, 1611, in OG, 11:75.
49. Francesco Sizzi devoted his *Dianoia* to Giovanni, as did Ludovico Delle Colombe with the *Discorso apologetico* aimed at refuting Galileo's hydrostatic theories. On Giovanni de' Medici's hostility toward Galileo, see Chapter 10. On relations between Morandi and Giovanni de' Medici, see Ernst, "Scienza, astrologia e politica nella Roma barocca," 224–227.
50. OG, 20:600.
51. OG, 10:400.
52. See Horky to Kepler, June 30, 1619, in OG, 10:386.
53. On Kepler's copy, see Hasdale to Galileo, August 9, 1610, in OG, 10:418. As Kepler informed Horky, likewise on August 9: "Tuam Peregrinationem ex concessu D. Mathei Welseri nactus, legi" (OG, 10:419). On June 30 Horky sent a copy of the *Peregrinatio* to Kepler (OG, 10:386), but it is likely that the gift never reached him, or the astronomer would not have needed Welser's copy. On Sarpi's copy, see Horky to Sarpi, July 10, 1610, in OG, 10:399–400.
54. See BL, MS Sloane 682, fols. 46–47. See also Antonio Favaro, "Un inglese a Padova al tempo di Galileo," *Atti e memorie della R. Accademia di scienze, lettere ed arti in Padova* 34 (1918): 12–14; Mordechai Feingold, "Galileo in England: The First Phase," in *Novità celesti e crisi del sapere*, ed. Paolo Galluzzi (Florence: Giunti, 1984), 411–420, especially 414.
55. Maestlin to Kepler, September 7, 1610, in OG, 10:429.
56. See OG, 10:386–387. In Florence, Ludovico Delle Colombe also had another copy. See OG, 10:398.

57. Horky to Kepler, May 24, 1610, in OG, 10:358.
58. See OG, 3:141.
59. Galileo to Giuliano de' Medici, October 1, 1610, in OG, 10:440.
60. OG, 10:414.
61. OG, 10:419.
62. We learn this from a letter from Kepler to Galileo dated October 25, 1610; see OG, 10:457.
63. See OG, 10:457–458.
64. OG, 10:422: Galileo to Kepler, August 19, 1610.
65. See above, note 14.
66. See above, note 11. This was drafted in August 1610 (the dedication is dated August 19) and the manuscript was sent to Galileo—provisionally in Italian—by the end of September. See OG, 10:437, 440.
67. Roffeni to Galileo, June 29, 1610, in OG, 10:385.
68. See BUB, Aula V, Tab. I, D. 1, vol. 319 (insert 5). In effect, there are elements in the work that show in-depth knowledge of some of the most recent solutions to astronomical problems, a skill mastered by Magini alone. Thus, the *Epistola apologetica* refers to the "catalogue of the fixed stars sent some time ago by Tycho to Magini, including a thousand observed stars, and far richer than the one published by Tycho in his *Progymnasmata*" (OG, 3:197). It is clear that only the recipient of the "gift"—Magini, of course—could have suggested such an argument.
69. See OG, 10:378–379.
70. Galileo to Kepler, August 19, 1610, in OG, 10:422–423. Criticism of the idea that "philosophy is a book and fantasy of man, like the *Iliad* and *Orlando Furioso,* books in which the least important things is that what is written is true," was proposed again in *Il Saggiatore* (see OG, 6:232).

6. The Battle of Prague

1. John Banville, *Kepler: A Novel* (London: Secker & Warburg, 1981), 124–125.
2. OG, 10:316; KGW, 16:302.
3. OG, 10:309.
4. See Antonio Favaro, *Amici e corrispondenti di Galileo,* ed. Paolo Galluzzi (Florence: Salimbeni, 1983), 1:365–366; Christoph Clavius, *Corrispondenza,* ed. Ugo Baldini and Pier Daniele Napolitani (Pisa: Università di Pisa, Dipartimento di Matematica, 1992), 6(1):111–112.
5. See Favaro, *Amici e corrispondenti,* 1:354, 360–363.
6. See Kepler, *Discussion avec le Messager céleste,* ed. Isabelle Pantin (Paris: Les Belles Lettres, 1993), xxxv–xxxvi.

7. Gloriosi to Schreck, May 29, 1610, in OG, 10:363–364.
8. See Chapter 10.
9. OG, 10:364.
10. On Hasdale, see Favaro, *Amici e corrispondenti,* 1:600–606; Galileo, *Le Messager céleste,* ed. Isabelle Pantin (Paris: Les Belles Lettres, 1992), xiii; Robert J. W. Evans, *Rudolf II and His World: A Study in Intellectual History, 1576–1612* (Oxford: Clarendon Press, 1973), 73.
11. Hasdale to Galileo, April 15, 1610, in OG, 10:314.
12. Fulgenzio Micanzio, *Vita del Padre Paolo dell'Ordine de' Servi e theologo della Serenissima* (Leiden 1646), in Paolo Sarpi, *Istoria del Concilio Tridentino, seguita dalla Vita del padre Paolo di Fulgenzio Micanzio,* ed. Corrado Vivanti (Turin: Einaudi, 1974), 2:1343–1344. See also Gaetano Cozzi, "Galileo Galilei, Paolo Sarpi e la società veneziana," in *Paolo Sarpi tra Venezia e l'Europa* (Turin: Einaudi, 1979), 155 n. 50.
13. ASVe, Consiglio dei Dieci, Segreta, register 15, fol. 59*r.* The information is cited in Cozzi, "Galileo Galilei, Paolo Sarpi e la società veneziana," 155 n. 50, with several inaccuracies in the transcription.
14. See Hasdale to Galileo, May 31, 1610, in OG, 10:366.
15. See Favaro, *Amici e corrispondenti,* 1:605, and the website http://documenta.rudolphina.org under the entry "Martin Hasdale"; see the letters from the years 1611 and 1612.
16. OG, 10:314.
17. OG, 10:315.
18. DNS, 12.
19. DNS, 12–13.
20. DNS, 22.
21. OG, 10:314.
22. Kepler, *Discussion avec le Messager céleste,* xi–xxii. On the *Dissertatio,* see also Robert S. Westman, *The Copernican Question: Prognostication, Skepticism, and Celestial Order* (Berkeley: University of California Press, 2011), 460–465.
23. See Klaus-Dieter Herbst, "Galilei's Astronomical Discoveries Using the Telescope and Their Evaluation Found in a Writing-Calendar from 1611," *Astronomische Nachrichten* 6 (2009): 537; Eileen Reeves, "Variable Stars: A Decade of Historiography on the *Sidereus nuncius,*" *Galilaeana* 8 (2011): 65–66.
24. Hasdale to Galileo, May 31, 1610, in OG, 10:365.
25. Hasdale to Galileo, April 28, 1610, in OG, 10:345.
26. Hasdale to Galileo, August 9, 1610, in OG, 10:418.
27. DNS, 13.

28. Roderigo Alidosi, *Relazione di Germania e della Corte di Rodolfo II imperatore*, ed. Giuseppe Campori (Modena: Tipografia e Litografia Cappelli, 1872), 6. Alidosi was ambassador to Prague from 1605 to 1607. On Rudolf, see above all Evans, *Rudolf II and His World*; Thomas Da Costa Kaufmann, *The School of Prague* (Chicago: University of Chicago Press, 1988); Kaufmann, *The Mastery of Nature: Aspects of Art, Science and Humanism in the Renaissance* (Princeton, NJ: Princeton University Press, 1993); and Eliška Fučíková et al., eds., *Rudolf II and Prague. The Court and the City* (London: Thames and Hudson, 1997).
29. Dispatch of October 22, 1600, in ASF, Mediceo del Principato 4356, fol. 505r. See Massimo Bucciantini, "Galileo e Praga," in *Toscana e Europa. Nuova scienza e filosofia tra '600 e '700*, ed. Ferdinando Abbri and Massimo Bucciantini (Milan: F. Angeli, 2006), 109–121.
30. Evans, *Rudolf II and His World*, 84.
31. Dispatch of October 22, 1600, in ASF, Mediceo del Principato 4356, fol. 505r.
32. Evans, *Rudolf II and His World*, 196.
33. Ibid.
34. See OG, 10:390, 418.
35. See Kepler, *Discussion avec le Messager céleste*, xi, 55 n. 30, 63 n. 57; Hasdale to Galileo, August 9, 1610, in OG, 10:418.
36. Letter of August 17, 1610, in OG, 10:420.
37. Hasdale to Galileo, August 24, 1610, in OG, 10:427. See Kepler, *Discussion avec le Messager céleste*, xi–xxxii.
38. Hasdale to Galileo, August 24, 1610, in OG, 10:427.
39. KGW, 16:333–334; Maestlin to Kepler, September 7, 1610, in OG, 10:429.
40. OG, 10:426–427.
41. OG, 10:358. On Matteo Carosio, see Favaro, *Amici e corrispondenti*, 3:1657–1667.
42. Galileo to Kepler, August 19, 1610, in OG, 10:421.
43. Hasdale to Galileo, June 7, 1610, in OG, 10:370.
44. BRF, MS 2446, letter nos. 215, 308, and 494.
45. Quoted in Marco Beretta, "Galileo in Sweden: Legend and Reality," in *Sidereus Nuncius and Stella Polaris: The Scientific Relations between Italy and Sweden in Early Modern History*, ed. Marco Beretta and Tore Frängsmyr (Canton, OH: Science History Publications, 1997), 12.
46. August 24, 1610, in OG, 10:426–427.
47. Segeth to Galileo, October 24, 1610, in OG, 10:455.
48. Kepler, *Narratio*; see OG, 3:185.
49. OG, 3:187.

50. The reference is to the letter Galileo sent Kepler on August 4, 1597, after receiving the *Mysterium cosmographicum*. See Massimo Bucciantini, "Agosto 1597: microstoria di una lettera," chapter 3 in *Galileo e Keplero. Filosofia, cosmologia e teologia nell'età della Controriforma* (Turin: Einaudi, 2003); Paolo Galluzzi, "Genesi e affermazione dell'universo macchina," in Galluzzi, ed., *Galileo. Immagini dell'universo dall'antichità al telescopio* (Florence: Giunti, 2009), 289–297.
51. Galileo to Giuliano de' Medici, December 11, 1610, in OG, 10:483.
52. OG, 3(2):876. See Antonio Favaro, "Elementi di un nuovo anagramma galileiano," in *Scampoli galileiani*, ed. Lucia Rossetti and Maria Laura Soppelsa (Trieste: LINT, 1992), 2:446–447.
53. Galileo to Vinta, July 30, 1610, in OG, 10:410.
54. See J. Kepler, *Dioptrice* (1611), in KGW, 4:344–354.
55. Kepler, *Narratio;* see OG, 3:184.
56. "Machinas nonnullas ad illa configuranda atque expolienda escogitavi": Galileo to Kepler, August 19, 1610, in OG, 10:421.
57. "Magno me desiderio incendisti videndi tui instrumenti, ut tandem iisdem tecum potiar caelestibus spectaculis": Kepler to Galileo, August 9, 1610, in OG, 10:413–414.
58. Kepler to Galileo, August 9, 1610, in OG, 10:414. See also Kepler, *Discussion avec le Messager céleste*, 140.
59. See OG, 10:382–383, 407–410.
60. See Maximilian of Bavaria to Galileo, July 8, 1610, in OG, 10:393.
61. See Cioli to Vinta, September 13, 1610, in OG, 10:430.
62. Giuliano de' Medici to Vinta, November 14 and 21, 1611, in OG, 11:234–235.
63. Giuliano de' Medici to Galileo, in OG, 10:493.
64. Galileo to Giuliano de' Medici, October 1, 1610, in OG, 10:440–441.
65. Giuliano de' Medici to Galileo, August 23, 1610, in OG, 10:426.
66. Galileo to Kepler, August 19, 1610, in OG, 10:422.
67. Kepler to Galileo, August 9, 1610, in OG, 10:416.
68. On the conflict between Maestlin and the Catholic Church, and above all Christoph Clavius regarding the calendar reform, see Bucciantini, *Galileo e Keplero*, 74–81.

7. Across the English Channel: Poets, Philosophers, and Astronomers

1. Lower to Harriot, June 11, 1610, in BL, Additional MS 6789, fol. 425r. Albeit with several transcription errors, the letter is also reproduced in Stephen Peter Rigaud, *Supplement to Dr. Bradley's Miscellaneous Works: With an Account of*

Harriot's Astronomical Papers (Oxford: Oxford University Press, 1833), 25. With the reference to "dutchmen that weare eaten by beares in Nova Zembla," Lower is referring to Willem Barents's 1594 and 1596 expeditions in the Arctic regions to seek a Northeast passage; see John Horace Parry, *The Age of Reconnaissance: Discovery, Exploration and Settlement, 1450–1650* (Berkeley: University of California Press, 1981), 205–206.

2. BL, Additional MS 6789, fol. 425*v*; Rigaud, *Supplement to Dr. Bradley's Miscellaneous Works*, 26. Regarding the significance the letter acquires also in the context of the circulation of Bruno's works in England, see Saverio Ricci, *La fortuna del pensiero di Giordano Bruno, 1600–1750* (Florence: Le Lettere, 1990), 76–78.
3. BL, Additional MS 6789, fol. 425*v*; Rigaud, *Supplement to Dr. Bradley's Miscellaneous Works*, 26.
4. See John W. Shirley, *Thomas Harriot: A Biography* (Oxford: Clarendon Press, 1983), 358–379.
5. See Jean Jacquot, "Harriot, Hill, Warner and the New Philosophy," in *Thomas Harriot: Renaissance Scientist*, ed. John W. Shirley (Oxford: Clarendon Press, 1974), 107–128; Stephen Clucas, "Corpuscular Matter Theory in the Northumberland Circle," in *Late Medieval and Early Modern Corpuscular Matter Theories*, ed. Cristoph Lüthy, John E. Murdoch, and William E. Newman (Leiden: Brill, 2001), 181–207.
6. See Robert H. Kargon, *Atomism in England from Hariot to Newton* (Oxford: Clarendon Press, 1966), 7. On the fact that, in the late sixteenth century, Harriot and others in the Northumberland Circle were among the first supporters in England of Copernican cosmology, see Robert S. Westman, "The Astronomer's Role in the Sixteenth Century: A Preliminary Study," *History of Science* 18 (1980): 136 n. 6.
7. See Mark Kishlansky, *A Monarchy Transformed: Britain 1603–1714* (London: Allen Lane, 1996), 65ff.
8. Shirley, *Thomas Harriot: A Biography*, 327–331.
9. Ibid., 340–341.
10. See Christopher Hill, *Intellectual Origins of the English Revolution* (Oxford: Clarendon Press, 1965), 131–132, 154.
11. John W. Shirley, "Sir Walter Ralegh and Thomas Harriot," in *Thomas Harriot: Renaissance Scientist*, ed. Shirley, 23–27.
12. See Jean Jacquot, "Thomas Harriot's Reputation for Impiety," *Notes and Records of the Royal Society of London* 9 (1952): 164–187. For a more recent analysis, see Scott Mandelbrote, "The Religion of Thomas Harriot," in *Thomas Harriot: An Elizabethan Man of Science*, ed. Robert Fox (Aldershot: Ashgate, 2000), 246–279.
13. See KGW, 16:172.

14. The only book Harriot published was a report of his trip to Virginia (*A Briefe and True Report of the New Found Land of Virginia,* 1588), where Walter Ralegh had sent him in 1585 with the expedition headed by Sir Richard Grenville, to survey and measure land. See David B. Quinn, "Thomas Harriot and the Problem of America," in *Thomas Harriot: An Elizabethan Man of Science,* ed. Robert Fox, 9–27.
15. See Chapter 6.
16. See KGW, 3:20.
17. PH, MS HMC, 241/III.2, fol. 12*r;* also in Rigaud, *Supplement to Dr. Bradley's Miscellaneous Works,* 27. On the astronomical calculations, see Giovanni Antonio Magini, *Ephemerides . . . ab anno Domini 1581 usque ad annum 1620 secundum Copernici hypotheses, Prutenicosque canones* (Venice: Damianum Zenerium, 1582), fol. 453*v.*
18. See DNS, 14.
19. For a detailed biography of Lower, see Paul M. Hunneyball, "Sir William Lower and the Harriot Circle," Durham Thomas Harriot Seminar, Occasional Paper no. 31, 2002. See also John J. Roche, "Lower, Sir William (ca. 1570–1615)," in *Oxford Dictionary of National Biography* (Oxford: Oxford University Press, 2004), under the specific entry.
20. In 1631, Warner published Harriot's *Artis analyticae praxis;* see Muriel Seltman, "Harriot's Algebra: Reputation and Reality," in *Thomas Harriot: An Elizabethan Man of Science,* ed. Fox, 153–154; Jacqueline A. Stedall, "Rob'd of Glories: The Posthumous Misfortunes of Thomas Harriot and His Algebra," *Archive for History of Exact Sciences* 54 (2000): 455–497. Torpoley wrote a confutation of Harriot's atomism, conserved at the BL (Birch MS 4458, fols. 6–89) and reproduced in the appendix to Jacquot, "Thomas Harriot's Reputation for Impiety," 183–187. See Kargon, *Atomism in England from Hariot to Newton,* 33–35; Stephen Clucas, "The Atomism of the Cavendish Circle: A Reappraisal," *The Seventeenth Century* 2 (1994): 249.
21. For a list of this correspondence, nine letters in all, see Shirley, *Thomas Harriot: A Biography,* 391 n. 33.
22. The original of this part of the letter has now been lost, whereas the second one is at the BL (Additional MS 6789, fols. 427–428). The entire letter is published in Rigaud, *Supplement to Dr. Bradley's Miscellaneous Works,* 42–45. Protheroe would later become one of the executors of Harriot's will; see Shirley, *Thomas Harriot: A Biography,* 412–414.
23. See Lower to Harriot, September 30, 1607, in PH, MS HMC, 241/VII, fols. 1–6.
24. That Harriot keenly assessed this information is demonstrated by the fact that he wrote long and detailed notes on the back and in the margins of the

sheets comprising Lower's letter. See Shirley, *Thomas Harriot: A Biography*, 395–396.

25. See Robert S. Westman, *The Copernican Question: Prognostication, Skepticism, and Celestial Order* (Berkeley: University of California Press, 2011), 411.
26. Kepler to Harriot, October 2, 1606, in KGW, 15:348–352.
27. Only five letters from this correspondence have survived: three of Kepler's, who started the exchange, and two of Harriot's. All five are reproduced in KGW, 15:348–352, 365–368; 16:31–32, 172–173, 250–251. See Shirley, *Thomas Harriot: A Biography*, 385–388.
28. Harriot to Kepler, December 2, 1606, in KGW, 15:368.
29. See Gordon R. Batho, "Thomas Harriot and the Northumberland Household," in *Thomas Harriot: An Elizabethan Man of Science*, ed. Fox, 39.
30. See Stephen Clucas, "Thomas Harriot and the Field of Knowledge in the English Renaissance," in *Thomas Harriot: An Elizabethan Man of Science*, ed. Fox, 93–135.
31. See BL, Additional MS 6789, fols. 1–538. See also Johannes A. Lohne, "Essays on Thomas Harriot. III. A Survey of Harriot's Scientific Writings," *Archive for History of Exact Sciences* 20 (1979): 265–312.
32. See BL, Additional MS 6789, fols. 266 and 268. See also John W. Shirley, "An Early Experimental Determination of Snell's Law," *American Journal of Physics* 19 (1951): 507–508; Johannes A. Lohne, "Thomas Harriot (1560–1621): The Tycho Brahe of Optics," *Centaurus* 6 (1959): 116–118.
33. See Batho, "Thomas Harriot and the Northumberland Household," 34.
34. See Clucas, "Thomas Harriot and the Field of Knowledge in the English Renaissance," 111–126.
35. See note 26. On Harriot's important mathematical discoveries and the introduction of an innovative symbolic language, see Jacqueline A. Stedall, "Symbolism, Combinations, and Visual Imagery in the Mathematics of Thomas Harriot," *Historia Mathematica* 34 (2007): 380–401.
36. The details about Harriot's home can be found in the will he drew up shortly before his death. See Rosalind Cecilia H. Tanner, "The Study of Thomas Harriot's Manuscripts: 1. Harriot's Will," *History of Science* 6 (1967): 1–16.
37. On this approach, in addition to the classic study by Eva Germaine Rimington Taylor, *The Mathematical Practitioners of Tudor and Stuart England* (Cambridge: Cambridge University Press, 1967), see also Stephen Johnston, "The Mathematical Practitioners and Instruments in Elizabethan England," *Annals of Science* 48 (1991): 319–344. Regarding Harriot, see Jim A. Bennett, "Instruments, Mathematics, and Natural Knowledge: Thomas Harriot's Place on the Map of Learning," in *Thomas Harriot: An Elizabethan Man of Science*, ed. Fox, 142–152.

38. See Shirley, *Thomas Harriot: A Biography*, 382–383.
39. See the letter from William Trumbull, English ambassador to Brussels, to the Earl of Salisbury dated November 30, 1609, and that of Captain Bruce to Trumbull dated February 14, 1610, in *Papers of William Trumbull the Elder, 1605–1610*, ed. Edward Kelly Purnell and Allen Banks Hinds (London: His Majesty's Stationery Office, 1924–1925), 2:186, 239.
40. See Chapter 1.
41. John W. Shirley, "Thomas Harriot's Lunar Observations," in *Science and History: Studies in Honor of Edward Rosen (Studia Copernicana*, XVI), ed. Erna Hilfstein, Pawel Czartoryski, and Frank. D. Grande (Wrocław: Polish Academy of Sciences Press, 1978), 283.
42. See Chapter 3.
43. See Terrie F. Bloom, "Borrowed Perceptions: Harriot's Maps of the Moon," *Journal for the History of Astronomy* 9 (1978): 117.
44. See Samuel Y. Edgerton Jr., "Galileo, Florentine 'Disegno', and the 'Strange Spottednesse' of the Moon," *Art Journal* 44 (1984): 225–232; Ewan A. Whitaker, "Selenography in the Seventeenth Century," in *The General History of Astronomy*, ed. René Taton and Curtis Wilson (Cambridge: Cambridge University Press, 1989), 2:122–124.
45. See Amir R. Alexander, "Lunar Maps and Coastal Outlines: Thomas Harriot's Mapping of the Moon," *Studies in History and Philosophy of Science* 29 (1998): 345–368; Stephen Pumfrey, "Harriot's Maps of the Moon: New Interpretations," *Notes and Records of the Royal Society* 63 (2009): 163–168.
46. Lower to Harriot, February 6, 1610, in Rigaud, *Supplement to Dr. Bradley's Miscellaneous Works*, 42 (our italics).
47. Lower to Harriot, June 11, 1610, in BL, Additional MS 6789, fol. 425*r* (our italics).
48. We will cite just one example: Allan Chapman, "A New Perceived Reality: Thomas Harriot's Moon Maps," *Astronomy and Geophysics* 50 (2009): 2–33.
49. Lower to Harriot, June 11, 1610, in BL, Additional MS 6789, fol. 425*v*; see also OG, 3:78; SN, 61.
50. Heydon's work, titled *A Defence of Judiciall Astrologie*, was a polemical response to John Chamber's *Treatise against Iudicial Astrologie* (1601); see Westman, *The Copernican Question*, 409–411. See also Mordechai Feingold, *The Mathematician's Apprenticeship: Science, Universities and Society in England, 1560–1640* (Cambridge: Cambridge University Press, 1984), 140–141.
51. Heydon to Camden, July 6, 1610, in William Camden, *Epistolae* (London: Richard Chiswell, 1691), 129–130.

52. See Shirley, "Thomas Harriot's Lunar Observations."
53. See Bloom, "Borrowed Perceptions: Harriot's Maps of the Moon," 121.
54. See Shirley, "Thomas Harriot's Lunar Observations," 303.
55. See note 2.
56. See Bryant Tuckerman, *Planetary, Lunar, and Solar Positions: A.D. 2 to A.D. 1649* (Philadelphia: American Philosophical Society, 1962), 519.
57. See PH, MS HMC, 241/IV, fols. 1–15.
58. PH, MS HMC, 241/IV, fol. 3*r*.
59. PH, MS HMC, 241/IV, fols. 1, 6, 8, 10; Shirley, *Thomas Harriot: A Biography*, 429.
60. Walter Ralegh, *The History of the World* (London: Printed by William Stansby for Walter Burre, 1614), 100.
61. PH, MS HMC, 241/IV, fol. 3*r*.
62. BL, Additional MS 6789, fol. 429*v*.
63. See John D. North, "Thomas Harriot and the First Telescopic Observations of Sunspots," *Thomas Harriot: Renaissance Scientist*, ed. Shirley, 141.
64. PH, MS HMC, 241/IV, fol. 2*v*.
65. See OG, 3:80; SN, 74.
66. See PH, MS HMC, 241/IV, fols. 16–44*v*.
67. See John J. Roche, "Harriot, Galileo, and Jupiter's Satellites," *Archives Internationales d'Histoire des Sciences* 32 (1982): 34.
68. See Gordon R. Batho, "Thomas Harriot's Manuscripts," in *Thomas Harriot: An Elizabethan Man of Science*, ed. Fox, 286–297.
69. See PH, MS HMC, 241/IV, fol. 26*r*.
70. See Roche, "Harriot, Galileo, and Jupiter's Satellites," 35–49. This was also a method that Harriot continued to apply later, as in the case of *Mundus Iovialis* (1614) by the German astronomer Simon Mayr, whose ephemerides of the satellites represented a further chance to compare his calculations (see PH, MS HMC, 241/IV, fols. 40–44*v*).
71. See North, "Thomas Harriot and the First Telescopic Observations of Sunspots," 132.
72. See PH, MS HMC, 241/IV, fol. 4*r*.
73. See Roche, "Harriot, Galileo, and Jupiter's Satellites," 19.
74. See PH, MS HMC, 241/IV, fol. 13*r*.
75. See Chapter 6.
76. See Chapter 6.
77. Francis Bacon, *Philosophical Studies, c. 1611–c. 1619*, ed. and with introduction, notes, and commentary by Graham Rees (Oxford: Clarendon Press, 1996), 6:132, 156, 174, 192. For a detailed analysis of these important writings by

Bacon, see Philippe Boulier, "Cosmologie et science de la nature chez Francis Bacon et Galilée," Ph.D. diss., Paris, 2010, 534–565.

78. See Marjorie Hope Nicolson, "The 'New Astronomy' and English Imagination," *Studies in Philology*, 32 (1935): 428–462; Nicolson, *Science and Imagination* (Ithaca, NY: Great Seal Books, 1956), 46–49, 53.
79. On the hostility and fear that the atomistic conception sparked in the minds of English poets, see Stephen Clucas, "Poetic Atomism in Seventeenth-Century England: Henry More, Thomas Traherne and 'Scientific Imagination,'" *Renaissance Studies* 3 (1991): 328.
80. See *The Poems of John Donne*, ed. H. J. C. Grierson (Oxford: Oxford University Press, 1912), 1:237–238.
81. See Massimo Bucciantini, *Galileo e Keplero. Filosofia, cosmologia e teologia nell'età della Controriforma* (Turin: Einaudi, 2003), 246–247.
82. See Marjorie Hope Nicolson, "Kepler, the Somnium and John Donne," *Journal of the History of Ideas* 3 (1940): 268.
83. On the date the *Conclave Ignati* was written, see Willem Heijting and Paul R. Sellin, "John Donne's 'Conclave Ignati': The Continental Quarto and Its Printing," *Huntington Library Quarterly* 62 (1999): 401; see also Robert Cecil Bald, *John Donne: A Life* (Oxford: Oxford University Press, 1970), 227–228.
84. See Evelyn Simpson, *A Study of the Prose Works of John Donne* (Oxford: Clarendon Press, 1948^2), 194–195. See also Howard Marchitello, *The Machine in the Text: Science and Literature in the Age of Shakespeare and Galileo* (New York: Oxford University Press, 2011), 116–122.
85. See John Donne, *Ignatius his Conclave: An Edition of the Latin and English Texts*, ed. Timothy Stafford Healy (Oxford: Clarendon Press, 1969), 112.
86. Regarding accusations against the Jesuits, whom most public opinion at the time considered the moral instigators of Ravaillac's murder, see Georges Minois, *Le couteau et le poison. L'assassinat politique en Europe* (Paris: Fayard, 1997), chapter 7. Regarding the image of the Jesuits as political conspirators, found also in *Ignatius his Conclave*, or the so-called Black Legend that accompanied them from their first missions in England, see William S. Maltby, *The Black Legend in England: The Development of Anti-Spanish Sentiment, 1558–1660* (Durham, NC: Duke University Press, 1971).
87. Donne, *Ignatius his Conclave*, 6–7.
88. See R. Chris Hassel Jr., "Donne's 'Ignatius His Conclave' and the New Astronomy," *Modern Philology* 68 (1971): 329–337.
89. Donne, *Ignatius his Conclave*, 9.
90. On the meaning of the inclusion of these figures in Donne's satirical pamphlet, see Denise Albanese, *New Science, New World* (Durham, NC: Duke

University Press, 1996), 39–58. Regarding Machiavelli, whose dialogue with Ignatius occupies one-third of the work, see Stefania Tutino, "Notes on Machiavelli and Ignatius Loyola in John Donne's 'Ignatius his Conclave' and 'Pseudo-Martyr,'" *English Historical Review* 119 (2004): 1308–1321.

91. Donne, *Ignatius his Conclave*, 13.
92. Ibid., 81.
93. John Webster, *The Duchess of Malfi*, ed. John Russell Brown (Manchester: Manchester University Press, 1997), Act II, Scene IV, 90. On the derivation of this passage from Donne's work, see Robert William Dent, *John Webster's Borrowing* (Berkeley: University of California Press, 1960), 200. Webster was also very familiar with *An Anatomy of the World*, the verses of which recur insistently in his drama; see Matthew Winston, "Gendered Nostalgia in *The Duchess of Malfi*," *The Renaissance Papers*, 1998, 103–113.
94. See Bald, *John Donne: A Life*, 119–123, 145–150.
95. On this hypothesis, see Nicolson, "Kepler, the Somnium, and John Donne," 269.
96. See Chapter 4.
97. On the important historical details regarding relations between Donne and Percy, see Dennis Thomas Flynn, *John Donne and the Ancient Catholic Nobility* (Bloomington: Indiana University Press, 1995), 16, 83, 159–159, 177. See also Bald, *John Donne: A Life*, 133–134.
98. See Jeffrey Johnson, "'One, Four, and Infinite': John Donne, Thomas Harriot, and 'Essayes in Divinity,'" *John Donne Journal* 22 (2003): 115.
99. See Gordon R. Batho, "The Library of the 'Wizard' Earl: Henry Percy Ninth Earl of Northumberland (1564–1632)," *The Library* 15 (1960): 246–261.
100. See John Donne, *Letters to Severall Persons of Honour* (London: J. Flesher, 1651), 152, 195.
101. See Batho, "Thomas Harriot and the Northumberland Household," 37–38, 41.
102. See *The Poems of John Donne*, 1:237.
103. See Lower to Harriot, February 7, 1610, in Rigaud, *Supplement to Dr. Bradley's Miscellaneous Works*, 43.
104. See Kargon, *Atomism in England from Hariot to Newton*, 20.
105. See Jacquot, "Harriot, Hill, Warner and the New Philosophy," 107–109.
106. See Johnson, "'One, Four, and Infinite,'" 115–116.
107. See Bald, *John Donne: A Life*, 389ff.

8. Conquering France

1. Christophe Justel, *Codex canonum Ecclesiae universae* (Paris: H. Beys, 1610). The work, which contained 207 canons developed by the various councils, was published in May. On the twenty-fifth of the month, Sarpi had already penned a letter to Castrino to say that he was "eagerly" awaiting it. At the end of August, he commented, "I am enjoying it and find choices made with exquisite judgment." Paolo Sarpi, *Lettere ai protestanti,* ed. Manlio Duilio Busnelli (Bari: Laterza, 1931), 2:86, 102.
2. See Chapter 2.
3. Pierre de L'Estoile, *Mémoires-Journaux,* ed. Gustave Brunet et al. (Paris: Librairie des Bibliophiles, 1881), 10:200–201. L'Estoile's words were "un livre fort curioux."
4. See Peiresc to Galileo, January 26, 1634, in OG, 16:27. See also OG, 12:405, 16:169. On his stay in Padua, see Cecilia Rizza, *Peiresc e l'Italia* (Turin: Giappichelli, 1965), 9–19, 51–67.
5. See Pierre Gassendi, *Viri illustris Nicolai Claudii Fabricii de Peiresc . . . vita* (The Hague: Vlacq, 1655), 77.
6. Peiresc to Pace, July 28, 1610, in BIC, MS 1875, fol. 93*r*: "des personnes qui peuvent tout sur moy." See Rizza, *Peiresc e l'Italia,* 191. On Pace, see Cesare Vasoli, "Giulio Pace e la diffusione europea di alcuni temi aristotelici padovani," in *Aristotelismo veneto e scienza moderna,* ed. Luigi Olivieri (Padua: Antenore, 1983), 2: 1009–1034.
7. Pace to Peiresc, late August 1610, in BNF, MS Fonds Français 9541, fol. 28*r*. Pace's son sent the book on August 23. See BNF, MS Fonds Français 9541, fol. 29*r*. A letter from Peiresc to Pace dated November 20 tells us that the volume, delivered to du Vair, was lost: "Il fault que je vous confesse que vostre [livre] s'est esgaré, je ne sçay comment, chez Mons.r le premier president" (BIC, MS 1875, fol. 96*v*).
8. See Rizza, *Peiresc e l'Italia,* 191.
9. On Gaultier, who was the member of the Provençal group with the greatest astronomical expertise, see Pierre Humbert, "Joseph Gaultier de La Valette, astronome provençal (1564–1647)," *Revue d'histoire des sciences et de leurs applications* 1 (1948): 314–322. On Peiresc's efforts, see Nicolas-Claude Fabri de Peiresc, *Lettres à Malherbe (1606–1628),* ed. Raymond Lebègue (Paris: CNRS, 1976), 36–37.
10. François de Malherbe, *Oeuvres,* ed. M. Ludovic Lalanne (Paris: Hachette, 1862–1869), 3:109. Peiresc claimed that his collection of telescopes included an instrument built by Jacob Metius; see OG, 16:27.

11. See Gassendi, *Viri illustris Nicolai Claudii Fabricii de Peiresc . . . vita*, 78.
12. Peiresc to Pace, November 7, 1610, in BIC, MS 1875, fol. 96*r*.
13. Ibid.; "[J]'ey faict mettre des petites pointes sur les extremités du dehors comme aux harquebuzes."
14. BIC, MS 1875, fol. 95*v*. See also Rizza, *Peiresc e l'Italia*, 191–192.
15. Peiresc to Pace, November 20, 1610, in BIC, MS 1875, fol. 96*v*. See Rizza, *Peiresc e l'Italia*, 192.
16. Peiresc to Pace, January 10, 1611, in BIC, MS 1875, fols. 105–106. See Rizza, *Peiresc e l'Italia*, 194–195.
17. Peiresc to Pace, June 21, 1611, in BIC, MS 1875, fol. 110*r*. See Rizza, *Peiresc e l'Italia*, 196.
18. See Seymour L. Chapin, "The Astronomical Activities of Nicolas Claude Fabri de Peiresc," *Isis* 48, no. 1 (1957): 19.
19. Peter N. Miller, "Description Terminable and Interminable: Looking at the Past, Nature and Peoples in Peiresc's Archive," in *Historia: Empiricism and Erudition in Early Modern Europe*, ed. Gianna Pomata and Nancy Siraisi (Cambridge, MA: MIT Press, 2005), 374. The material is conserved in BIC, MS 1803.
20. See Miller, "Description Terminable and Interminable," 375–376.
21. Peiresc to Pace, September 29, 1611, in BIC, MS 1875, fol. 113*r*.
22. See Miller, "Description Terminable and Interminable," 375. In his letter to Pace dated September 29, 1611, Peiresc offered a detailed description of the observations made with two different telescopes; see BIC, MS 1875, fol. 113*r*.
23. Peiresc to Pace, January 10, 1611, in BIC, MS 1875, fols. 105–106. See Rizza, *Peiresc e l'Italia*, 194–195.
24. Peiresc to Pace, June 21, 1611, in BIC, MS 1875, fol. 110*r*.
25. BIC, MS 1875, fol. 124*v*.
26. Peiresc to Pace, September 20, 1611, in BIC, MS 1875, fol. 111*r*.
27. Peiresc to Pace, January 10, 1611, in BIC, MS 1875, fols. 105–106.
28. Peiresc to Pace, June 21, 1611, in BIC, MS 1875, fol. 110*r*. See Rizza, *Peiresc e l'Italia*, 195.
29. Peiresc to Pace, January 10, 1611, in BIC, MS 1875, fols. 105–106.
30. "Pour avoyr moien de faire ma dedication a la rayne": Peiresc to Pace, June 21, 1611, in BIC, MS 1875, fol. 110*r*.
31. See Malherbe, *Oeuvres*, 3:241–242.
32. On this, see M. l'abbé Auguste (Alphonse Auguste), "Un dessin inédit de Chalette," *Bulletin de la Societé archéologique du Midi de la France* (1912–1914), 185–187.
33. See Gassendi, *Viri illustris Nicolai Claudii Fabricii de Peiresc*, 79.

34. The limitations of the activities of Peiresc and his group are emphasized by Antonio Favaro, "Niccolò Fabri de Peiresc," in Favaro, *Amici e corrispondenti di Galileo,* ed. Paolo Galluzzi (Florence: Salimbeni, 1983), 3:1552–1553. Chapin, "The Astronomical Activities of Nicolas Claude Fabri de Peiresc," 17, instead deems the data to be "remarkably accurate considering the short time spent in observation and the rudimentary equipment used."
35. Peiresc to Gualdo, January 2, 1615, in OG, 12:125 (our italics).
36. Peiresc to Galileo, January 26, 1634, in OG, 16:27 (our italics).
37. See Gassendi, *Viri illustris Nicolai Claudii Fabricii de Peiresc,* 79–80; Rizza, *Peiresc e l'Italia,* 198–199; Chapin, "The Astronomical Activities of Nicolas Claude Fabri de Peiresc," 17–18.
38. Quoted in Cecilia Rizza, "Galileo nella corrispondenza di Peiresc," *Studi francesi* 15 (1961): 433–451, esp. 437.
39. Camille de Rochemonteix, *Un Collège de Jésuites aux XVIIe et XVIIIe siècles. Le Collège Henry IV de la Flèche* (Le Mans: Leguicheux, 1889), 1:52–53.
40. For a detailed description of the event, see Rochemonteix, *Un Collège de Jésuites,* 139–143; see also *Le convoy du coeur de tres-auguste, tres-clément et tres-victorieux Henry le Grand . . . depuis la ville de Paris jusques au Collège Royal de La Flèche* (Lyon: Morillon, 1610).
41. See Rochemonteix, *Un Collège de Jésuites,* 144–146.
42. "Sur la mort du Roy Henry le Grand et sur la decouverte de quelques nouvelles planettes ou estoiles errantes autour de Jupiter, faicte l'année d'icelle par Galilée, celèbre mathématicien du duc de Florence." The sonnet was published in the collection titled *In Anniversarium Henrici Magni obitus diem Lacrymae Collegii Flexiensis Regii S. J.* (La Flèche: Rezé, 1611).
43. Quoted in Stephen Toulmin, *Cosmopolis: The Hidden Agenda of Modernity* (New York: Free Press, 1990), 60. The original text reads:

> La France avait deja repandu tant de pleurs
> Pour la mort de son Roy, que l'empire de l'onde
> Gros de flots ravageait à la terre ses fleurs,
> D'un déluge second menaçant tout le monde;
>
> Lorsque l'astre de jour, qui va faisant la ronde
> Autour de l'Univers, meu des proches malheurs
> Qui hastaient devers nous leur course vagabonde
> Lui parle de la sorte, au fort de ses douleurs:
>
> France de qui les pleurs, pour l'amour de ton Prince
> Nuisent par leur excès à toute autre province,
> Cesse de t'affliger sur son vide tombeau;

Car Dieu l'ayant tiré tout entier de la terre,
Au ciel de Jupiter maintenant il esclaire
pour servir aux mortels de céleste flambeau.

See Rochemonteix, *Un Collège de Jésuites*, 147.

44. Ibid., 148.
45. See Toulmin, *Cosmopolis*, 60–61. According to Geneviève Rodis-Lewis, "We can believe that, in the humanities class of 1610–11, he contributed to a collection of poems written by the students of La Flèche for the anniversary of the death of their founder." Geneviève Rodis-Lewis, *Descartes: His Life and Thought*, trans. Jane Marie Todd (Ithaca, NY: Cornell University Press, 1998), 14. Originally published as *Descartes, Biographie* (Paris: Calmann-Levy, 1995).
46. See Maurizio Torrini, "'Et vidi coelum novum et terram novam.' A proposito di rivoluzione scientifica e libertinismo," *Nuncius* 1 (1986), 51.
47. This is the hypothesis of Antonella Romano, *La contre-réforme mathématique. Constitution et diffusion d'une culture mathématique jésuite à la Renaissance* (Rome: École Française de Rome, 1999), 490–491.
48. Giuliano Giraldi, *Esequie d'Arrigo Quarto . . . celebrate in Firenze dal Serenissimo Don Cosimo II* (Florence: Bartolommeo Sermartelli et fratelli, 1610), 8, quoted in *"Parigi val bene una messa!" L'omaggio dei Medici a Enrico IV re di Francia e di Navarra*, ed. Monica Bietti, Francesca Fiorelli Malesci, and Paul Mironneau (Livorno: Sillabe, 2010), 59, 65.
49. Sara Mamone, "Il re è morto, viva la regina," in *"Parigi val bene una messa!,"* 33–34.
50. Sarpi, *Lettere ai protestanti*, 1:130.
51. Galileo to Giugni, June 25, 1610, in OG, 10:381.
52. Traiano Boccalini, *Ragguagli di Parnaso e pietra del paragone politico* ("Per segno di intimo dolore con una oscurissima nube si velò subito la faccia, dalla quale per tre giorni continui versò pioggia di abbondantissime lacrime; e i letterati tutti, spagnuoli, inglesi, fiamminghi, tedeschi e italiani, con abbondanza maggiore di lacrime si son veduti pianger il caso infelicissimo di tanto re che gli stessi franzesi"). The English translation is by Henry, Earl of Monmouth, *I Ragguagli di Parnasso: or Advertisements from Parnassus in Two Centuries with the Politick Touch-stone* (London: Humphrey Mosely and Thomas Heath, 1556). On this point, see Frances Amelia Yates, *Astraea: The Imperial Theme in the Sixteenth Century* (London: Routledge & Kegan Paul, 1975), 213. See also Torrini, "Et vidi coelum novum et terram novam."
53. Vinta to Pannocchieschi d'Elci, May 23, 1610, in ASF, Mediceo del Principato 302, fols. 106*v*–107*r*. For an almost journalistic chronicle of the event, see Vincent J. Pitts, *Henri IV of France: His Reign and Age* (Baltimore: Johns Hopkins University Press, 2009), 317–331.

54. ASF, Mediceo del Principato 302, fol. 107*r*. Note: this passage was published by Antonio Favaro in OG, 10:356, but the preceding one regarding the death of Henry IV is missing. According to the publishing standards followed in the Edizione Nazionale, only passages discussing Galileo are included in the correspondence between other figures (i.e., when Galileo is neither the sender nor the recipient). In this case, however, the general sense of the letter is lost.
55. See Botti to Vinta, July 6, 1610, in OG, 10:392.
56. Letter from Vinta to Cioli, August 23, 1610, in ASF, Mediceo del Principato 4872, unnumbered fol., with date.
57. Cioli to Vinta, September 13, 1610, in OG, 10:430.
58. Fontanelli to Ruggeri, April 1610, in OG, 10:347.
59. Botti to Vinta, September 19, 1610, in OG, 10:433.
60. Botti to Galileo, August 18, 1611, in OG, 11:173.
61. Ibid.

9. Milan: At the Court of "King" Federico

1. Quoted in Federico Borromeo, *Il Museo,* Italian trans. L. Grasselli, preface and notes Luca Beltrami (Milan: Tip. U. Allegretti, 1909), 57 n. 1.
2. Federico Borromeo, *Musaeum. La Pinacoteca ambrosiana nelle memorie del suo fondatore,* commentary Gianfranco Ravasi, Italian trans. and notes Piero Cigada (Milan: Claudio Gallone Editore, 1997), 27. Regarding *Air,* he observed: "He surrounded Air, seen as a plain of light, with all sorts of pleasant things. If one must add something for the purpose of comparison, for this last painting the artist seems to have lavished all care, as it was the completion of the entire work" (29).
3. ABIB, filza L.III.20, "Minute del cardinale F. Borromeo," 1616–1617.
4. Stefania Bedoni, *Jan Brueghel in Italia e il collezionismo del Seicento,* foreword by Pierluigi De Vecchi and introduction by Bert W. Meijer (Florence: Litografia Rotoffset, 1983), 193. On Borromeo's artistic interests, see also Irene Baldriga, *L'occhio della lince. I primi Lincei tra arte, scienza e collezionismo (1603–1630)* (Rome: Accademia Nazionale dei Lincei, 2002), 15–31; Lucy C. Cutler, "Virtue and Diligence: Jan Brueghel I and Federico Borromeo," in *Virtus: virtuositeit en kunstliefhebbers in de Nederlanden 1500–1700,* ed. Jan de Jong, Dulcia Meijers, and Mariët Westermann (Zwolle: Waanders Nederlands kunsthistorisch jaarboek, 2004), 203–227.
5. Federico Borromeo, *Della pittura sacra libri due,* ed. Barbara Agosti (Pisa: Scuola Normale Superiore, 1994), 30. The work, in Latin, was published in 1624. We should also consider Ferdinando Bologna's comments regarding the

"disorder" and lack of decorum that many of Caravaggio's paintings must have had in Borromeo's eyes: "Federico Borromeo . . . could never have appreciated a work by Caravaggio like the London *Emmaus*, in which still life, set in a dazzling foreground, competes in importance with the sacred story 'cuius causa [tabula] instituta erat.'" Bologna, *L'incredulita del Caravaggio e l'esperienza delle "cose naturali"* (Turin: Bollati Boringhieri, 1992), 131.

6. Borromeo, *Della pittura sacra*, 30.
7. ASDM, Mensa, 19, 357b. On this subject, see Isabella Balestrieri, *Le fabbriche del Cardinale Federico Borromeo, 1595–1631. L'Arcivescovado e l'Ambrosiana* (Benevento: Hevelius, 2005), 41, which mentions the wrong date of January 13, 1614; the exact transcription is: "May 2, 1614 . . . as credit to Messer Gio. Battista Velate for the cost of a spyglass sent to His Excellency from Venice."
8. BAM, G 264 inf., fol. 41*r*, about which see Carlo Marcora, "La biografia del cardinal Federico Borromeo scritta dal suo medico personale Giovanni Battista Mongilardi," *Memorie storiche della Diocesi di Milano* 15 (1968): 134–136. Identification of the site of Pobigo is uncertain, but it might be the modern Pobiga, in the municipality of Besana Brianza; on this, see Pierino Boselli, *Toponimi lombardi* (Milan: SugarCo, 1977), 217.
9. Francesco Rivola, *Vita di Federico Borromeo* (Milan: Gariboldi, 1656), 635.
10. Letter dated September 3, 1616, in OG, 12:276. See Favaro, *Amici e corrispondenti di Galileo*; Giuseppe Gabrieli, "Federico Borromeo e gli accademici Lincei," in Gabrieli, *Contributi alla storia dell'Accademia dei Lincei* (Rome: Accademia dei Lincei, 1989), 2:1465–1486.
11. On Schreck, see Giuseppe Gabrieli, "Giovanni Schreck linceo, gesuita e missionario in Cina e le sue lettere dall'Asia," in Gabrieli, *Contributi alla storia dell'Accademia dei Lincei*, 2:1011–1051; Isaia Iannaccone, *Iohann Schreck Terrentius: le scienze rinascimentali e lo spirito dell'Accademia dei Lincei nella Cina dei Ming* (Naples: Istituto Universitario Orientale, 1998).
12. "While Father Terenzio was passing through here, he said he would like a spyglass, and I promised to send him one. Since I do not know where he is now, I will send it to you that you might safely deliver it to him." Borromeo to Faber, March 30, 1616, quoted in Fabrizio Cortesi, "Lettere inedite del cardinale Federico Borromeo a Giovan Battista Faber segretario dei primi Lincei," *Aevum* 6 (1932): 516. See also Gabrieli, "Federico Borromeo e gli accademici Lincei," in Gabrieli, *Contributi alla storia dell'Accademia dei Lincei*, 2:1473.
13. OG, 12:275.
14. Ibid.
15. BAM, G 253 inf., fol. 154*r*. The texts referred to by Welser are the three *Epistolae de maculis solaribus* and the *De maculis solaribus et stellis circa Iovem errantibus*

accuratior disquisitio, respectively published in January and September 1612 by the Jesuit Christoph Scheiner.

16. Cavalieri to Galileo, May 13, 1621, in OG, 13:55.
17. Christoph Scheiner, *De maculis solaribus et stellis circa Iovem errantibus accuratior disquisitio* (Augsburg: Ad insigne pinus, 1612); see OG, 5:62.
18. Federico Borromeo, *Pro suis studiis,* in BAM, G 310 inf., ins. 8, 273–276. The text should also be read for other considerations on natural philosophy, and particularly for several optical experiments citing the use of the *camera obscura,* conducted personally by Borromeo. "We will also mention what seemed to be a great secret of nature, that test I have conducted many times and always with great admiration, as one makes a hole in a panel, and then after placing a piece of glass over the hole, everything outside is represented on a sheet of paper, and if the glass is good it is represented with great proportion and such lively colors that no painter can rival it. So that in one point it is possible to paint a beautiful picture, on a sheet of paper or something else, that will have nothing comparable in painters' workshops. I believe that art has not found this secret, but that chance and the will of God led to people discovering it accidentally: from here, I would like to make a consideration, and I think to myself how many other things of the marvel of nature, and of God who produced it, are still hidden and unknown, and perhaps may never be known" (276–278).
19. Pamela M. Jones, *Federico Borromeo and the Ambrosiana: Art Patronage and Reform in Seventeenth-Century Milan* (Cambridge: Cambridge University Press, 1993), 83–84.
20. Borromeo, *Occhiale celeste,* in BAM, I 52 ff., 42, now in Giliola Barbero, Massimo Bucciantini, and Michele Camerota, "Uno scritto inedito di Federico Borromeo: l'"Occhiale celeste,'" *Galilaeana* 4 (2007): 322. On this, see the letter from Gerolamo Settala to Borromeo dated December 1, 1618: "Two days ago at ten o'clock, we saw a star rise in the east with rays towards the sky resembling a comet in everything else, but before there was another dense body of air shaped like a long beam, quite curved, and it seems to show the colors of the rainbow but without blue" (BAM, G 313a inf., fol. 171*r*).
21. Borromeo, *Occhiale celeste,* 42, in Barbero, Bucciantini and Camerota, "Uno scritto inedito," 322.
22. Ibid., 327.
23. As can be noted from the page numbers that have been cited, Federico used the Frankfurt reprint of the *Sidereus.* See ibid., 322 and 332–333. However, we must note that the seventeenth-century catalogues of the Ambrosiana also record the presence of a copy of the first Venetian edition. See BAM, Z 38 inf., fol. 82*v*; MS Z 27 inf., 144.

24. Borromeo, *Occhiale celeste,* 45, in Barbero, Bucciantini, and Camerota, "Uno scritto inedito," 324.
25. Ibid., 325–326.
26. Federico Borromeo, *I tre libri delle laudi divine* (Milan, 1632), 24.
27. "*De veritate Scripturae Sacrae.* Sumitur occasio a Galilaeo. Examinantur illius rationes. Confutantur," in BAM, G 310 inf. (ins. 4), fol. 17*r*.
28. "Quaestio de motu terrae ex Gallileo confutatur, et [iuxta] illud Psalmi 135, *qui firmavit terram super aquas*": BAM, G 72 ff., 125.
29. BAM, I 58 inf. The full title is *Prima pars introductoriae constructionis astronomiae duas in partes distributae. In eam videlicet quae de primo mobili et eam quae de secundis mobilibus est.* The second part is missing. On Casati and his *Constructio astronomiae,* see Michele Camerota, "Galileo e il Parnaso tychonico," in *Galileo e il Parnaso tychonico. Un capitolo inedito del dibattito sulle comete tra finzione letteraria e trattazione scientifica,* ed. Ottavio Besomi and Michele Camerota (Florence: Olschki, 2000), 112–118.
30. "Opinionem de Terrae motu secundum hypothesim Copernici, ita exacte ipsis caelestibus phaenomenis congruere, ut nihil hac in re verisimilius adduci posse": BAM, I 58 inf., fol. 22*v*.
31. "Cum ipsis caelesti numine praefatis magis consentiens": BAM, I 58 inf., fol. I*r*.
32. "Nam etiam si exploratum habeam fore ut primo haereas talibus attonitus veris atque etiam perhorrescas nimium insolitae novitatis aspectu, ordine praepostero universi orbis naturam ac dispositionem ob oculos ponentis et omnium philosophorum (licet non ipsius verae philosophiae) dogmata longo usu et inveterata omnium opinione sancita prorsus abrogantis; nihilominus, si utroque, ut dicitur, oculo rem ipsam introspicere ac diligenti animadversione perpendere volueris in consilium adhibitis rationibus quae ibi ad hanc rem insinuandam copiose satis subministrantur, persuadere mihi nullo modo possum te tantum humana philosophorum auctoritati daturum, ut propter illam rationes ipsas divinis oraculis confirmatas continuo tibi esse abiciendas putes": BAM, I 58, inf., fol. I*v*.
33. "Si quis tentaret supradictam de terrae motu opinionem sacrarum litterarum testimonio corroborare, id ei forte ex voto felicius contingeret, ac multo magis verisimilibus coniecturis confirmare posset opinionem suam, quam reliqui alterius sententiae suam": BAM, I 58, inf., fols. 17*v*–18*r*.
34. Cavalieri to Galileo, April 28, 1621, in OG, 13:62.
35. See Chapter 1. In August 1609, a message from Milan noted that "a Frenchman gave His Excellency [Governor Pedro Enríquez de Acevedo, Count of Fuentes] an instrument shaped like a trumpet, with which he could see faraway things

as if they were near, and since they provided good recognition, he ordered that two more be made to send to Spain": ASV, Fondo Borghese, IV.52, fol. 305*r*.

36. Vinta to Beccheria, June 21, 1611, in ASF, Mediceo del Principato 3137, fol. 383*v*.
37. See Cristoforo Borri, *Tractatus astrologiae*, BAM, A 83 ff., fols. 1–104. A copy of the work, dated 1615, is at the BNCR, Fondo Gesuitico, MS 587. On Borri and his *Tractatus*, see Camerota, "Galileo e il Parnaso tychonico," 119–126.
38. See Borri, *Tractatus astrologiae*, in BAM, A 83 ff., fols. 17*r*–*v* and 52*v*–53*r*.
39. BAM, A 83 ff., fols. 63*v*–67*v*.
40. "Septem ab hinc annis ex quo animum mathematicis scientiis applicare coepi, cum, communem coelestium orbium descriptionem et distributionem ingressus, animadvertissem confusionem illam tot tantorumque epiciclorum, eccentricorumque a Ptolomaeis fictorum, ita animus meus ab illis abhorruit, ut nunquam adduci potuerim, ut illos crederem": BAM, A 83 ff., fol. 63*r*–*v*.
41. "Certissimum esse Lunam non esse perfectissime rotundam, sed multis vallibus ac montibus reddi inaequalem; probabilius etiam est caetera astra esse montibus referta sicut Luna. De Luna non indiget probatione cum sensu pateat benefitio perspicilli, qui nuper a Galileo de Galileis Florentino, Patavii Gymnasii publico mathematico repertum esse dicitur, quem quidem perspicillum statim ac praemanibus habui, observavi hos montes et valles in Luna, sed non sum ausus antea proferre, ne alicuius temeraritatis notam subirem, sed postquam haec et alia, quae suo loco dicemus, iam video ab ipso Galileo observata, immo et in lucem edita, veritatem hanc confirmare et propagare non dubitabo": BAM, A 83 ff., fols. 31*v*–32*r*.
42. "Ut sic videmus in Terra circa Solis ortum, cum valles terrae nondum lumine perfusas, montes vero illas ex adverso Solis circundantes iam iam splendorem fulgentes intuemur, ac veluti terrestrium cavitatum umbrae, Sole sublimiora petente, imminuuntur, ita ut lunares istae cavitates, crescente parte luminosa, tenebras amittunt": BAM, A 83 ff., fol. 32*r*. The passage is identical to what we find in the *Sidereus nuncius*; see OG, 3:63–64. "And we have an almost entirely similar sight on Earth, around sunrise, when the valleys are not yet bathed in light but the surrounding mountains facing the Sun are already seen shining with light. And just as the shadows of the earthly valleys are diminished as the Sun climbs higher, so those lunar spots lose their darkness as the luminous part grows." Galileo Galilei, *Sidereus Nuncius or the Sidereal Messenger*, trans. with introduction, conclusion, and notes by Albert Van Helden (Chicago: University of Chicago Press, 1989), 41.
43. "Addit Cheplerus, in sua *Dioptrica*, Saturnum summa cum admiratione deprehendi, non unam solum esse stellam, sed tres inter se proximas, adeo ut sese mutuo quasi contingant" (In his *Dioptrice*, Kepler said he had learned with

great admiration that Saturn is not a single planet but three separate ones, so close to each other that they almost touch): Borri, *Tractatus astrologiae*, in BAM, A 83 ff., fol. 39*v*. On Venus, see BAM, A 83 ff., fol. 40*r*.

44. "Hoc non indiget probatione, ut demum est de montibus in Luna, cum sensu pateat, ope eiusdem perspicilli, quae quidem stellae ab ipso Galileo ante omnes observatae sunt, et ego ipse quam diligenter [diligentius] potui, observavi. Horum vero veri motus et periodi nondum determinari potuerunt": BAM, A 83 ff., fol. 39*r*.

45. Barbavara left a number of manuscripts, including trigonometric studies in BAM, G 84 ff. On Muzio Oddi and his mathematical activity in Milan, see Alexander Marr, *Between Raphael and Galileo: Mutio Oddi and the Mathematical Culture of Late Renaissance Italy* (Chicago: University of Chicago Press, 2011), 57–105. Rho went to China as a Jesuit missionary in 1618, translating and composing a number of works on astronomy into Chinese, including one discussing Galileo's discoveries. The *provveditore* of the University of Pisa, Girolamo da Sommaia, had this to say about him: "Messer Alessandro Ro [*sic*], who studied in Pisa and has a musician son in Milan and another who is a Jesuit [Giacomo], and who today is in the Indies and is a great mathematician, particularly of instruments and mechanical things, which he makes himself, would have been a very great mathematician if his fathers had let him study mathematics alone, but they diverted his attention to many other things" (BNCF, MS Magl. XI, 57, fol. 7*v*). And Baranzano's *Uranoscopia* was published in 1617. Baranzano took his vows in Monza in 1609 and stayed in Milan until 1615.

46. BAM, G 309 inf., 7; quoted in Eraldo Bellini, *Stili di pensiero nel Seicento italiano. Galileo, i Lincei, i Barberini* (Pisa: Edizioni ETS, 2009), 97. The work in question is *Salomon, sive Opus Regium*, printed in Latin in 1617.

10. The Dark Skies of Florence

1. Piccolomini of Aragon to Galileo, March 27, 1610, in OG, 10:305.
2. See Chapter 4 (letter from Sertini, March 27, 1610, in OG, 10:305).
3. ASF, Guidi, 542, unnumbered fol., under the date. The recipient is anonymous.
4. BRF, MS 2446, letter no. 134.
5. These supplies are detailed in the ledgers of his workshop-foundry at San Marco. Among the payment orders for 1610, under the date of November 30 we find "fourteen tubes [*cannoni*] for spyglasses ordered from Venice," as well as "thirteen tubes for spyglasses" in August and "120 lenses for spyglasses . . . [and]

eight tubes for spyglasses ordered from Venice" on March 31, 1611 (ASF, Mediceo del Principato 5132/A, payment orders for 1610, unnumbered fols., under the respective dates). On Antonio de' Medici, see Paolo Galluzzi, "Motivi paracelsiani nella Toscana di Cosimo II e di Don Antonio dei Medici: alchimia, medicina chimica e riforma del sapere," in *Scienze, credenze occulte, livelli di cultura* (Florence: Olschki, 1982), 31–62.

6. OG, 10:423.
7. See Antonio Santucci, *Trattato nuovo delle comete, che le siano prodotte in cielo, e non nella regione dell'aria, come alcuni dicono* (Florence: Caneo, 1611), 102–116.
8. Quoted in Massimo Bucciantini, *Contro Galileo. Alle origini dell'"affaire"* (Florence: Olschki, 1995), 27 n. 1.
9. Bonifacio Vannozzi, *Delle lettere miscellanee . . . volume terzo* (Bologna: Bartolomeo Cochi, 1617), 407. For bibliographical information about Vannozzi and Baldinotti, see Vittorio Capponi, *Biografia pistoiese* (Pistoia: Tipografia Rossetti, 1878), 30, 384–386.
10. William Shakespeare, *Hamlet*, Act II, Scene 2. As we know, *Hamlet* was first staged between the end of 1598 and 1601.
11. The timing can be determined from the fact that pages 403–405 of the *Lettere miscellanee* include a letter dated August 17, 1610, in which Baldinotti thanks Vannozzi for sending him a spyglass. In his answer, Vannozzi observed that "Galileo's spyglass deserves to be held in great esteem because it was a gift given to me by His Excellency, but the invention is not considered new, as some say that Archimedes also made them, and [Della] Porta mentions it in his treatise on *specilli*, which resemble the lens. Nevertheless, being its renewer is certainly praiseworthy" (405). Immediately after this (406–407) is Vannozzi's letter to Baldinotti, cited in the text but without the date.
12. Undated letter, but ascribed to 1610, in Vannozzi, *Delle lettere miscellanee*, 182.
13. Ibid., 181 (part of the aforementioned letter).
14. See Luigi Guerrini, *Galileo e la polemica anticopernicana a Firenze* (Florence: Edizioni Polistampa, 2009), 25–26, 30–32; Eileen Reeves, "Variable Stars: A Decade of Historiography on the *Sidereus nuncius*," *Galilaeana* 8 (2011): 41–42.
15. See Massimo Bucciantini, "Reazioni alla condanna di Copernico: nuovi documenti e nuove ipotesi di ricerca," *Galilaeana* 1 (2004): 12.
16. Ibid., 11 n. 29.
17. On Don Giovanni de' Medici, see ibid., 3–19; Brendan Dooley, "Narrazione e verità: don Giovanni de' Medici e Galileo," *Bruniana & Campanelliana* 14, no. 2 (2008): 389–403; Paola Volpini in DBI, 73, under the specific entry. See also Edward Goldberg's studies on Benedetto Blanis, a Jewish theologian and Don Giovanni's librarian: *Jews and Magic in Medici Florence: The Secret World of*

Benedetto Blanis (Toronto: University of Toronto Press, 2011) and *A Jew at the Medici Court: The Letters of Benedetto Blanis Hebreo (1615–1621)* (Toronto: University of Toronto Press, 2011).

18. See Chapter 5.
19. See M. Galilei to Galileo, April 14, 1610, in OG, 10:312–313; Hasdale to Galileo, April 15, 1610, in OG, 10:314–315; Altobelli to Galileo, April 17, 1610, in OG, 10:317–318.
20. See Kepler to Galileo, April 19, 1610, in OG, 10:319–340.
21. Galileo to Vinta, May 7, 1610, in OG, 10:349 (our italics).
22. OG, 10:351.
23. OG, 10:351–352.
24. See Chapter 6.
25. See Carlo Carabba and Giuliano Gasparri, "La vita e le opere di Girolamo Magagnati," *Nouvelles de la République des Lettres* 2 (2005): 61–85; Girolamo Magagnati, *Lettere a diversi*, ed. and with an introduction by Laura Salvetti Firpo, contained in "La vita di Girolamo Magagnati," ed. Carlo Carabba and Giuliano Gasparri (Florence: Olschki, 2006), vii–xx.
26. Girolamo Magagnati, *Meditazione poetica sopra i pianeti medicei* (Venice: Heredi d'Altobello Salicato, 1610), fol. [4*v*].
27. Vinta to Galileo, May 22, 1610, in OG, 10:356.
28. See Giuliano de' Medici to Galileo, July 19, 1610, in OG, 10:403–404.
29. Galileo to Vinta, June 18, 1610, in OG, 10:373. For the date of July 25, when Galileo resumed his observation of Jupiter, see OG, 10:373 n. 2.
30. Galileo to Vinta, June 18, 1610, in OG, 10:374.
31. Dal Monte to Galileo, June 4, 1610, in OG, 10:367: "I was especially pleased to learn what you are thinking about doing with rock crystal, so I hope that, through your erudition and ingenuity, other marvelous things may be found."
32. Galileo to Vinta, July 30, 1610, in OG, 10:410.
33. Barbolani to Vinta, June 26, 1610, in OG, 10:384.
34. See Sagredo to Galileo, August 18, 1611, in OG, 11:172.
35. Ibid., 172.
36. Ibid., 170–172.
37. See, for example, Gaetano Cozzi, "Galileo Galilei, Paolo Sarpi e la società veneziana," in *Paolo Sarpi tra Venezia e l'Europa* (Turin: Einaudi, 1979), 216–217 n. 163; Paolo Sarpi, *Istoria del Concilio Tridentino, seguita dalla Vita del padre Paolo di Fulgenzio Micanzio*, ed. Corrado Vivanti (Turin: Einaudi, 1974), 2:lv–lvi.
38. [Francesco Griselini], *Del genio di P. Paolo Sarpi* (Venice: Leonardo Bossaglia, 1785), 2: 70–71.
39. Galileo to Sarpi, February 12, 1611, in OG, 11:48.

40. Micanzio to Galileo, February 26, 1611, in OG, 11:57.
41. Galileo to Sarpi, February 12, 1611, in OG, 11:48.
42. We are grateful to Corrado Pin for providing his detailed considerations regarding Sarpi's language and this note in particular.
43. Galileo to Sarpi, February 12, 1611, in OG, 11:47 (our italics).
44. Galileo to Vinta, August 20, 1610, in OG, 10:425.
45. Galileo to Vinta, March 19, 1610, in OG, 10:299.
46. Galileo to Vinta, June 18, 1610, in OG, 10:373.
47. Galileo to Vinta, July 30, 1610, in OG, 10:410.
48. Sertini to Galileo, August 7, 1610, in OG, 10:412.
49. See Galileo to Kepler, August 19, 1610, in OG, 10:421.
50. See Galileo to Giuliano de' Medici, October 1, 1610, in OG, 10:440.
51. Cesi to Galileo, August 20, 1611, in OG, 11:175.
52. See OG, 10:431–432, 11:46–50.
53. OG, 10:431.
54. Cigoli to Galileo, October 1, 1610, in OG, 10:442.
55. Santini to Galileo, October 9, 1610, in OG, 10:445.
56. OG, 10:431.
57. OG, 10:432.
58. Pignoria to Gualdo, September 26, 1610, in OG, 10:436.
59. Clavius to Galileo, December 17, 1610, in OG, 10:485.
60. Galileo to Clavius, December 30, 1610, in OG, 10:500 (our italics).

11. The Roman Mission

1. See OG, 3:442.
2. Galileo to Vinta, April 1, 1611, in OG, 11:80.
3. See Niccolini to Cosimo II, March 30, 1611, in OG, 11:78–79. On Galileo's trip to Rome, see also John L. Heilbron, *Galileo* (Oxford: Oxford University Press, 2010), 170–77.
4. See Elena Fasano Guarini, "'Roma officina di tutte le pratiche del mondo': dalle lettere del cardinale Ferdinando de' Medici a Cosimo I e a Francesco I," in *La corte di Roma tra Cinque e Seicento. "Teatro" della politica europea*, ed. Gianvittorio Signorotto and Maria Antonietta Visceglia (Rome: Bulzoni, 1998), 265–297.
5. Vinta to Galileo, January 20, 1611, in OG, 11:28–29.
6. "Tu purgasti oculos hominum, et novum ostendis caelum, et novam terram in luna": Campanella to Galileo, January 13, 1611, in OG, 11:23. The passage alludes to verses from *Apocalypse*, 21:1.

7. See OG, 11:60–61, 80–81.
8. Niccolini to Cosimo II, March 30, 1611, in OG, 11:79.
9. Cosimo II to Dal Monte, February 27, 1611, in OG, 11:61.
10. Galileo to Vinta, April 1, 1611, in OG, 11:79.
11. "Galileus saepius ad Collegium nostrum et mathematicos venit": SBD, MS XV.247, fol. 208*v*.
12. See BNCF, MS Gal. 49, fols. 4*r* and 5*r*; see also OG, 3:863–864. In a letter to Vinta dated April 1, Galileo emphasized that his data agreed with the Jesuits'; see OG, 11:80.
13. See BNCF, MS Gal. 13, fol. 2*r*; see also OG, 11:87–88. The autograph signature seems to suggest that Galileo may have received the paper directly from its author or from one of those who had been contacted. However, we should note that in a letter dated August 11, 1611, Ludovico Cigoli announced, "From Cardinal dal Monte's secretary I received the note with the question His Excellency Bellarmine asked the Jesuits": OG, 11:168. Therefore, perhaps he was the one who gave Bellarmine's note to Galileo.
14. OG, 11:92–93.
15. See Welser to Galileo, January 7, 1611, in OG, 11:13–14. In reality, the idea went back to the Middle Ages and was discussed above all starting with Averroës's commentary on *De caelo*. From this standpoint, it was thought that the Moon reflected sunlight after having been "drenched" in it and that this light was uneven due to the different density of its surface. This would explain the appearance of "moonspots," meaning the shadows visible when the Moon is full. Above all, however, by denying that the smooth and perfectly uniform lunar surface reflected the Sun's rays like a mirror, the theory of "imbibition" explained why the light of the Moon would be visible in the same way from every corner of the Earth (whereas reflection would have polarized it to specific places on Earth and would vary depending on the position of the celestial body). On this, see Roger Ariew, "Galileo's Lunar Observations in the Context of Medieval Lunar Theory," *Studies in History and Philosophy of Science* 15, no. 3 (1984): 213–226; Isabelle Pantin, "Galilée, la lune et les Jesuites," in *Galilaeana* 2 (2005), 22–23.
16. Vincenzo Figliucci, *Stanze sopra le stelle e macchie solari scoperte col nuovo occhiale* (Rome: Mascardi, 1615), 19. The first part of the work was composed in spring 1611, when Figliucci was in Rome. The original reads: "Ma più stimar si dee conforme al vero / Ciò che sempre creduto hanno i più dotti, / Ch'in cotal corpo impuro e non sincero / Sian veramente i rai del Sole indotti, / Onde poi non rimandi il lume intiero, / Ma siano i suoi splendor difformi e rotti./ E come il denso fa splendor più chiaro, / Così oscuro lo fa quel ch'è più raro."

See Luigi Guerrini, "Le 'Stanze sopra le stelle e macchie solari scoperte col nuovo occhiale' di Vincenzo Figliucci. Un episodio poco noto della visita di Galileo Galilei a Roma nel 1611," *Lettere Italiane* 50, no. 3 (1998): 387–415.

17. See Galileo to Gallanzoni, July 16, 1611, in OG, 11:151.
18. On the reply, see BNCF, MS Gal. 13, fol. 2bis*r*. A copy of the document is now in the Archivum Romanum Societatis Iesu: Opp. NN 245, fol. 430. Ugo Baldini later identified another copy at the Biblioteca Oliveriana in Pesaro (MS1582, fasc. II, 1); see Ugo Baldini, *Saggi sulla cultura della Compagnia di Gesù* (Rome: Bulzoni, 1995), 238 n. 60. This is probably the document mentioned in a letter from Giovanni Colle to Pier Matteo Giordani dated May 28, 1611: "Raised here are those doubts stated by His Excellency Cardinal Bellarmine regarding the things of Galileo, also with the answers containing the affirmation of several Jesuit fathers." See Enrico Gamba, "Galilei e l'ambiente scientifico urbinate: Testimonianze epistolari," *Galilaeana* 4 (2007): 354. On the assembly at the Collegio Romano, see Welser to Galileo, June 17, 1611, in OG, 11:127.
19. Johannes Albertus Franciscus Orbaan, *Documenti sul Barocco in Roma* (Rome: Società Romana di Storia Patria, 1920), 284.
20. "Presentibus 4 cardinalibus et ipso Galileo Galileo": Guldin to Lanz, May 21, 1611, in SBD, MS XV.247, fol. 234*r*. On the identity of the speaker, see OG, 11:162–163. For the conference on the *nova* of 1604, see Ugo Baldini, *Legem impone subactis. Studi su filosofia e scienza dei Gesuiti in Italia. 1540–1632* (Rome: Bulzoni, 1992, 155–182). Regarding the meeting at the Collegio Romano, see in particular Pantin, "Galilée, la lune et les Jesuites"; Eileen Reeves and Albert Van Helden, "Verifying Galileo's Discoveries: Telescope-making at the Collegio Romano," in *Der Meister und die Fernrohre: das Wechselspiel zwischen Astronomie und Optik in der Geschichte*, ed. Jürgen Hamel and Inge Keil (Frankfurt am Main: H. Deutsch, 2007), 127–141.
21. See OG, 3:293–298.
22. See OG, 3:295.
23. See OG, 3:298.
24. "Non absque Philosophorum murmure": Grégoire de Saint-Vincent to Huygens, October 4, 1659, in Christian Huygens, *Oeuvres completes, Correspondance*, II, 1657–1659 (The Hague: M. Nijhoff, 1889), 490.
25. See OG, 11:274.
26. Tamburelli to Grienberger, November 11, 1611, in OG, 11:233.
27. See Christoph Grienberger, *Catalogus veteres affixarum longitudines conferens cum novis* (Rome: B. Zanettum, 1612), ii [unnumbered fol.].
28. Figliucci, *Stanze sopra le stelle e macchie solari*, 9. The original reads: "Tu Galileo sopra il terrestre limo / Il sentier chiuso a noi primiero apristi. / Tu co' i cristalli

ch'io ne' carmi esprimo, / Di nuove stelle il ciel ricco scopristi. / Mentr'altri al terreo suol, tu il cor alzasti / A merci eterne, e il mar del ciel solcasti."

29. Grienberger to Galileo, January 22, 1611, in OG, 11:33 (our italics).
30. OG, 11:34 (our italics).
31. Ibid. (our italics).
32. Colonna to Galileo, May 16, 1614, in OG, 12:63–64.
33. In his dedication to Cosimo II, Galileo noted that Jupiter and its satellites "all together, in mutual harmony, complete their great revolutions . . . about the center of the world, that is, about the Sun itself": SN, 31. Further ahead, this statement is repeated as a specific reference to the validity of the Copernican system: see SN, 84.
34. DNS, 42–43.
35. Galileo to Vinta, January 15, 1611, in OG, 11:27.
36. Gualdo to Galileo, May 6, 1611, in OG, 11:100.
37. *Ratio atque Institutio Studiorum Societatis Iesu. Ordinamento degli studi della Compagnia di Gesù,* edited by Angelo Bianchi (Milan: Rizzoli, 2002), 194–195.
38. "Quaeso rogas P. Clavio et P. Grinbergero an pro salvandis motibus novorum illorum planetarum Iovis, Saturni, Martis, sufficiat ponere epicyclos, quorum centra sint eadem cum centris Jovis, Saturni, Martis, an vero alia theoria sit excogitanda": UBG, MS 159, env. no. 17, fol. 5*r*. The satellites attributed to Saturn and Mars can be explained through the singular appearances of the two planets viewed using the telescope; in a previous letter Lanz clarified that the confreres at the Collegio Romano had seen two stars on either side of Saturn and others "circa Martem": UBG, MS 159, env. no. 17, fol. 2*r*. He also thought it likely that "non Jovem tantum sed et alios planetas suos habere satellites, cum Sol, Venerem et Mercurium habent" (not only Jupiter, but also the Sun, Venus, and Mercury have theirs satellites): UBG, MS 159, env. no. 17, fol. 7a*r*.
39. See Chapter 12.
40. See Chapter 9.
41. "Confessemos que Venus e Mercurio se movem a o redor do Sol, e que hora abaixo ou assima de lle, hora antes hora depois de lle fazem seu curso, como tambem se pode collegir das varias oppinioines dos antiguos, dos quaes hunos poserâo estes 2 plametas assima outros abaixo do Sol. . . . Ambas estes oppinioines conciliou Ticho Brahe, dilligentissimo e mais moderno observador do curso dos planetas e estrellas, o quoal determinâo que se moviâo a o redor do corpo Sollar no lib. 2 *De mundi aetherei recentioribus phenomenis,* cap. 8, por estas pallavras: Em Venus e Mercurio os mesmos circuitos menores a o redor

do Sol e que nao cercâo a terra, pois parece que tem hum certo modo de epiciclos; e tinha dito assima que todos os planetas tirando a Luna se moviâo a o redor do Sol, como a o redor de capitâo ou Rey seu." ANTT, Livraria, MS 1770, fols. 33*v*–34*r*. For the passage quoted from Brahe, see Tycho Brahe, *Opera Omnia*, ed. John Louis Emil Dreyer (Haunia: In Libraria Gyldendaliana, 1913; repr. Amsterdam: Swets & Zeitlinger, 1972), 4:157.

42. Figliucci, *Stanze sopra le stelle e macchie solari*, 21–22. The original reads: "Ma comunque si sia, Venere e il Sole / In un istesso Ciel convien riporre. / E se ciò che il gran Dano Ticon vuole, / Non ricusiam per ver noi di supporre, / Che Marte sotto al Sol talhor si suole / Trovar, e col suo moto in giù ricorre; / Un Orbe havran comune tre Pianeti, / Quinci e quindi da due fien poi discreti."

43. See Claudio Acquaviva, "De soliditate et uniformitate doctrinae," in *Epistolae selectae praepositorum generalium ad superiores Societatis* (Rome: Typis Poliglottis Vaticani, 1911), 207–209.

44. Galileo to Vinta, April 27, 1611, in OG, 11:94.

45. Galileo to Salviati, April 22, 1611, in OG, 11:89.

46. On the Garden at Trinità dei Monti, see Orbaan, *Documenti sul Barocco*, 197. We can assume that Galileo must have met Joyeuse, based on the letters from Gallanzoni, Borsacchi, and Joyeuse himself, who in September 1611 received a telescope from Galileo as a gift; see OG, 11:132, 137, 208, 211. See also Galileo's letter to Gallanzoni dated July 16, 1611, in OG, 11:141. On Dal Monte's residence, see OG, 11:78–79, and Orbaan, *Documenti sul Barocco*, 163, 168. On Barberini's residence, see OG, 11:80–81, and Orbaan, *Documenti sul Barocco*, nn. 101, 103. Regarding Bandini's residence, see OG, 11:86, and Orbaan, *Documenti sul Barocco*, n. 26. On the Palazzo Farnese, see OG, 11:132.

47. See Eraldo Bellini, *Umanisti e lincei. Letteratura e scienza a Roma nell'età di Galileo* (Padua: Antenore, 1997), 13–14; Federica Favino, "Le ragioni del 'patronage.' I Farnese di Roma e Galileo," in *Il caso Galileo. Una rilettura storica, filosofica, teologica*, ed. Massimo Bucciantini, Michele Camerota, and Franco Giudice (Florence: Olschki, 2011), 174–175.

48. See OG, 11:205. On Agucchi's astronomical interests, see Massimo Bucciantini, "Teologia e nuova filosofia. Galileo, Federico Cesi, Giovambattista Agucchi e la discussione sulla fluidità e corruttibilità del cielo," in *Sciences et Religions. De Copernic à Galilée* (Rome: École Française de Rome, 1999), 411–442.

49. Galileo to Orsini, April 8, 1611, in OG, 11:82–83; Giovan Battista Strozzi, "Lettione in biasmo della superbia malvagia e fino a che segno riprensibile," in Strozzi, *Orazioni et altre prose* (Rome: Grignani, 1635), 206. In his monograph

on Strozzi, Silvio Barbi dated the reading of the speech on pride to 1624. See Silvio Adrasto Barbi, *Un accademico mecenate e poeta. Giovan Battista Strozzi il giovane* (Florence: Sansoni, 1900), 55. In reality, as confirmed by Galileo's aforesaid letter, this discussion was held in April 1611. According to information in a letter from Salvatore Luci to Vinta dated April 9, 1611, the speech went on for an hour and was followed by the reading of a *canzone* by Ciampoli (see ASF, Mediceo del Principato 972, fol. 117*r*).

50. See OG, 11:87, 92.
51. OG, 11:89. See also the report of Ambassador Niccolini to Vinta, OG, 11:92, and William R. Shea and Mariano Artigas, *Galileo in Rome: The Rise and Fall of a Troublesome Genius* (Oxford: Oxford University Press, 2003), 37–39.
52. See OG, 19:265.
53. Orbaan, *Documenti sul Barocco*, 283.
54. See Giulio Cesare Lagalla, *De phoenomenis in orbe Lunae novi telescopii usu a D. Gallileo Gallileo nunc iterum suscitatis Physica disputatio* (Venice: Baglioni, 1612); see OG, 3:330. See also Girolamo Sirtori, *Telescopium, sive ars perficiendi* (Frankfurt: Iacobi, 1618), 27. For further information about the meeting, see Johannes Faber, *Animalia Mexicana descriptionibus scholiisque exposita* (Rome: I. Mascardum, 1628), 473, and the biography of Persio penned by his secretary Giovanni Bartolini; see Gabrieli, *Contributi alla storia dell'Accademia dei Lincei*, 1:877.
55. In the spring of 1611, from the "garden of Monte Cavallo"—the Quirinal Hill—Galileo also conducted some of his earliest observations of sunspots; see OG, 5:81–82; 11:305, 335, 329, 418, 424; 12:175; 18:297.
56. Cesi to Stelluti, April 30, 1611, in OG, 11: 99.
57. Ibid., 94.
58. Dal Monte to Cosimo II, May 31, 1611, in OG, 11:119.
59. Ibid. On Galileo's departure, see Guicciardini to Vinta, June 4, 1611, in OG, 11:121.
60. OG, 19:275; *I documenti vaticani del processo di Galileo Galilei (1611–1741)*, new expanded edition revised and annotated by Sergio Pagano (Vatican City: Archivio Segreto Vaticano, 2009), 171–172.
61. See Antonino Poppi, *Cremonini e Galilei inquisiti a Padova nel 1604. Nuovi documenti d'archivio* (Padua: Antenore, 1992).
62. For a reconstruction of the events, see Massimo Bucciantini, *Contro Galileo. All'origine dell' "affaire"* (Florence: Olschki, 1995); Michele Camerota, *Galileo Galilei e la cultura scientifica nell'età della Controriforma* (Rome: Salerno Editrice, 2004), 272–332.

63. These are the words in Caccini's deposition at the Holy Office: "They say you are very close to that Brother Paolo, the Servite, who is so famous in Venice for his impiety." Galileo Galilei, *Scienza e religione. Scritti copernicani,* ed. Massimo Bucciantini and Michele Camerota (Rome: Donzelli, 2009), 255.
64. Guicciardini to Picchena, December 5, 1615, in OG, 12:207.
65. On the subject, see Adriano Prosperi, *Tribunali della coscienza. Inquisitori, confessori, missionari* (Turin: Einaudi, 1996), 194–210.
66. SN, 40 (for the Latin text, see OG, 3:62–63).
67. Erwin Panofsky, *Galileo as a Critic of the Arts* (The Hague: Nijhoff, 1954), 5. Before Panofsky, the ties between this work and Galileo's telescopic discoveries were highlighted by Guido Anichini, "La cupola del Cigoli a S. Maria Maggiore e un cimelio galileiano," *L'Illustrazione Vaticana* 3, no. 16 (1932): 814.
68. See Giovan Battista Cardi, *Vita di Lodovico Cardi da Cigoli,* ed. Guido Battelli (Florence: Barbèra, 1913), 38–39; Filippo Baldinucci, *Notizie de' Professori del Disegno da Cimabue in qua* (Milan: Società Tipografica de' Classici Italiani, 1812), 9:134–135.
69. The first payment order for the work is dated September 25, 1610. See Anna Maria Corbo, "I pittori della cappella Paolina in S. Maria Maggiore," *Palatino* 11, no. 3 (1967): 307. In a letter to Galileo dated October 19, 1612, Cigoli informs him that he is "at the end of the dome": OG, 11:418.
70. See Cigoli to Galileo, April 13, 1612, in OG, 11:291. As to criticism, Cigoli relates that some of his "enemies" accused him of not knowing how to paint, pointing out that the figures in the fresco "seem to be painted in oils": OG, 11:425.
71. On this, see David M. Stone, "Bad Habit: Scipione Borghese, Wignacourt and the Problem of Cigoli's Knightood," in *Celebratio Amicitiae. Essays in Honour of Giovanni Bonello,* ed. Maroma Camilleri and Theresa Vella (Malta: Fondazzjoni Patrimonju Malti, 2006), 207–229.
72. Steven F. Ostrow, *Art and Spirituality in Counter-Reformation Rome* (Cambridge: Cambridge University Press, 1996), 241. See also Eileen Reeves, *Painting the Heavens: Art and Science in the Age of Galileo* (Princeton, NJ: Princeton University Press, 1997), 141–154; Steven F. Ostrow, "Cigoli's 'Immacolata' and Galileo's Moon: Astronomy and the Virgin in Early Seicento Rome," *The Art Bulletin* 78, no. 2 (1996): 225–229.
73. Francisco Pacheco, *Arte de la pintura. Su antiguedad y grandezas* (Seville: Faxardo, 1649), 483. See Reeves, *Painting the Heavens,* 194.
74. Two brothers, both members of the Oratorians, devised the pictorial cycle: Tommaso and Francesco Bozio. The former has been credited with most of the cycle. See Ostrow, *Art and Spirituality in Counter-Reformation Rome,* 175–190.

75. BVR, MS O 57, fol. 377*r*. See Ostrow, "Cigoli's 'Immacolata' and Galileo's Moon," 219 n. 13. See also the virtually identical version in the archive at the basilica (Archivio capitolare di Santa Maria Maggiore, Collezione Bianchini, tomo X, fol. 539*r*); see Corbo, "I pittori della cappella Paolina," 304.
76. Blas Viegas, *Commentarii exegetici in Apocalypsim Ioannis Apostoli* (Paris: D. Binet, 1606), 520.
77. Jean de La Haye, *Commentarii literales et conceptuales in Apocalypsim Sancti Ioannis Evangelistae* (Paris: N. Buon et D. Thierry, 1644), 424.
78. João da Sylveira, *Commentarii in Apocalypsim D. Joannis Apostoli* (Lyon: Nissonios et J. Posuel, 1681), II, 23. See Reeves, *Painting the Heavens*, 140.
79. Johann Heinrich Bullinger, *In Apocalypsim Iesu Christi revelatam* (Zurich: Officina Froschoviana, 1590), 75.
80. For a similar conclusion (but not backed by exegetic references), see Sara Elizabeth Booth and Albert Van Helden, "The Virgin and the Telescope: The Moons of Cigoli and Galileo," in *Galileo in Context*, ed. Jürgen Renn (Cambridge: Cambridge University Press, 2001), 193–216.
81. Andrea Vittorelli, *Gloriose memorie della Beatissima Vergine Madre di Dio* (Rome: Facciotto, 1616), 223 and 225.
82. Andrea Vittorelli, *Dei ministerii et operationi angeliche libri sei* (Vicenza: Tozzi, 1611), 150.
83. SN, 58. See OG, 3:76.
84. Vittorelli, *Dei ministerii et operationi angeliche*, 233–234. As opposed to the previous passage, this one is reported in Ostrow, *Art and Spirituality in Counter-Reformation Rome*, 243.
85. OG, 11:449.
86. Giovanni Severato, *Memorie delle sette Chiese di Roma* (Rome: Mascardi, 1630), 1:710. The Latin quotation is from the *Diario del cerimoniere papale*, which details the inaugural event; see Orbaan, *Documenti sul Barocco*, 12. In the handwritten notes of G. B. Costaguti, majordomo of Paul V, we also find that the pontiff outfitted the chapel with "very beautiful, rich, and charming furnishings" that included "six candlesticks . . . two large silver torchères, and a lamp, and similar things in brass." See Ludwig von Pastor, *Storia dei Papi dalla fine del Medio Evo* (Rome: Desclée, 1908–1934), 12:705.
87. OG, 11:89.
88. See Chapters 1 and 6.
89. Samuel Y. Edgerton Jr., "Galileo, Florentine 'Disegno' and the 'Strange Spottednesse' of the Moon," *Art Journal* 44, no. 3 (1984): 230.
90. See Elisa Acanfora, "Cigoli, Galileo e le prime riflessioni sulla cupola barocca," *Paragone* 51, no. 31 (2000): 32.

91. On this, see Filippo Camerota, *Linear Perspective in the Age of Galileo: Ludovico Cigoli's "Prospettiva Pratica"* (Florence: Olschki, 2010).
92. Guicciardini to Cosimo II, March 4, 1616, in OG, 12:242 (our italics).
93. Guicciardini to Picchena, December 5, 1615, in OG, 12:217.

12. In Motion: Portugal, India, China

1. APUG, 534, fol. 55*r*.
2. APUG, 534, fol. 56*v*.
3. See Ugo Baldini, "The Portuguese Assistancy of the Society of Jesus and Scientific Activities in Its Asian Missions until 1640," in *História das Ciências Matemáticas, Portugal e o Oriente*, ed. Luís Saraiva (Lisbon: Fundaçao Oriente, 2000), 49–104; *Breve memoria della vita del V. P. Giovanni Antonio Rubino della Compagnia di Gesù, martire del Giappone* (Turin: Tipografia G. Derossi, 1898), 6–7. For short biographical information on Rubino, see also Joseph Dehergne, *Répertoire des Jésuites de Chine de 1552 à 1800* (Rome: Institutum Historicum S. I., 1973), 729.
4. On the activity of the *aula da esfera*, see Luís de Albuquerque, "A 'Aula da Esfera' do Colégio de Santo Antão no século XVII," in *Anais da Academia Portuguesa de História* 21, no. 2 (1972): 337–391; Baldini, "The Portuguese Assistancy of the Society of Jesus"; Ugo Baldini, "L'insegnamento della matematica nel Collegio di S. Antão a Lisbona, 1590–1640," in *A Companhia de Jesús e a Missionação no Oriente* (Lisbon: Brotéria e Fundação Oriente, 2000), 275–310; Baldini, "The Teaching of Mathematics in the Jesuit Colleges of Portugal, from 1640 to Pombal," in *The Practice of Mathematics in Portugal: Papers from the International Meeting Organized by the Portuguese Mathematical Society*, ed. Luís Saraiva and Henrique Leitão (Coimbra: Actas Universitatis Conimbrigensis, 2004), 293–465.
5. See Baldini, "L'insegnamento della matematica nel Collegio di S. Antão a Lisbona," 284 n. 31.
6. On the route taken by the Jesuit missionaries to reach the Far East, see Pasquale M. D'Elia, *Galileo in China: Relations through the Roman College between Galileo and the Jesuit Scientist-Missionaries (1610–1640)* (Cambridge, MA: Harvard University Press, 1960), 19; and, above all, Liam Brockey, "Largos Caminhos e Vastos Mares: Jesuit Missionaries and the Journey to China in the Sixteenth and Seventeenth Centuries," in *Bulletin of Portuguese/Japanese Studies* (2001), 1, 45–72, offering a detailed analysis of the stages of the journey but also the actual conditions.
7. Manuel Dias was known at the time as "junior" to distinguish him from his older contemporary by the same name, Manuel Dias Sr.; see Joseph Sebes,

"Dias (o Novo), Manuel," in *Diccionario Histórico de la Compañía de Jesús,* ed. Charles E. O'Neil and Joaquín M. Dominguez (Madrid: Universidad Pontificia Comillas, 2001), 2:1113. See also Dehergne, *Répertoire des Jésuites de Chine de 1552 à 1800,* 76–77.

8. See Nicolas Standaert, "Jesuits in China," in *The Cambridge Companion to the Jesuits,* ed. Thomas Worcester (Cambridge: Cambridge University Press, 2008), 169–185, especially 172–175.
9. See Baldini, "The Portuguese Assistancy of the Society of Jesus," 60–61.
10. See José Wicki, "Liste der Jesuiten-Indienfahrer, 1541–1758," in *Aufsätze zur Portugiesischen Kulturgeschichte,* 7 (1967): 283.
11. See Baldini, "The Portuguese Assistancy of the Society of Jesus," 79 and 85.
12. See Ugo Baldini, "The Jesuit College in Macao as a Meeting Point of the European, Chinese and Japanese Mathematical Traditions. Some Remarks on the Present State of Research, Mainly Concerning Sources (16th–17th Centuries)," in *The Jesuits, the Padroado and East Asian Science (1552–1773),* ed. Luís Saraiva and Catherine Jami (Singapore: World Scientific Publishing, 2008), 33–79, especially 51.
13. See Giuliano Bertuccioli in DBI, 39, under the specific entry.
14. Rubino took the *Nossa Senhora da Paz,* while De Ursis sailed on the *Nossa Senhora da Bigonha.* See Wicki, *Liste der Jesuiten-Indienfahrer,* 283–284.
15. See Erik Zürcher, "Giulio Aleni's Chinese Biography," in *Scholar from the West: Giulio Aleni S. J. (1582–1649) and the Dialogue between Christianity and China,* ed. Tiziana Lippiello and Roman Malek (Brescia–Sankt Augustin: Fondazione Civiltà Bresciana-Monumenta Serica Institute, 1997), 101; Favaro, *Carteggio inedito,* 347–349; D'Elia, *Galileo in China,* 21.
16. See Henrique Leitão, "The Contents and Context of Manuel Dias' Tianwenlüe," in *The Jesuits, the Padroado and East Asian Science,* ed. Saraiva and Jami, 100 n. 5 (which is also the source of the quotation).
17. The copy at the Biblioteca Apostolica Vaticana, Borgia Cinese, 324, is considered the *editio princeps.* On the problems related to the different editions of Dias's text, see Rui Magone, "The Textual Tradition of Manuel Dias' Tianwenlüe," in *The Jesuits, the Padroado and East Asian Science,* ed. Saraiva and Jami, 123–138. See also D'Elia, *Galileo in China,* 90, note 61.
18. See Isaia Iannaccone, "From N. Longobardo's Explanation of Earthquakes as Divine Punishment to F. Verbiest's Systematic Instrumental Observations: The Evolution of European Science in China in the Seventeenth Century," in *Western Humanistic Culture Presented to China by Jesuit Missionaries (XVII–XVIII Centuries),* ed. Federico Masini (Rome: Institutum Historicum S. I., 1996), 160.
19. For a summary of the contents of the *Tian Wen Lüe,* see Leitão, "The Contents and Context of Manuel Dias' Tianwenlüe," 110–113.

20. We are quoting from D'Elia, *Galileo in China*, 18–19, making a few changes thanks to the revision of Eugenio Menegon, to whom we are grateful.
21. Ibid.
22. See Leitão, "The Contents and Context of Manuel Dias' Tianwenlüe," 117.
23. See James M. Lattis, *Between Copernicus and Galileo: Christoph Clavius and the Collapse of Ptolemaic Cosmology* (Chicago: University of Chicago Press, 1994), 180.
24. Christoph Clavius, *Commentarium in Sphaeram Ioannis de Sacrobosco*, in *Opera Mathematica V Tomis distributa* (Mainz: sumptibus A. Hierat excudebat R. Eltz, 1611–1612), 3:75.
25. See Leitão, "The Contents and Context of Manuel Dias' Tianwenlüe," 116.
26. News of the celestial discoveries reached Lisbon between 1611 and 1612, and any missionary going to Asia could easily have been its messenger; see Henrique Leitão, "Galileo's Telescopic Observations in Portugal," in *Largo campo di filosofare*, ed. José Montesinos and Carlos Solís (La Orotava: Fundación Canaria Orotava de Historia de la Ciencia, 2001), 903–913.
27. APUG, 534, fol. 56*v*. On this hypothesis, see D'Elia, *Galileo in China*, 17.
28. See D'Elia, *Galileo in China*, 33; Zhishan Zhang, "Johann Adam Schall von Bell and his Book *On the Telescope*," in *Western Learning and Christianity in China: The Contribution and Impact of Johann Adam Schall von Bell, S. J. (1592–1666)*, ed. Roman Malek (Sankt Augustin: Monumenta Serica Institute, 1998), 2:681–690.
29. We are quoting from D'Elia, *Galileo in China*, 50, with a few changes.

Epilogue

1. OG, 12:171. See Maurice Clavelin, *Galilée copernicien* (Paris: Albin Michel, 2004), 54–68.
2. OG, 12:172 (our italics).
3. Lorini to Cardinal Sfondrati, February 7, 1615, and Galileo to Castelli, December 21, 1613, in Galileo Galilei, *Scienza e religione. Scritti copernicani*, ed. Massimo Bucciantini and Michele Camerota (Rome: Donzelli, 2009), 249, 8. See also Massimo Bucciantini, "'L'affaire' Galileo," in *Atlante della letteratura italiana*, ed. Sergio Luzzatto and Gabriele Pedullà (Turin: Einaudi, 2011), 2:338–343.
4. Galileo Galilei and Christoph Scheiner, *On Sunspots*, trans. Eileen Reeves and Albert Van Helden (Chicago: University of Chicago Press, 2010), 296.
5. One of the exceptions is the text composed between 1546 and 1547 by the Florentine Dominican Giovanni Maria Tolosani but never published. See Eu-

genio Garin, "Alle origini della polemica anticopernicana," in *Rinascite e rivoluzioni. Movimenti culturali dal XIV al XVIII secolo* (Rome: Laterza, 1975), 283–295.

6. See Adriano Prosperi, "Intellettuali e Chiesa all'inizio dell'età moderna," in *Storia d'Italia. Annali 4. Intellettuali e potere*, ed. Corrado Vivanti (Turin: Einaudi, 1981), 166.
7. Decree of General Congregation of the Index, March 5, 1616, in *The Galileo Affair: A Documentary History*, ed. and trans. Maurice A. Finocchiaro (Berkeley: University of California Press, 1989), 149.
8. Francesco De' Vieri, *Trattato nel quale si contengono i tre primi libri delle metheore* (1582), quoted in Elide Casali, *Le spie del cielo. Oroscopi, lunari e almanacchi nell'Italia moderna* (Turin: Einaudi, 2003), 102.
9. Angelo Grillo, *Delle lettere* (Venice: Evangelista Deuchino, 1616), 2:305. The letter, undated, was sent to Tommaso Arigucci. On Grillo, see Luigi Matt in DBI, 59, under the specific entry; Maria Cristina Farro, "Un 'libro di lettere' da riscoprire. Angelo Grillo e il suo epistolario," *Esperienze letterarie* 18 (1993): 69–81; Nick Wilding, "Instrumental Angels," in *Conversations with Angels: Essays Towards a History of Spiritual Communication, 1100–1700*, ed. Joad Raymond (Basingstoke: Palgrave Macmillan, 2011), 67–89.
10. Grillo, *Delle lettere*, 2:305.
11. Lagalla, *De phoenomenis in orbe Lunae physica disputatio*, in OG, 3(1): 347–349.
12. Tommaso Campanella, *Apologia pro Galileo*, ed. Michel-Pierre Lerner, trans. Germana Ernst (Pisa: Scuola Normale Superiore, 2006), 21.
13. Ibid., xlii–xliii.
14. Wentzel von Meroschwa, *Epistola ad Ioannem Traut Noribergensem de statu praesentis belli, et urbium imperialium* (1620). See Massimo Bucciantini, *Galileo e Keplero. Filosofia, cosmologia e teologia nell'età della Controriforma* (Turin: Einaudi, 2003), 247–251.
15. Adam Tanner, *Dioptra Fidei, Das ist: Allgemeiner, Catholischer und Gründtlicher Religions-Discurs von dem Richter und Richtschnur in Glaubenssachen* (Ingolstadt: Angermayr, 1617).
16. Leonard Lessius, *De providentia numinis et animi immortalitate* (Antwerp: Officina Plantiniana, 1613), 18–19. On Lessius and Tanner, see Bucciantini, "Novità celesti e teologia," in *Largo campo di filosofare*, ed. José Montesinos and Carlos Solís (La Orotava: Fundación Canaria Orotava de Historia de la Ciencia, 2001), 795–808.
17. Letter to Arigucci (n.d.), in Grillo, *Delle lettere*, 2:306.
18. Letter to Chiocco (n.d.), in Grillo, *Delle lettere*, 2:149 (our italics).

19. John H. Elliott, *The Old World and the New: 1492–1650* (Cambridge: Cambridge University Press, 1970), 15–16. Elliot's considerations are worth reading in full: "In some respects the Renaissance involved, at least in its earlier stages, a closing rather than an opening of the mind. The veneration of antiquity became more slavish; authority staked fresh claims against experience. Both the boundaries and the content of traditional disciplines such as cosmography or social philosophy had been clearly determined by reference to the texts of classical antiquity, which acquired an extra degree of definitiveness when for the first time they were fixed on the printed page. Fresh information from alien sources was therefore liable to seem at worst incredible, at best irrelevant, when set against the accumulated knowledge of the centuries. Given this deference to authority, there was unlikely to be any undue precipitation, least of all in academic circles, to accept the New World into consciousness."
20. Ibid., 30.
21. See Marguerite Yourcenar, *En pèlerin et en étranger* (Paris: Gallimard, 1989).

References

Primary Sources

Acquaviva, Claudio. "De soliditate et uniformitate doctrinae." In *Epistolae selectae praepositorum generalium ad superiores Societatis*. Rome: Typis Poliglottis Vaticani, 1911.

Alidosi, Roderigo. *Relazione di Germania e della Corte di Rodolfo II imperatore*. Edited by Giuseppe Campori. Modena: Tipografia e Litografia Cappelli, 1872.

Bacon, Francis. *Philosophical Studies, c. 1611–c. 1619*. Edited and with introduction, notes, and commentary by Graham Rees. Oxford: Clarendon Press, 1996.

Berchet, Guglielmo. *Relazioni dei Consoli Veneti nella Siria*. Turin: Paravia, 1886.

Boccalini, Traiano. *Ragguagli di Parnaso e pietra del paragone politico*. Edited by Giuseppe Rua. Bari: Laterza, 1934.

Borromeo, Federico. *Della pittura sacra libri due*. Edited by Barbara Agosti. Pisa: Scuola Normale Superiore, 1994.

Brahe, Tycho. *Opera Omnia*. Edited by John Louis Emil Dreyer. Haunia: In Libraria Gyldendaliana, 1913; reprinted, Amsterdam: Swets & Zeitlinger, 1972, 15 vols.

Breve memoria della vita del V. P. Giovanni Antonio Rubino S.J., martire del Giappone. Turin: Tipografia Derossi, 1898.

Bullinger, Johann Heinrich. *In Apocalypsim Iesu Christi rivelata*. Zurich: Officina Froschoviana 1590.

Cairasco de Figueroa, Bartolomé. *Templo militante. Flos sanctorum, y triumphos de sus virtudes*. Lisbon: Pedro Crasbeeck, 1613.

Camden, William. *Epistolae*. London: Richard Chiswell, 1691.

Campanella, Tommaso. *Apologia pro Galileo*. Edited by Michel-Pierre Lerner, translated by Germana Ernst. Pisa: Scuola Normale Superiore, 2006.

Cavalieri, Giovanni Michele. *Galleria de' sommi Pontefici, Patriarchi, Arcivescovi e Vescovi dell'Ordine de' Predicatori*. Benevento: Stamperia Arcivescovile, 1696.

Clavius, Christoph. "Commentarium in Sphaeram Ioannis de Sacrobosco." In id., *Opera Mathematica V Tomis distribute*. Mainz: R. Eltz, 1611–1612.

———. *Corrispondenza*. Edited by Ugo Baldini and Pier Daniele Napolitani. Pisa: Università di Pisa, Dipartimento di matematica, 1992.

Considerazioni d'Alimberto Mauri sopra alcuni luoghi del Discorso di Lodovico delle Colombe intorno alla stella apparita nel 1604. Florence: Gio. Antonio Caneo, 1606.

Le convoy du coeur de tres-auguste, tres-clément et tres-victorieux Henry le Grand . . . depuis la ville de Paris jusques au College Royal de La Flèche. Lyon: Morillon, 1610.

de La Haye, Jean. *Commentarii literales et conceptuales in Apocalypsim Sancti Ioannis Evangelistae*. Paris: N. Buon et D. Thierry, 1644.

Della Porta, Giovan Battista. *Magia naturalis libri XX*. Naples: Horatium Salvianum, 1589.

I documenti vaticani del processo di Galileo Galilei (1611–1741). New edition by Sergio Pagano. Città del Vaticano: Archivio Segreto Vaticano, 2009.

Donne, John. *Letters to severall persons of honour*. London: J. Flesher, 1651.

———. *Ignatius his Conclave: An Edition of the Latin and English Texts*. Edited by Timothy Stafford. Oxford: Clarendon Press, 1969.

———. "An Anatomy of the World." In *The Poems of John Donne*, edited by H. J. C. Grierson, 2 vols. Oxford: Oxford University Press, 1912.

Faber, Johannes. *Animalia Mexicana descriptionibus scholiisque exposita*. Rome: Mascardi, 1628.

Figliucci, Vincenzo. *Stanze sopra le stelle e macchie solari scoperte col nuovo occhiale*. Rome: Mascardi, 1615.

Galilei, Galileo. *Opere*. Edited by Antonio Favaro. Florence: Barbèra, 1890–1909; reprinted 1968.

———. *Sidereus Nuncius. Le messager céleste*. Edited by Isabelle Pantin. Paris: Les Belles Lettres, 1992.

———. *Sidereus Nuncius or The Sidereal Messenger*. Translated with introduction, conclusions, and notes by Albert Van Helden. Chicago and London: University of Chicago Press, 1989.

———. *Scienza e religione. Scritti copernicani*. Edited by Massimo Bucciantini and Michele Camerota. Rome: Donzelli, 2009.

———. *Sidereus nuncius: o mensageiro das estrema*. Edited by Henrique Leitão. Lisbon: Fundaçao Calouste Gulbenkian, 2010.

Galilei, Galileo, and Christoph Scheiner. *On Sunspots*. Translated by Eileen Reeves and Albert Van Helden. Chicago: University of Chicago Press, 2010.

Garzoni, Tommaso. *Dello specchio di scientia universale*. Venice: Andrea Ravenoldo, 1567.

Gassendi, Pierre. *Viri Illustri Nicolai Claudij Fabricij de Peiresc . . . vita*. The Hague: Vlacq, 1655.

Gilbert, William. *De Magnete*. London: Petrus Short, 1600.

Giraldi, Giuliano. *Esequie d'Arrigo Quarto . . . celebrate in Firenze dal Serenissimo don Cosimo II*. Florence: Bartolommeo Sermartelli e fratelli, 1610.

Grienberger, Christoph. *Catalogus veteres affixarum longitudines conferens cum novis*. Rome: Zanetti, 1612.

Grillo, Angelo. *Delle lettere*. Venice: Evangelista Deuchino, 1616.

[Griselini, Francesco]. *Del genio di P. Paolo Sarpi*. Venice: Leonardo Bossaglia, 1785.

Horky, Martin. *Disputatio ethica de beatitudine politica*. Liegnitz: N. Sartorius, 1605.

———. *Brevissima peregrinatio contra Nuncium sidereum.* Modena: Iulianum Cassianum, 1610.

———. *Ein richtiger und sehr nützlicher Wegweiser, wie man sich für der Pestilentz bewahren sole.* Rostock: Sachs, 1624.

———. *Eine Neue Diania Astromantica, oder gewisser Beweiß, Was zu halten sey von den schrecklichen Göttlichen Wunderwerck, so diß jetzige 1629 Jahr Christi an der Sonnen ist gesehen worden.* N.p., 1629.

———. *Talentum astromanticum Oder Natuerliche Weissagung und Verkuendigung aus des Himmels Lauff, vom Zustand und Beschaffenheit des Schalt-Jahrs nach Christii Geburt 1632.* Leipzig: Ritzsch, 1632.

———. *Das grosse Prognosticon, oder Astrologische Wunder Schrifft . . . auff das Jahr Christi MDC.XXXIII.* Hamburg: n.p., 1633.

———. *Chrismologium physico-astromanticum, oder Natürliche Weissagung und Erkundigung auss dem Gestirn und Himmelslauff, vom Zustand und Beschaffenheit dess 1639 Jahrs Christi.* Nuremberg: n.p., 1639.

———. *Alter und Newer SchreibKalender, sampt der Planeten Aspecten Lauff und derselben Influentzen, Auffs Jahr . . . MDCL . . . gestellet.* Nuremberg: Endters, 1649 (?).

Huygens, Christian. *Oeuvres completes, Correspondance,* II, 1657–1657. The Hague: M. Nijhoff, 1889.

Justel, Christophe. *Codex canonum Ecclesiae universae.* Paris: H. Beys, 1610.

Kepler, Johannes. *Gesammelte Werke.* Edited by Walther von Dyck, Max Caspar, and Franz Hammer. Munich: Beck, 1937.

———. *Dissertatio cum Nuncio sidereo. Discussion avec le Messager céleste; Narratio de observatis Jovis satellitibus. Rapport sur l'observation des satellites de Jupiter.* Edited by Isabelle Pantin. Paris: Les Belles Lettres, 1993.

———. *Kepler's Conversation with Galileo's Sidereal Messenger.* First complete translation, with an introduction and notes by Edward Rosen. New York: Johnson Reprint, 1965.

———. *Opera Omnia.* Edited by Carl Frisch. Frankfurt: Heyder & Zimmer, 1858–1871.

Lagalla, Giulio Cesare. *De phoenomenis in orbe Lunae novi telescopii usu a D. Gallileo Gallileo nunc iterum suscitatis Physica disputatio.* Venice: Baglioni, 1612.

L'Estoile, Pierre Taisan de. *Mémoires-Journaux.* Edited by Gustave Brunet et al. Paris: Librairie des Bibliophiles, 1881.

Lessius, Leonard. *De providentia numinis et animi immortalitate.* Antwerp: ex Officina Plantiniana, apud Viduam & Filios Io. Moreti, 1613.

Lipsius, Justus. *Epistolarum selectarum centuria prima.* Antwerp: C. Plantinum, 1586.

Magagnati, Girolamo. *Meditazione poetica sopra i pianeti medicei.* Venice: Heredi d'Altobello Salicato, 1610.

———. *Lettere a diversi*. Edited by Laura Salvetti Firpo, with introduction. Foreward "La vita di Girolamo Magagnati" edited by Carlo Carabba and Giuliano Gasparri. Florence: Olschki, 2006.

Magini, Giovanni Antonio. *Ephemerides . . . ab anno Domini 1581 usque ad annum 1620 secundum Copernici hypotheses, Prutenicosque canones*. Venice: Damianum Zenerium, 1582.

———. *Supplementum Ephemeridum ac Tabularum Secundorum Mobilium*. Venice: Haeredem Damiani Zenarii, 1614.

Malherbe, François de. *Oeuvres*. Edited by M. Ludovic Lalanne. Paris: Hachette, 1862.

Manzini, Carlo Antonio. *L'occhiale all'occhio*. Bologna: Herede del Benacci, 1660.

Masini, Antonio. *Bologna perlustrata*. Third edition. Bologna: Erede di Vittorio Benacci, 1666.

Mayr, Simon. *Mundus Iovialis*. Nuremburg: Typis I. Lauri, 1614.

Meroschwa, Wentzel von. *Epistola ad Ioannem Traut Noribergensem de statu praesentis belli, et urbium imperialium*. N.p., 1620.

Micanzio, Fulgenzio. *Vita del Padre Paolo dell'Ordine de' Servi e theologo della Serenissima (Leida 1646)*. In Paolo Sarpi, *Istoria del Concilio Tridentino, seguita dalla "Vita del padre Paolo" di Fulgenzio Micanzio*. Edited by Corrado Vivanti. Turin: Einaudi 1974.

Monmouth, Henry, Earl of. *I Ragguagli di Parnasso: or Advertisements from Parnassus in Two Centuries with the Politick Touch-stone*. London: Humphrey Mosely and Thomas Heath, 1556.

Les Négotiations de Monsieur le President Jeannin. Paris: Pierre le Petit, 1656.

Pacheco del Río, Francisco. *Arte de la pintura. Su antiguedad y grandezza*. Seville: Faxardo, 1649.

Pasquali Alidosi, Giovanni Niccolò, *Diario. Overo raccolta delle cose che nella Città di Bologna giornalmente occorrono per l'Anno MDCXIV*. Bologna: Bartolomeo Cochi, 1614.

Peiresc, Nicolas-Claude Fabri de. *Lettres à Malherbe (1606–1628)*. Edited by Raymond Lebègue. Paris: CNRS, 1976.

Ralegh, Walter. *The History of the World*. London: William Stansby for Walter Burre, 1614.

Ratio atque Institutio Studiorum Societatis Iesu. Ordinamento degli studi della Compagnia di Gesù. Edited by Angelo Bianchi. Milan: Rizzoli, 2002.

Rivola, Francesco. *Vita di Federico Borromeo*. Milan: Gariboldi, 1656.

Roffeni, Giovanni Antonio. *Epistola apologetica*. Bologna: Heredes Joannis Rossij, 1611.

Sampieri, Vincenzo. *Origine e fondatione di tutte le Chiese che di presente si trovano nella Città di Bologna*. Bologna: Clemente Ferroni, 1633.

Santucci, Antonio. *Trattato nuovo delle comete, che le siano prodotte in cielo, e non nella regione dell'aria, come alcuni dicono*. Florence: Caneo, 1611.

Sanudo, Marin. *I diarii:. MCCCCXCVI–MDXXXIII) dall'autografo Marciano ital. cl. VII codd. CDXIX–CDLXXVII.* Venice: F. Visentini, 1882.

Sarpi, Paolo. *Lettere.* Edited by Filippo Luigi Polidori. Florence: Barbèra, 1863.

———. *Lettere ai protestanti.* Edited by Manlio Duilio Busnelli. Bari: Laterza, 1931.

———. *Lettere ai gallicani.* Edited by Boris Ulianich. Wiesbaden: Steiner, 1961.

———. *Opere.* Edited by Gaetano Cozzi and Luisa Cozzi. Milan-Naples: Ricciardi, 1969.

———. *Pensieri naturali, metafisici e matematici.* Edited by Luisa Cozzi and Libero Sosio. Milan-Naples: Ricciardi, 1996.

Scheiner, Christoph. *De maculis solaribus et stellis circa Iovem errantibus accuratior disquisitio.* Augsburg: Ad insigne pinus, 1612.

Severato, Giovanni. *Memorie delle sette Chiese di Roma.* Rome: Mascardi, 1630.

Sirtori, Girolamo. *Telescopium, sive ars perficiendi.* Frankfurt: Iacobi, 1618.

Sizzi, Francesco. *Dianoia astronomica, optica, physica, qua Syderei nuncii rumor de quatuor planetis a Galilaeo . . . vanus redditur.* Venice: P. M. Bertanum, 1611.

Smith, Logan Pearsall. *The Life and Letters of Sir Henry Wotton.* Oxford: Clarendon Press, 1907.

Strozzi, Giovan Battista. "Lettione in biasmo della superbia malvagia e fino a che segno riprensibile," in id., *Orazioni et altre prose.* Rome: Grignani, 1635.

Sur la mort du Roy Henry le Grand et sur la decouverte de quelques nouvelles planettes ou estoiles errantes autour de Jupiter, faicte l'année d'icelle par Galilée, celèbre mathématicien du duc de Florence, In *Anniversarium Henrici Magni obitus diem Lacrymae Collegii Flexiensis Regii S. J., Rezé.* La Flèche: Rezé, 1611.

Sylveira, João da. *Commentarii in Apocalypsim D. Joannis Apostoli.* Lyon: A. Nissonios et J. Posuel, 1681.

Tanner, Adam. *Dioptra Fidei, Das ist: Allgemeiner, Catholischer und Gründtlicher Religions-Discurs von dem Richter und Richtschnur in Glaubenssachen.* Ingolstadt: Angermayr, 1617.

Trumbull the Elder, William. *Papers, 1605–1610.* Edited by Edward Kelly Purnell and Allen Banks Hinds. London: His Majesty's Stationery Office, 1924–1995; 6 vols.

Ugurgieri Azzolini, Isidoro. *Le pompe sanesi o' vero relazione delli huomini, e donne illustri di Siena e suo Stato.* Pistoia: Stamperia di Pier'Antonio Fortunati, 1649.

Vannozzi, Bonifacio. *Della suppellettile degli avvertimenti politici, morali, et cristiani.* Bologna: Heredi di Giovanni Rossi, 1609–1613.

———. *Delle lettere miscellanee . . . volume terzo.* Bologna: Bartolomeo Cochi, 1617.

Viegas, Blas. *Commentarii exegetici in Apocalypsim Ioannis Apostoli.* Paris: D. Binet, 1606.

Vittorelli, Andrea. *Dei ministerii et operationi angeliche libri sei.* Vicenza: Tozzi, 1611.

———. *Gloriose memorie della Beatissima Vergine Madre di Dio.* Rome: Facciotto, 1616.

Walchius, Johannes Georgius. *Decas fabularum humani generis.* Strasbourg: Lazar Zetzner, 1609.

Webster, John, *The Duchess of Malfi.* Edited by John Russell Brown. Manchester: Manchester University Press, 1997.

Secondary Sources

Acanfora, Elisa. "Cigoli, Galileo e le prime riflessioni sulla cupola barocca," *Paragone* 51, no. 31 (2000): 29–52.

Albanese, Denise. *New Science, New World.* Durham, NC: Duke University Press, 1996.

Albert & Isabella, 1598–1621: Catalogue, edited by Werner Thomas and Luc Duerloo. Turnhout: Brepols, 1998.

Albuquerque, Luís de. "A 'Aula da Esfera' do Colégio de Santo Antão no século XVII," *Anais da Academia Portuguesa de História* 21, no. 2 (1972): 337–391.

Alexander, Amir R. "Lunar Maps and Coastal Outlines: Thomas Harriot's Mapping of the Moon," *Studies in History and Philosophy of Science* 29 (1998): 345–368.

Anichini, Guido. "La cupola del Cigoli a S. Maria Maggiore e un cimelio galileiano," *L'Illustrazione Vaticana* 3, no. 16 (1932): 814.

Aricò, Denise. "Giovanni Antonio Roffeni: un astrologo bolognese amico di Galileo," *Il Carrobbio* 24 (1998): 67–96.

Ariew, Roger. "Galileo's Lunar Observations in the Context of Medieval Lunar Theory," *Studies in History and Philosophy of Science* 15, no. 3 (1984): 213–226.

Auguste A., Abbé. "Un dessin inédit de Chalette," *Bulletin de la Société archéologique du Midi de la France* (1912–1914): 185–187.

Baart, Jan M. "Una vetreria di tradizione italiana ad Amsterdam." In *Archeologia e storia della produzione del vetro preindustriale,* edited by Marja Mendera, 423–437. Florence: All'Insegna del Giglio, 1991.

Balbiani, Laura. "La ricezione della Magia naturalis di Giovan Battista Della Porta. Cultura e scienza dall'Italia all'Europa," *Bruniana & Campanelliana* 5 (1999): 277–303.

Bald, Robert Cecil. *John Donne: A Life.* Oxford: Oxford University Press 1970.

Baldini, Ugo. *Legem impone subactis. Studi su filosofia e scienza dei Gesuiti in Italia 1540–1632.* Rome: Bulzoni, 1992.

———. "L'insegnamento della matematica nel Collegio di S. Antão a Lisbona, 1590–1640," in *A Companhia de Jesús e a Missionação no Oriente,* 275–310. Lisbon: Brotéria e Fundação Oriente, 2000.

———. *Saggi sulla cultura della Compagnia di Gesú*. Rome: Bulzoni, 1995.

———. "The Portuguese Assistancy of the Society of Jesus and Scientific Activities in Its Asian Missions until 1640." In *História das Ciências Matemáticas, Portugal e o Oriente*, edited by Luís Saraiva, 49–104. Lisbon: Fundação Oriente, 2000.

———. "The Teaching of Mathematics in the Jesuit Colleges of Portugal, from 1640 to Pombal." In *The Practice of Mathematics in Portugal. Papers from the International Meeting Organized by the Portuguese Mathematical Society*, edited by Luís Saraiva and Heinrique Leitão, 293–465. Coimbra: Actas Universitatis Conimbrigensis, 2004.

———. "The Jesuit College in Macao as a Meeting Point of the European, Chinese and Japanese Mathematical Traditions. Some Remarks on the Present State of Research, Mainly Concerning Sources (16th–17th Centuries)." In *The Jesuits, the Padroado and East Asian Science (1552–1773)*, edited by Luís Saraiva and Catherine Jami, 33–79. Singapore: World Scientific Publishing, 2008.

Baldinucci, Filippo. *Notizie de' Professori del Disegno da Cimabue in qua*. Milan: Società Tipografica de' Classici Italiani, 1812.

Baldriga, Irene. *L'occhio della lince. I primi Lincei tra arte, scienza e collezionismo (1603–1630)*. Rome: Accademia Nazionale dei Lincei, 2002.

Balestrieri, Isabella. *Le fabbriche del Cardinale Federico Borromeo, 1595–1631. L'Arcivescovado e l'Ambrosiana*. Benevento: Hevelius, 2005.

Baltrušaitis, Jurgis. *Le miroir: révélations, science-fiction et fallacies*. Paris: Éditions du Seuil, 1979.

Banville, John. *Kepler: A Novel*. London: Secker & Warburg, 1981.

Barbero, Giliola, Massimo Bucciantini, and Michele Camerota. "Uno scritto inedito di Federico Borromeo: l'"Occhiale celeste,'" *Galilaeana* 4 (2007): 309–341.

Barbi, Silvio Adrasto. *Un accademico mecenate e poeta. Giovan Battista Strozzi il giovane*. Florence: Sansoni, 1900.

Batho, Gordon R. "The Library of the 'Wizard' Earl: Henry Percy Ninth Earl of Northumberland (1564–1632)," *The Library* 15 (1960): 246–261.

———. "Thomas Harriot and the Northumberland Household." In *Thomas Harriot: An Elizabethan Man of Science*, edited by Robert Fox, 28–46. London: Ashgate, 2000.

———. "Thomas Harriot's Manuscripts." In *Thomas Harriot: An Elizabethan Man of Science*, edited by Robert Fox, 286–297. London: Ashgate, 2000.

Battisti, Eugenio, and Giuseppa Saccaro Battisti. *Le macchine cifrate di Giovanni Fontana*. Milan: Arcadia Edizioni, 1984.

Bedini, Silvio A. "The Instruments of Galileo Galilei." In *Galileo Man of Science*, edited by Ernst McMullin, 272–280. New York: Basic Books, 1967.

Bedoni, Stefania. *Jan Brueghel in Italia e il collezionismo del Seicento*. Florence-Milan: Litografia Rotoffset, 1983.

Bellini, Eraldo. *Umanisti e lincei. Letteratura e scienza a Roma nell'età di Galileo*. Padua: Antenore, 1997.

———. *Stili di pensiero nel Seicento italiano. Galileo, i Lincei, i Barberini*. Pisa: Edizioni ETS, 2009.

Bennett, Jim A. "Instruments, Mathematics, and Natural Knowledge: Thomas Harriot's Place on the Map of Learning." In *Thomas Harriot: An Elizabethan Man of Science*, edited by Robert Fox, 137–152. London: Ashgate, 2000.

Beretta, Marco. "Galileo in Sweden: Legend and Reality." In *Sidereus Nuncius and Stella Polaris: The Scientific Relations between Italy and Sweden in Early Modern History*, edited by Marco Beretta and Tore Frängsmyr, 5–23. Canton, OH: Science History Publications, 1997.

Bertuccioli, Giuliano. "De Ursis, Sabatino," in *Dizionario biografico degli italiani*, 39, under the specific entry. Rome: Istituto della Enciclopedia Italiana, 1960–.

Biagioli, Mario. *Galileo, Courtier. The Practice of Science in the Culture of Absolutism*. Chicago: University of Chicago Press, 1993.

———. "Replication or Monopoly? The Economies of Invention and Discovery in Galileo's Observations of 1610," *Science in Context* 13, nos. 3–4 (2000): 547–590.

———. *Galileo's Instruments of Credit: Telescopes, Images, Secrecy*. Chicago: University of Chicago Press, 2006.

———. "Did Galileo Copy the Telescope? A 'New' Letter by Paolo Sarpi." In *The Origins of the Telescope*, edited by Albert Van Helden, Sven Dupré, Rob Van Gent, and Huib Zuidervaart, 203–230. Amsterdam: Knaw Press, 2010.

Bigourdan, Guillaume. *Histoire de l'astronomie d'observation et des observatoires en France*. Paris: Gauthier-Villars, 1918.

Bloom, Terrie F. "Borrowed Perceptions: Harriot's Maps of the Moon," *Journal for the History of Astronomy* 9 (1978): 117–122.

Bologna, Ferdinando. *L'incredulità del Caravaggio e l'esperienza delle "cose naturali."* Turin: Bollati Boringhieri, 1992.

Booth, Sara Elizabeth, and Albert Van Helden. "The Virgin and the Telescope: The Moons of Cigoli and Galileo." In *Galileo in Context*, edited by Jürgen Renn, 193–216. Cambridge: Cambridge University Press, 2001.

Borromeo, Federico. *Il Museo*. Italian translation by L. Grasselli, preface and notes Luca Beltrami. Milan: Tip. U. Allegretti, 1909.

———. *I tre libri delle laudi divine*. Milan, 1632.

———. *Musaeum. La Pinacoteca ambrosiana nelle memorie del suo fondatore*. Commentary Gianfranco Ravasi, Italian translation and notes Piero Cigada. Milan: Claudio Gallone Editore, 1997.

Boselli, Pierino. *Toponimi lombardi.* Milan: SugarCo, 1977.

Boulier, Philippe. "Cosmologie et science de la natura chez Francis Bacon et Galilée." Ph.D. diss., Université de Paris IV—Sorbonne, 2010.

Braudel, Fernand-Paul. *The Mediterranean and the Mediterranean World in the Age of Philip II,* 2 vols. Translated by Siân Reynolds. Berkeley: University of California Press, 1995.

Bredekamp, Horst. "Gazing Hands and Blind Spots: Galileo as Draftsman," *Science in Context* 13 (2000): 423–462.

———. *Galilei der Künstler. Der Mond. Die Sonne. Die Hand.* Berlin: Akademie Verlag, 2007.

Brockey, Liam. "Largos Caminhos e Vastos Mares: Jesuit Missionaries and the Journey to China in the Sixteenth and Seventeenth Centuries," *Bulletin of Portuguese/Japanese Studies* 1 (2001): 45–72.

Bucciantini, Massimo. *Contro Galileo. Alle origini dell' "affaire."* Florence: Olschki, 1995.

———. "Teologia e nuova filosofia. Galileo, Federico Cesi, Giovambattista Agucchi e la discussione sulla fluidità e corruttibilità del cielo." In *Sciences et Religions. De Copernic à Galilée,* 411–442. Rome: École Française de Rome, 1999.

———. "Novità celesti e teologia." In *Largo campo di filosofare,* edited by José Montesinos and Carlos Solís. La Orotava: Fundación Canaria Orotava de Historia de la Ciencia, 2001.

———. *Galileo e Keplero. Filosofia, cosmologia e teologia nell'età della Controriforma.* Turin: Einaudi, 2003.

———. "Reazioni alla condanna di Copernico: nuovi documenti e nuove ipotesi di ricerca," *Galilaeana* 1 (2004): 3–19.

———. "Galileo e Praga." In *Toscana e Europa. Nuova scienza e filosofia tra '600 e '700,* edited by Ferdinando Abbri and Massimo Bucciantini, 109–121. Milan: F. Angeli, 2006.

———. "L''affaire' Galileo." In *Atlante della letteratura italiana,* edited by Sergio Luzzatto and Gabriele Pedullà, 338–343- Turin: Einaudi, 2011.

Busnelli, Manlio Duilio. *Études sur Fra Paolo Sarpi: Et autres essais italiens et français.* Geneva: Editions Slatkine, 1986.

Calore, Marina, and Gian Luigi Betti. "'Il molto illustre Cavaliere Hercole Bottrigari.' Contributi per la biografia di un eclettico intellettuale bolognese del Cinquecento," *Il Carrobbio* 35 (2009): 93–120.

Camerota, Filippo. *Linear Perspective in the Age of Galileo: Ludovico Cigoli's Prospettiva Pratica.* Florence: Olschki, 2010.

Camerota, Michele. "Flaminio Papazzoni: un aristotelico bolognese maestro di Federico Borromeo e corrispondente di Galileo." In *Method and Order in Renaissance Philosophy of Nature: The Aristotle Commentary Tradition,* edited by Daniel

Di Liscia, Eckhard Kessler, and Charlotte Methuen, 271–300. Aldershot: Ashgate, 1997.

———. "Galileo e il Parnaso tychonico." In *Galileo e il Parnaso tychonico. Un capitolo inedito del dibattito sulle comete tra finzione letteraria e trattazione scientifica*, edited by Ottavio Besomi and Michele Camerota, 1–158. Florence: Olschki, 2000.

———. *Galileo Galilei e la cultura scientifica nell'età della Controriforma*. Rome: Salerno Editrice, 2004.

———. "Francesco Sizzi. Un oppositore di Galileo tra Firenze e Parigi." In *Toscana e Europa. Nuova scienza e filosofia tra '600 e '700*, edited by Ferdinando Abbri and Massimo Bucciantini, 83–107. Milan: F. Angeli, 2006.

Capponi, Vittorio. *Biografia pistoiese*. Pistoia: Tipografia Rossetti, 1878.

Carabba, Carlo, and Giuliano Gasparri. "La vita e le opere di Girolamo Magagnati," *Nouvelles de la Republique des Lettres* 11 (2005): 61–85.

Cardi, Giovan Battista. *Vita di Lodovico Cardi da Cigoli*. Edited by Guido Battelli. Florence: Barbèra, 1913.

Casali, Elide. *Le spie del cielo. Oroscopi, lunari e almanacchi nell'Italia moderna*. Turin: Einaudi, 2003.

Cavicchi, Elizabeth. "Painting the Moon," *Sky and Telescope* 83 (1991): 313–315.

Chapin, Seymour L. "The Astronomical Activities of Nicolas Claude Fabri de Peiresc," *Isis* 48, no. 1 (1957): 13–29.

Chapman, Allan. "A New Perceived Reality: Thomas Harriot's Moon Maps," *Astronomy and Geophysics* 50 (2009): 2–33.

Clavelin, Maurice. *Galilée copernicien*. Paris: Albin Michel, 2004.

Clerici, Alberto. "Ragion di Stato e politica internazionale. Guido Bentivoglio e altri interpreti italiani della Tregua dei Dodici Anni (1609)," *Dimensioni e problemi della ricerca storica* 2 (2009): 187–223.

Clucas, Stephen. "Corpuscular Matter Theory in the Northumberland Circle." In *Late Medieval and Early Modern Corpuscular Matter Theories*, edited by Christoph Lüthy, John E. Murdoch, and William E. Newman, 181–207. Leiden: Brill, 2001.

———. "Poetic Atomism in Seventeenth-Century England: Henry More, Thomas Traherne and 'Scientific Imagination,'" *Renaissance Studies* 3 (1991): 327–340.

———. "The Atomism of the Cavendish Circle: A Reappraisal," *The Seventeenth Century* 2 (1994): 247–273.

———. "Thomas Harriot and the Field of Knowledge in the English Renaissance." In *Thomas Harriot: An Elizabethan Man of Science*, edited by Robert Fox, 93–135. London: Ashgate, 2000.

Corbo, Anna Maria. "I pittori della cappella Paolina in S. Maria Maggiore," *Palatino* 11, no. 3 (1967): 301–313.

Cortesi, Fabrizio. "Lettere inedite del cardinale Federico Borromeo a Giovan Battista Faber segretario dei primi Lincei," *Aevum* 6 (1932): 514–518.

Cozzi, Gaetano. *Paolo Sarpi tra Venezia e l'Europa*. Turin: Einaudi, 1979.

Cozzi, Gaetano, and Luisa Cozzi. "Da Mula, Agostino," in *Dizionario biografico degli italiani*, 32, under the specific entry. Rome: Istituto della Enciclopedia Italiana, 1960–.

Cutler, Lucy C. "Virtue and Diligence: Jan Brueghel I and Federico Borromeo." In *Virtus: virtuositeit en kunstliefhebbers in de nederlanden 1500–1700*, edited by Jan de Jong, Dulcia Meijers, and Mariët Westermann, 203–227. Zwolle: Waanders Nederlands kunsthistorisch jaarboek, 2004.

D'Elia, Pasquale M. *Galileo in China. Relations through the Roman College between Galileo and the Jesuit Scientist-Missionaries (1610–1640)*. Cambridge, MA: Harvard University Press, 1960.

De Mas, Enrico. "Il 'De radiis visus et lucis.' Un trattato scientifico pubblicato a Venezia nel 1611 dallo stesso editore del 'Sidereus Nuncius.'" In *Novità celesti e crisi del sapere*, edited by Paolo Galluzzi, 159–166. Florence: Giunti, 1984.

De Renzi, Silvia, and Donatella L. Sparti. "Mancini Giulio." In *Dizionario biografico degli italiani*, 68, under the specific entry. Rome: Istituto della Enciclopedia Italiana, 1960–.

De Waard, Cornelis. *De uitvinding der verrekijkers: eene bijdrage tot de beschavingsgeschiedenis*. Rotterdam: W. L. & J. Brusse 1906.

De Zanche, Luciano. "I vettori dei dispacci diplomatici veneziani da e per Costantinopoli," *Archivio per la storia postale* 2 (1999): 19–43.

———. *Tra Costantinopoli e Venezia. Dispacci di Stato e lettere di mercanti dal Basso Medioevo alla caduta della Serenissima*. Prato: Istituto di studi storici postali, 2000.

Dehergne, Joseph. *Répertoire des Jésuites de Chine de 1552 à 1800*. Rome: Institutum Historicum S. I., 1973.

Demeester, Joëlle. "Le domaine de Mariemont sous Albert et Isabelle (1598–1621)," *Annales du Cercle archéologique de Mons* 71 (1978–1981): 181–291.

Dent, Robert William. *John Webster's Borrowing*. Berkeley: University of California Press, 1960.

Díaz Padrón, Matías. *Museo del Prado: Catálogo de Pinturas I, Escuela Flamenca Siglo XVII*. Madrid: Museo del Prado y Patronato Nacional de Museos, 1975.

Dooley, Brendan. *Morandi's Last Prophecy and the End of Renaissance Politics*. Princeton, NJ: Princeton University Press, 2002.

———. "Narrazione e verità: don Giovanni de' Medici e Galileo," *Bruniana & Campanelliana* 14, no. 2 (2008): 389–403.

Drake, Stillman. "Galileo's First Telescopes at Padua and Venice," *Isis* 50 (1959): 345–354.

———. "Galileo Gleanings XIII: An Unpublished Fragment Relating to the Telescope and Medicean Stars," *Physis* 4 (1962): 342–344.

———. "Galileo's First Telescopic Observations," *Journal for the History of Astronomy* 7 (1976): 153–168.

———. *Galileo against Philosophers in His Dialogue of Cecco di Ronchitti (1605) and Considerations of Alimberto Mauri*. Los Angeles: Zeitlin & Ver Brugge, 1976.

Dupré, Sven. "Galileo, the Telescope, and the Science of Optics in the Sixteenth Century: A Case Study of Instrumental Practice in Art and Science," Ph.D. diss., Université de Gent, 2002.

———. "Ausonio's Mirrors and Galileo's Lenses: The Telescope and Sixteenth-Century Optical Knowledge," *Galilaeana* 2 (2005): 145–180.

Dursteller, Eric R. "Power and Information: The Venetian Postal System in the Early Modern Eastern Mediterranean." In *From Florence to the Mediterranean and Beyond: Studies in Honor of Anthony Molho*, edited by Diogo Ramada Curto, Eric R. Dursteller, Julius Kirshner, and Francesca Trivellato, 1:601–623. Florence: Olschki, 2009.

Duyvendak, Jan Julius Lodewijk. "The First Siamese Embassy to Holland," *T'oung Pao* 32 (1936): 285–292.

Eamon, William. *Science and the Secrets of Nature: Books of Secrets in Medieval and Early Modern Culture*. Princeton, NJ: Princeton University Press, 1994.

Eastwood, Bruce S. "Alhazen, Leonardo, and Late-Medieval Speculation on the Inversion of Images in the Eye," *Annals of Science* 43 (1986): 413–446.

Edgerton, Samuel Y. Jr. "Galileo, Florentine 'Disegno,' and the 'Strange Spottednesse' of the Moon," *Art Journal* 9 (1984): 225–232.

———. *The Mirror, the Window and the Telescope: How Renaissance Linear Perspective Changed Our Vision of the Universe*. Ithaca, NY: Cornell University Press, 2009.

Elliott, John H. *The Old World and the New: 1492–1650*. Cambridge: Cambridge University Press, 1970.

Ernst, Germana. "Scienza, astrologia e politica nella Roma barocca. La biblioteca di Orazio Morandi." In *Bibliothecae selectae. Da Cusano a Leopardi*, edited by Eugenio Canone, 217–252. Florence: Olschki, 1993.

Ertz, Klaus. *Jan Brueghel der Ältere (1568–1625). Die Gemälde mit kritischem Oeuvrekatalog*. Cologne: DuMont, 1979.

Evans, Robert J. W. *Rudolph II and His World: A Study in Intellectual History (1576–1612)*. Oxford: Clarendon Press, 1973.

Farro, Maria Cristina. "Un 'libro di lettere' da riscoprire. Angelo Grillo e il suo epistolario," *Esperienze letterarie* 18 (1993): 69–81.

Fasano Guarini, Elena. "'Roma officina di tutte le pratiche del mondo': dalle lettere del cardinale Ferdinando de' Medici a Cosimo I e a Francesco I." In *La corte di Roma tra Cinque e Seicento. "Teatro" della politica europea*, edited by Gianvittorio Signorotto and Maria Antonietta Visceglia, 265–297. Rome: Bulzoni, 1998.

Favaro, Antonio. *Galileo Galilei e lo Studio di Padova*, 2 vols. Florence: Le Monnier, 1883; reprinted, Padua: Antenore, 1966.

———. *Carteggio di Ticone Brahe, Giovanni Keplero e di altri celebri astronomi e matematici dei secoli XVI e XVII con Giovanni Antonio Magini*. Bologna: Zanichelli, 1886.

———. "La libreria di Galileo Galilei," *Bullettino di bibliografia e di storia delle scienze matematiche e fisiche* 19 (1886): 219–293.

———. "Le osservazioni di Galileo circa i Pianeti Medicei dal 7 gennaio al 23 febbraio 1613," *Atti del Reale Istituto veneto di scienze, lettere ed arti* 59 (1900): 519–526.

———. "Un inglese a Padova al tempo di Galileo," *Atti e Memorie della R. Accademia di scienze, lettere ed arti in Padova* 34 (1918): 12–14.

———. *Amici e corrispondenti di Galileo*. Edited by Paolo Galluzzi. Florence: Salimbeni, 1983.

———. "Elementi di un nuovo anagramma galileiano." In id., *Scampoli galileiani*, 2, edited by Lucia Rossetti and Maria Laura Soppelsa, 446–447. Trieste: LINT, 1992.

Favino, Federica. "Le ragioni del 'patronage.' I Farnese di Roma e Galileo." In *Il caso Galileo. Una rilettura storica, filosofica, teologica*, edited Massimo Bucciantini, Michele Camerota, and Franco Giudice, 163–185. Florence: Olschki, 2011.

Fedele, Umberto. "Le prime osservazioni di stelle doppie," *Coelum* 17 (1949): 65–69.

Feingold, Mordechai. "Galileo in England: The First Phase." In *Novità celesti e crisi del sapere*, edited by Paolo Galluzzi, 411–420. Florence: Giunti, 1984.

———. *The Mathematician's Apprenticeship: Science, Universities and Society in England, 1560–1640*. Cambridge: Cambridge University Press, 1984.

Finocchiaro, Maurice A., ed. and trans. *The Galileo Affair: A Documentary History*. Berkeley: University of California Press, 1989.

Flynn, Dennis Thomas. *John Donne and the Ancient Catholic Nobility*. Bloomington: Indiana University Press, 1995.

Frajese, Vittorio. *Sarpi scettico. Stato e Chiesa a Venezia tra Cinque e Seicento*. Bologna: Il Mulino, 1994.

Fučiková, Eliška, James M. Bradburne, Beket Bukovinska, Jaroslava Hausenblasova, Lumomir Konecny, Ivan Muchka, and Michal Sronek, eds. *Rudolf II and Prague: The Court and the City*. London: Thames and Hudson, 1997.

Gabrieli, Giuseppe. "Federico Borromeo e gli accademici Lincei." In *Contributi alla storia dell'Accademia dei Lincei*, 2:1465–1486. Rome: Accademia dei Lincei, 1989.

———. "Giovanni Schreck linceo, gesuita e missionario in Cina e le sue lettere dall'Asia." In *Contributi alla storia dell'Accademia dei Lincei,* 2:1011–1051. Rome: Accademia dei Lincei, 1989.

Galluzzi, Paolo. "Motivi paracelsiani nella Toscana di Cosimo II e di Don Antonio dei Medici: alchimia, medicina chimica e riforma del sapere." In *Scienze, credenze occulte, livelli di cultura,* 31–62. Florence: Olschki, 1982.

———. "Genesi e affermazione dell'universo macchina." In *Galileo. Immagini dell'universo dall'antichità al telescopio,* edited by Paolo Galluzzi, 289–297. Florence: Giunti, 2009.

Gamba, Enrico. "Galilei e l'ambiente scientifico urbinate: testimonianze epistolari," *Galilaeana* 4 (2007): 343–360.

Garin, Eugenio. "Alle origini della polemica anticopernicana." In *Rinascite e rivoluzioni. Movimenti culturali dal XIV al XVIII secolo,* 283–295. Rome-Bari: Laterza, 1975.

Gingerich, Owen, and Albert Van Helden. "From 'Occhiale' to Printed Page: The Making of Galileo's 'Sidereus Nuncius,'" *Journal for the History of Astronomy* 34 (2003): 251–267.

———. "How Galileo Constructed the Moons of Jupiter," *Journal for the History of Astronomy* 42 (2011): 259–264.

Giudice, Franco. "Only a Matter of Credit? Galileo, the Telescopic Discoveries, and the Copernican System," *Galilaeana* 4 (2007): 391–413.

Goldberg, Edward. *A Jew at the Medici Court: The Letters of Benedetto Blanis Hebreo (1615–1621).* Toronto: University of Toronto Press, 2011.

———. *Jews and Magic in Medici Florence: The Secret World of Benedetto Blanis.* Toronto: University of Toronto Press, 2011.

Guerrini, Luigi. "Le 'Stanze sopra le stelle e macchie solari scoperte col nuovo occhiale' di Vincenzo Figliucci. Un episodio poco noto della visita di Galileo Galilei a Roma nel 1611," *Lettere Italiane* 50, no. 3 (1998): 387–415.

———. *Galileo e la polemica anticopernicana a Firenze.* Florence: Edizioni Polistampa, 2009.

Hahn, Juergen. *The Origins of the Baroque Concept of Peregrinatio.* Chapel Hill: University of North Carolina Press, 1973.

Hamou, Philippe. *La mutation du visible. Essai sur la portée épistemologique des instruments d'optique au XVIIe siècle.* Villeneuve d'Ascq [Nord]: Presses Universitaires du Septentrion, 1999.

Hassel, R. Chris, Jr. "Donne's 'Ignatius His Conclave' and the New Astronomy," *Modern Philology* 68 (1971): 329–337.

Heijting, Willem, and Paul R. Sellin. "John Donne's 'Conclave Ignati': The Continental Quarto and Its Printing," *Huntington Library Quarterly* 62 (2001), 401–21.

Heilbron, J. L. *Galileo.* Oxford: Oxford University Press, 2010.

Hensen, Antonius Hubertus Leonardus. "De Verrekijkers van Prins Maurits en van Aartshertog Albertus," *Mededeelingen van het Nederlansh Historisch Institut te Rome* 3 (1923): 199–204.

Herbst, Klaus-Dieter. "Galilei's Astronomical Discoveries Using the Telescope and Their Evaluation Found in a Writing-calendar from 1611," *Astronomische Nachrichten* 6 (2009): 536–539.

Hill, Christopher. *Intellectual Origins of the English Revolution*. Oxford: Clarendon Press, 1965.

Houzeau, Jean-Charles. "Le téléscope à Bruxelles, au printemps de 1609," *Ciel et terre* 3 (1882): 25–28.

Humbert, Pierre. *Un amateur: Peiresc (1580–1637)*. Paris: Desclée de Brouwer, 1933.

———. "Joseph Gaultier de La Valette, astronome provençal (1564–1647)," *Revue d'histoire des sciences et de leurs applications*, 1 (1948): 314–322.

Hunneyball, Paul M. "Sir William Lower and the Harriot Circle," The Durham Thomas Harriot Seminar, Occasional Paper no. 31, 2002.

Iannaccone, Isaia. "From N. Longobardo's Explanation of Earthquakes as Divine Punishment to F. Verbiest's Systematic Instrumental Observations: The Evolution of European Science in China in the Seventeenth Century." In *Western Humanistic Culture Presented to China by Jesuit Missionaries (XVII–XVIII Centuries)*, edited by Federico Masini, 159–174. Rome: Institutum Historicum S. I., 1996.

———. *Iohann Schreck Terrentius: le scienze rinascimentali e lo spirito dell'Accademia dei Lincei nella Cina dei Ming*. Naples: Istituto Universitario Orientale, 1998.

Ilardi, Vincent. *Renaissance Vision from Spectacles to Telescopes*. Philadelphia: American Philosophical Society, 2007.

Infelise, Mario. "Ricerche sulla fortuna editoriale di Paolo Sarpi." In *Ripensando Paolo Sarpi*, edited by Corrado Pin, 519–546. Venice: Ateneo Veneto, 2006.

Israel, Jonathan I. *The Dutch Republic: Its Rise, Greatness, and Fall, 1477–1806*. Oxford: Oxford University Press, 1995.

———. *Conflicts of Empires: Spain, the Low Countries and the Struggle for World Supremacy, 1585–1713*. London: Hambledon Press, 1997.

Jacquot, Jean. "Thomas Harriot's Reputation for Impiety," *Notes and Records of the Royal Society of London* 9 (1952): 164–187.

———. "Harriot, Hill, Warner and the New Philosophy." In *Thomas Harriot: Renaissance Scientist*, edited by John W. Shirley, 107–128. Oxford: Clarendon Press, 1974.

Johnson, Jeffrey. "'One, Four, and Infinite': John Donne, Thomas Harriot, and 'Essayes in Divinity,'" *John Donne Journal* 22 (2003): 109–143.

Johnston, Stephen. "The Mathematical Practitioners and Instruments in Elizabethan England," *Annals of Science* 48 (1991): 319–344.

Jones, Pamela M. *Federico Borromeo and the Ambrosiana: Art Patronage and Reform in Seventeenth-Century Milan*. Cambridge: Cambridge University Press, 1993.

Kargon, Robert H. *Atomism in England from Hariot to Newton*. Oxford: Clarendon Press, 1966.

Kaufmann, Thomas DaCosta. *The Mastery of Nature: Aspects of Art, Science and Humanism in the Renaissance*. Princeton, NJ: Princeton University Press, 1993.

———. *The School of Prague*. Chicago: University of Chicago Press, 1988.

Keil, Inge. *Augustanus Opticus. Johann Wiesel (1583–1662) und 200 Jahre optisches Handwerk in Augsburg*. Berlin: Akademie Verlag, 2000.

Kishlansky, Mark. *A Monarchy Transformed: Britain 1603–1714*. London: Allen Lane, 1996.

Lattis, James M. *Between Copernicus and Galileo: Christoph Clavius and the Collapse of Ptolemaic Cosmology*. Chicago: University of Chicago Press, 1994.

Leitão, Henrique. "Galileo's Telescopic Observations in Portugal." In *Largo campo di filosofare*, edited by José Montesinos and Carlos Solís, 903–913. La Orotava: Fundación Canaria Orotava de Historia de la Ciencia, 2001.

———. "The Contents and Context of Manuel Dias' Tianwenlüe." In *The Jesuits, the Padroado and East Asian Science (1552–1773)*, edited by Luís Saraiva and Catherine Jami, 99–112. Singapore: World Scientific Publishing, 2008.

Lindberg, David C. *Theories of Vision from Al-Kindi to Kepler*. Chicago: University of Chicago Press, 1976.

Lohne, Johannes A. "Thomas Harriot (1560–1621): The Tycho Brahe of Optics," *Centaurus* 6 (1959) 113–21.

———. "Essays on Thomas Harriot. III. A Survey of Harriot's Scientific Writings," *Archive for History of Exact Sciences* 20 (1979): 265–312.

Magone, Rui. "The Textual Tradition of Manuel Dias' Tianwenlüe." In *The Jesuits, the Padroado and East Asian Science (1552–1773)*, edited by Luís Saraiva and Catherine Jami, 123–138. Singapore: World Scientific Publishing, 2008.

Malet, Antoni. "Kepler and the Telescope," *Annals of Science* 60 (2003): 107–136.

Maltby, William S. *The Black Legend in England: The Development of Anti-Spanish Sentiment, 1558–1660*. Durham, NC: Duke University Press, 1971.

Mamone, Sara. "Il re è morto, viva la regina." In *"Parigi val bene una messa!" 1610: l'omaggio dei Medici a Enrico IV re di Francia e di Navarra*, edited by Monica Bietti, Francesca Fiorelli Malesci, and Paul Mironneau, 32–39. Livorno: Sillabe, 2010.

Mandelbrote, Scott. "The Religion of Thomas Harriot." In *Thomas Harriot: An Elisabethan Man of Science*, edited by Robert Fox, 246–279. Aldeshot: Ashgate, 2000.

Marchitello, Howard. *The Machine in the Text: Science and Literature in the Age of Shakespeare and Galileo*. New York: Oxford University Press, 2011.

Marcora, Carlo. "La biografia del cardinal Federico Borromeo scritta dal suo medico personale Giovanni Battista Mongilardi," *Memorie storiche della Diocesi di Milano* 15 (1968): 125–232.

Marr, Alexander. *Between Raphael and Galileo: Mutio Oddi and the Mathematical Culture of Late Renaissance Italy*. Chicago: University of Chicago Press, 2011.

Matt, Luigi. "Grillo, Angelo," in *Dizionario biografico degli italiani*, 59, under the specific entry. Rome: Istituto della Enciclopedia Italiana, 1960–.

Miller, Peter N. "Description Terminable and Interminable: Looking at the Past, Nature and Peoples in Peiresc's Archive." In *Historia: Empiricism and Erudition in Early Modern Europe*, edited by Gianna Pomata and Nancy Siraisi, 355–397. Cambridge, MA: MIT Press, 2005.

Minois, Georges. *Le couteau et le poison. L'assassinat politique en Europe*. Paris: Fayard, 1997.

Molaro, Paolo, and Perluigi Selvelli. "The Mystery of the Telescopes in Jan Brueghel the Elder's Paintings," *Memorie della Società Astronomica Italiana* 75 (2008): 282–285.

Needham, Paul. *Galileo Makes a Book: The First Edition of "Sidereus Nuncius" Venice 1610*. Berlin: Akademie Verlag, 2011.

Nicolson, Marjorie Hope. "The 'New Astronomy' and English Imagination," *Studies in Philology* 32 (1935): 428–462.

———. "Kepler, the Somnium, and John Donne," *Journal of the History of Ideas* 3 (1940): 259–280.

———. *Science and Imagination*. Ithaca, NY: Great Seal Books, 1956.

North, John D. "Thomas Harriot and the First Telescopic Observations of Sunspots." In *Thomas Harriot: Renaissance Scientist*, edited by John W. Shirley, 130–165. Oxford: Clarendon Press, 1974.

Nováček, Vojtěch Jaromír. "Martin Horký, český hvězdář [*Martin Horky astronomo ceco*]," *Časopis Musea královstvý českého* 63 (1889): 389–400.

Orbaan, Johannes Albertus Franciscus. *Documenti sul Barocco in Roma*. Rome: Società Romana di Storia Patria, 1920.

Ostrow, Steven F. *Art and Spirituality in Counter-Reformation Rome*. Cambridge: Cambridge University Press, 1996.

———. "Cigoli's Immacolata and Galileo's Moon: Astronomy and the Virgin in Early Seicento Rome," *The Art Bulletin* 78, no. 2 (1996): 218–235.

Palladino, Franco. "Un trattato sulla costruzione del cannocchiale ai tempi di Galilei. Principi matematici e problemi tecnologici," *Nouvelles de la République des Lettres* 1 (1987): 83–102.

Panofsky, Erwin. *Galileo as a Critic of the Arts*. The Hague: Nijhoff, 1954.

Pantin, Isabelle. "La lunette astronomique: une invention en quête d'auteurs." In *Inventions et découvertes au temps de la Renaissance*, edited by Marie Thérèse Jones-Davies, 159–174. Paris: Klincksieck, 1994.

———. "Galilée, la lune et les Jésuites," *Galilaeana* 2 (2005): 19–42.

Parry, John Horace. *The Age of Reconnaissance: Discovery, Exploration and Settlement, 1450–1650*. Berkeley: University of California Press, 1981.

Pastor, Ludwig von. *Storia dei Papi dalla fine del Medio Evo*, Rome: Desclée, 1908–1934.

Pedersen, Olaf. "Sagredo's Optical Researches," *Centaurus* 13 (1968): 139–150.

Pin, Corrado. "'Qui si vive con esempi, non con ragione': Paolo Sarpi e la committenza di stato nel dopo-Interdetto." In *Ripensando Paolo Sarpi*, edited by Corrado Pin, 343–394. Venice: Ateneo Veneto, 2006.

Pitts, Vincent J. *Henri IV of France: His Reign and Age*. Baltimore: Johns Hopkins University Press, 2009.

Poppi, Antonino. *Cremonini e Galilei inquisiti a Padova nel 1604. Nuovi documenti d'archivio*. Padua: Antenore, 1992.

Prosperi, Adriano. "Intellettuali e Chiesa all'inizio dell'età moderna." In *Storia d'Italia. Annali 4. Intellettuali e potere*, edited by Corrado Vivanti, 159–252. Turin: Einaudi, 1981.

———. *Tribunali della coscienza. Inquisitori, confessori, missionari*. Turin: Einaudi, 1996.

Pumfrey, Stephen. "Harriot's Maps of the Moon: New Interpretations," *Notes and Records of the Royal Society* 63 (2009): 163–168.

Quinn, David B. "Thomas Harriot and the Problem of America." In *Thomas Harriot: An Elizabethan Man of Science*, edited by Robert Fox, 9–27. London: Ashgate, 2000.

Reeves, Eileen. *Painting the Heavens. Art and Science in the Age of Galileo*. Princeton, NJ: Princeton University Press, 1997.

———. *Galileo's Glassworks: The Telescope and the Mirror*. Cambridge, MA: Harvard University Press, 2008.

———. "Kingdoms of Heavens: Galileo and Sarpi on the Celestial," *Representations* 105 (2009): 61–84.

———. "Variable Stars: A Decade of Historiography on the *Sidereus nuncius*," *Galilaeana* 8 (2011): 37–72.

Reeves, Eileen, and Albert Van Helden. "Verifying Galileo's Discoveries: Telescope-Making at the Collegio Romano." In *Der Meister und die Fernrohre: das Wechselspiel zwischen Astronomie und Optik in der Geschichte*, edited by Jürgen Hamel and Inge Keil, 127–141. Frankfurt am Main: H. Deutsch, 2007.

Ricci, Saverio. *La fortuna del pensiero di Giordano Bruno, 1600–1750*. Florence: Le Lettere, 1990.

Rigaud, Stephen Peter. *Supplement to Dr. Bradley's Miscellaneous Works: With an Account of Harriot's Astronomical Papers.* Oxford: Oxford University Press, 1833.

Rizza, Cecilia. "Galileo nella corrispondenza di Peiresc," *Studi francesi* 15 (1961): 433–451.

———. *Peiresc e l'Italia.* Turin: Giappichelli, 1965.

Roche, John J. "Harriot, Galileo, and Jupiter's Satellites," *Archives Internationales d'Histoire des Sciences* 32 (1982): 9–51.

———. "Lower, Sir William (ca. 1570–1615)." In *Oxford Dictionary of National Biography.* Oxford: Oxford University Press, 2004.

Rochemonteix, Camille de. *Un Collège de Jèsuites aux XVIIe et XVIIIe siècles. Le Collège Henry IV de la Flèche.* Le Mans: Leguicheux, 1889.

Rodis-Lewis, Geneviève. *Descartes: His Life and Thought.* Translated by Jane Marie Todd. Ithaca, NY: Cornell University Press, 1998.

Romagnoli, Ettore. *Biografia cronologica de' Bellartisti Senesi, 1200–1800.* Florence: Edizioni S.P.E.S., 1976.

Romano, Antonella. *La contre-réforme mathématique. Constitution et diffusion d'une culture mathématique jésuite à la Renaissance.* Rome: École Française de Rome, 1999.

Ronchi, Vasco. *Galileo e il cannocchiale.* Udine: Idea, 1942.

Rosen, Edward. "When Did Galileo Make His First Telescope?," *Centaurus* 2 (1951): 44–51.

Savio, Paolo. "Per l'epistolario di Paolo Sarpi," *Aevum* 10 (1936): 3–104.

Sebes, Joseph. "Dias (o Novo), Manuel." In *Diccionario Histórico de la Compañia de Jesus,* edited by Charles E. O'Neil and Joaquín M. Domíniguez, 2:1113. Madrid: Universidad Pontificia Comillas, 2001.

Seltman, Muriel. "Harriot's Algebra: Reputation and Reality." In *Thomas Harriot: An Elizabethan Man of Science,* edited by Robert Fox, 153–185. London: Ashgate, 2000.

Settle, Thomas B. "Ostilio Ricci, a Bridge between Alberti and Galileo." In *Actes du XIIe Congrès International d'Histoire des Sciences,* 3. B: 121–126. Paris: Blanchard, 1971.

Shea, William R., and Mariano Artigas. *Galileo in Rome: The Rise and Fall of a Troublesome Genius.* Oxford: Oxford University Press, 2003.

Shirley, John W. "An Early Experimental Determination of Snell's Law," *American Journal of Physics* 19 (1951): 507–508.

———. "Thomas Harriot's Lunar Observations." In *Science and History: Studies in Honor of Edward Rosen* (*Studia Copernicana,* XVI), edited by Erna Hilfstein, Pawel Czartoryski, and Frank. D. Grande, 283–308. Wrocław: Polish Academy of Sciences Press, 1978.

———. "Sir Walter Ralegh and Thomas Harriot." In *Thomas Harriot: Renaissance Scientist,* edited by John W. Shirley, 16–35. Oxford: Oxford University Press, 1983.

———. *Thomas Harriot: A Biography*. Oxford: Clarendon Press, 1983.

Simon, Gérard. *Archéologie de la vision. L'optique, le corps, la peinture*. Paris: Éditions du Seuil, 2003.

Simpson, Evelyn. *A Study of the Prose Works of John Donne*. Oxford: Clarendon Press, 1948^2.

Sluiter, Engel. "The Telescope before Galileo," *Journal for the History of Astronomy* 28 (1997): 223–234.

Smolka, Josef. "Böhmen und die Annhame der Galileischen astronomischen Entdeckungen," *Acta historiae rerum naturalium necnon technicarum* 1 (1997): 41–69.

———. "Martin Horký a jeho kalendáře," in *Miscellanea. Oddělení rukopisů a starých tisků* (Prague: Národní Knihovna, 2005), 145–160.

Solaini, Piero. "Storia del cannocchiale," *Atti della Fondazione Giorgio Ronchi* 51 (1996): 805–872.

Sosio, Libero. "Galileo Galilei e Paolo Sarpi." In *Galileo Galilei e la cultura veneziana*, 269–311. Venice: Istituto veneto di scienze, lettere ed arti, 1995.

———. "Paolo Sarpi, un frate nella rivoluzione scientifica." In *Ripensando Paolo Sarpi*, edited by Corrado Pin, 183–236. Venice: Ateneo Veneto, 2006.

Standaert, Nicolas. "Jesuits in China." In *The Cambridge Companion to the Jesuits*, edited by Thomas Worcester, 169–185. Cambridge: Cambridge University Press, 2008.

Stedall, Jacqueline A. "Rob'd of Glories: The Posthumous Misfortunes of Thomas Harriot and his Algebra," *Archive for History of Exact Sciences* 54 (2000): 455–497.

———. "Symbolism, Combinations, and Visual Imagery in the Mathematics of Thomas Harriot," *Historia Mathematica* 34 (2007): 380–401.

Stone, David M. "Bad Habit: Scipione Borghese, Wignacourt and the Problem of Cigoli's Knightood." In *Celebratio Amicitiae. Studies in Honour of Giovanni Bonello*, edited by Maroma Camilleri and Theresa Vella, 207–229. Malta: Fondazzjoni Patrimonju Malti, 2006.

Straker, Stephen M. "Kepler, Tycho and 'The Optical Part of Astronomy': The Genesis of Kepler's Theory of Pinhole Images," *Archive for History of Exact Sciences* 24 (1981): 267–293.

Strano, Giorgio. "La lista della spesa di Galileo: un documento poco noto sul telescopio," *Galilaeana* 6 (2009): 197–211.

Strebel, Christoph. "Martinus Horky und das Fernrohr Galileis," *Sudhoffs Archiv* 90 (2006): 11–28.

Tanner, Rosalind Cecilia H. "The Study of Thomas Harriot's Manuscripts: 1. Harriot's Will," *History of Science* 6 (1967): 1–16.

Taylor, Eva Germaine Rimington. *The Mathematical Practitioners of Tudor and Stuart England*. Cambridge: Cambridge University Press, 1967.

Thomas, Werner. "Andromeda Unbound: The Reign of Albert and Isabella in the Southern Netherlands, 1598–1621." In *Albert & Isabella, 1598–1621: Essays*, edited by Werner Thomas and Luc Duerloo. Turnhout: Brepols, 1998.

Torrini, Maurizio. "'Et vidi coelum novum et terram novam.' A proposito di rivoluzione scientifica e libertinismo," *Nuncius* 1 (1986): 49–77.

———. "Il Rinascimento nell'orizzonte della nuova scienza." In *Nuovi maestri e antichi testi. Umanesimo e Rinascimento alle origini del pensiero moderno*, edited by Stefano Caroti and Vittoria Perrone Compagni, 315–330. Florence: Olschki, 2012.

Toulmin, Stephen. *Cosmopolis: The Hidden Agenda of Modernity*. New York: Free Press, 1990.

Trevor-Roper, Hugh. *Princes and Artists: Patronage and Ideology at Four Habsburg Courts, 1517–1633*. London: Thames and Hudson, 1976.

Trivellato, Francesca. *Fondamenta dei vetrai. Lavoro, tecnologia e mercato a Venezia tra Sei e Settecento*. Rome: Donzelli, 2000.

Tuckerman, Bryant. *Planetary, Lunar, and Solar Positions, A.D. 2 to A.D. 1649*. Philadelphia: American Philosophical Society, 1962.

Tutino, Stefania. "Notes on Machiavelli and Ignatius Loyola in John Donne's 'Ignatius his Conclave' and 'Pseudo-Martyr,'" *English Historical Review* 119 (2004): 1308–1321.

Ulianich, Boris. "Badoer, Giacomo," in *Dizionario biografico degli italiani*, 5, under the specific entry. Rome: Istituto della Enciclopedia Italiana, 1960–.

Valleriani, Matteo. *Galileo Engineer*. Dordrecht: Springer, 2010.

Van der Cruysse, Dirk. *Louis XIV et le Siam*. Paris: Fayard, 1991.

Van Helden, Albert. *The Invention of the Telescope*. Philadelphia: American Philosophical Society, 1977.

———. "Galileo and the Telescope." In *The Origins of the Telescope*, edited by Albert Van Helden, Sven Dupré, Rob Van Gent, and Huib Zuidervaart, 183–201. Amsterdam: Knaw Press 2010.

Vasoli, Cesare. "Giulio Pace e la diffusione europea di alcuni temi aristotelici padovani." In *Aristotelismo veneto e scienza moderna*, edited by Luigi Olivieri, 2:1009–1034. Padua: Antenore, 1983.

Vermij, Rienk H. "The Telescope at the Court of the Stadtholder Maurits." In *The Origins of the Telescope*, edited by Albert Van Helden, Sven Dupré, Rob Van Gent, and Huib Zuidervaart, 73–92. Amsterdam: Knaw Press 2010.

Volpini, Paola. "Medici, Giovanni de'," in *Dizionario biografico degli italiani*, 73, under the specific entry. Rome: Istituto della Enciclopedia Italiana, 1960–.

Westman, Robert S. "The Astronomer's Role in the Sixteenth Century: A Preliminary Study," *History of Science* 18 (1980): 105–147.

———. *The Copernican Question: Prognostication, Skepticism, and Celestial Order.* Berkeley: University of California Press, 2011.

Whitaker, Ewan A. "Galileo's Lunar Observations and the Dating of the Composition of 'Sidereus Nuncius,'" *Journal for the History of Astronomy* 9 (1978): 155–169.

———. "Selenography in the Seventeenth Century." In *The General History of Astronomy*, edited by René Taton and Curtis Wilson, 2:119–143. Cambridge: Cambridge University Press, 1989.

———. "Identificazione e datazione delle osservazioni lunari di Galileo." In *Galileo. Immagini dell'universo dall'antichità al telescopio*, edited by Paolo Galluzzi, 262–267. Florence: Giunti, 2009.

Wicki, José. "Liste der Jesuiten-Indienfahrer, 1541–1758," *Aufsätze zur Portugiesichen Kulturgeschichte* 7 (1967): 252–450.

Wilding, Nick. "Instrumental Angels." In *Conversations with Angels: Essays Towards a History of Spiritual Communication, 1100–1700*, edited by Joad Raymond, 67–89. Basingstoke: Palgrave Macmillan, 2011.

Willach, Rolf. "The Development of Lens Grinding and Polishing Techniques in the First Half of the 17th Century," *Bulletin of the Scientific Instrument Society* 68 (2001): 10–15.

———. *The Long Route of the Telescope*. Philadelphia: American Philosophical Society, 2008.

Winston, Matthew. "Gendered Nostalgia in *The Duchess of Malfi*," *Renaissance Papers*, 1998, 103–113.

Wootton, David. "New Light on the Composition and Publication of the 'Sidereus Nuncius,'" *Galilaeana* 6 (2009): 123–140.

———. *Galileo: Watcher of the Skies*. New Haven, CT: Yale University Press, 2010.

Yates, Frances Amelia. *Astraea: The Imperial Theme in the Sixteenth Century*. Boston: Routledge & Kegan Paul, 1975.

Yourcenar, Marguerite. *En pèlerin et en étranger*. Paris: Gallimard, 1989.

Zecchin, Luigi. "I cannocchiali di Galilei e gli 'occhialeri' veneziani." In *Vetro e vetrai di Murano: studi sulla storia del vetro*, 2:255–265. Venice: Arsenale, 1987–1990.

Zhang, Zhishan. "Johann Adam Schall von Bell and his Book 'On the Telescope.'" In *Western Learning and Christianity in China: The Contribution and Impact of Johann Adam Schall von Bell, S. J., (1592–1666)*, edited by Roman Malek, 2:681–690. Sankt Augustin: Monumenta Serica Institute, 1998.

Zürcher, Erik. "Giulio Aleni's Chinese Biography." In *Scholar from the West: Giulio Aleni S. J. (1582–1649) and the Dialogue between Christianity and China*, edited by Tiziana Lippiello and Roman Malek, 85–127. Brescia–Sankt Augustin: Fondazione Civiltà Bresciana/Monumenta Serica Institute, 1997.

Credits

Illustrations

Fig. 1. Galileo, objective lens. Florence, Museo Galileo.

Fig. 2. Galileo Galilei, objective lens. Florence, Museo Galileo. Photo: Giuseppe Molesini.

Figs. 3–4. Galileo's telescopes. Florence, Museo Galileo.

Figs. 5–6. Eyepiece and objective lens. Florence, Museo Galileo.

Fig. 7. Brueghel the Elder, *Landscape with View of the Castle of Mariemont.* Richmond, Virginia Museum of Fine Arts. Adolph D. and Wilkins C. Williams Fund. Photo: Katherine Wetzel. Copyright © Virginia Museum of Fine Arts.

Fig. 8. Brueghel the Elder, *Landscape with View of the Castle of Mariemont,* detail. Richmond, Virginia Museum of Fine Arts. Adolph D. and Wilkins C. Williams Fund. Photo: Katherine Wetzel. Copyright © Virginia Museum of Fine Arts.

Fig. 9. Galileo's shopping list. Florence, Biblioteca Nazionale Centrale, MS Gal. 88, fol. 107*r.*

Fig. 10. Galileo's first representation of Jupiter's satellites. Florence, Biblioteca Nazionale Centrale, MS Gal. 88, fol. 105*r.*

Fig. 11. Galileo's sketch of Jupiter's satellites. Archivio della Pontificia Università Gregoriana. From Galileo Galilei, *Opere,* national edition, ed. Antonio Favaro, 10:277. Florence: Barbèra, 1890–1909.

Fig. 12. Galileo, report of observations of Jupiter's satellites. Ann Arbor, University of Michigan Library, Special Collections.

Fig. 13. Galileo, first page of the autograph record of observations of Jupiter's satellites. Florence, Biblioteca Nazionale Centrale, MS Gal. 48, fol. 30*r.*

Fig. 14. Galileo, diagram of the positions of Jupiter's satellites, from MS Gal. 48, fol. 34*v.* From Galileo Galilei, *Opere,* national edition, ed. Antonio Favaro, 3:436. Florence: Barbèra, 1890–1909.

Fig. 15. Galileo, autograph transcription of Kepler's solution of the anagram of Saturn. Florence, Biblioteca Nazionale Centrale, MS Gal. 28, fol. 3*r.*

Fig. 16. Galileo, autograph experiments with anagrams. Florence, Biblioteca Nazionale Centrale, MS Gal. 70, fol. 11*v*.

Fig. 17. Harriot, drawing of the Moon. Sussex, Petworth House, MSS T. Harriot, MS HMC, 241/IX, fol. 26.

Fig. 18. Harriot, drawing of the Moon. Sussex, Petworth House, MSS T. Harriot, MS HMC, 241/IX, fol. 20.

Fig. 19. Galileo, drawing of the Moon. From Galileo Galilei, *Sidereus nuncius* (Venice, 1610).

Fig. 20. Chalette, *bozzetto* for the title page of *Astra medicea*. Carpentras, Bibliothèque d'Inguimbert, MS 803, fol. 285.

Fig. 21. Brueghel the Elder, *Allegory of Air*. Paris, Musée du Louvre (photo RMN—Musée du Louvre / Jean-Gilles Berizzi / distr. Alinari).

Fig. 22. Title page with the Medici arms. From Girolamo Magagnati, *Meditazione poetica sopra i pianeti medicei*. Venice: Heredi d'Altobello Salicato, 1610.

Fig. 23. Cigoli, *Assumption of the Virgin*, detail. Rome, Basilica of Santa Maria Maggiore, Pauline Chapel (courtesy Ministry for Cultural Heritage and Activities).

Fig. 24. Galileo, drawing of the Moon. Florence, Biblioteca Nazionale Centrale, MS Gal. 48, fol. 29*v*.

Fig. 25. Dias, *Tian Wen Lüe*. Vatican City, by permission of Biblioteca Apostolica Vaticana, with all rights reserved. Borgia Cinese, 324, fol. 43*a–b*.

Plates

Plate 1. Brueghel the Elder, *Landscape with View of the Castle of Mariemont*. Richmond, Virginia Museum of Fine Arts. Adolph D. and Wilkins C. Williams Fund. Photo: Katherine Wetzel. Copyright © Virginia Museum of Fine Arts.

Plate 2. Brueghel the Elder and Rubens, *Allegory of Sight*. Madrid, Museo Nazionale del Prado (photo Scala, Florence).

Plate 3. Brueghel the Elder, *Allegory of Air*. Paris, Musée du Louvre (photo RMN—Musée du Louvre / Jean-Gilles Berizzi / distr. Alinari).

Plate 4. Cigoli, *Assumption of the Virgin,* detail. Rome, Basilica of Santa Maria Maggiore, Pauline Chapel (courtesy Ministry for Cultural Heritage and Activities).

Plate 5. de Ribera, *Allegory of Sight.* Mexico City, Museo Franz Mayer.

Plate 6. Galileo, drawings of the Moon, detail. Florence, Biblioteca Nazionale Centrale, MS Gal. 48, fol. 28*r.*

Plate 7. Cassini, drawing of the Moon. Paris, Observatoire de Paris, MS D-VI-40, fol. 52.

Plate 8. English school, *Royal Observatory from Crooms Hill.* Greenwich, National Maritime Museum (photo National Maritime Museum / Archivi Alinari).

Acknowledgments

This book is not a collection of essays, but the outcome of collective work that, like any true collaboration, took an immense amount of time, but was always shared from the first to the last page. It reflects the awareness that we could never have done this alone. The subjects we examine here could not be broader in scope, but they are also quite specific and relied on the convergence of different strengths and skills. Massimo Bucciantini wrote and oversaw the final revision of Chapters 2, 4, 6, and 10, as well as parts of Chapters 8 and 9; Michele Camerota, Chapters 5 and 11, as well as parts of Chapters 8 and 9; Franco Giudice, Chapters 1, 3, 7, and 12. Many people assisted us with this project. First of all, we are especially grateful to our first readers, those who read all or part of our book during the writing process, improving it with their suggestions and advice: Pietro Corsi, Maria Pia Donato, Federica Favino, Enrico Giannetto, Isabelle Pantin, Corrado Pin, Patrizia Ruffo, Francesca Sandrini, Domenico Scarpa, Marta Stefani, and Maurizio Torrini. Along with them, we are also grateful to Antonella Barzazi, Gian Luigi Betti, Marco Ciardi, Anna C. Citernesi, Stephen Clucas, Vera Costantini, Krista De Jonge, Robert Ellis, Michele Maccherini, Eugenio Menegon, Maria Pia Pedani, Eileen Reeves, Jingjing Song, Giovanni Sordini, Clay Valeri, and Charles Webster.

Special thanks go to the entire staff at the Museo Galileo, as this work would never have been possible without their kindness and professional advice.

Index